AF838540

Student Solutions Manual for Cochener, Hodge, and Gustafson's

Beginning and Intermediate Algebra with Graphing Calculators

AN INTEGRATED APPROACH

David J. Cochener
Austin Peay State University

Brooks/Cole
Thomson Learning™

Australia • Canada • Mexico • Singapore • Spain • United Kingdom • United States

Project Development Editor: Michelle Paolucci
Marketing Manager: Leah Thomson
Editorial Assistant: Erin Wickersham
Production Editor: Stephanie Andersen
Cover Design: Laurie Albrecht
Cover Photo: Ed Young
Print Buyer: Tracy Brown
Printing and Binding: Globus Printing Company

For more information, contact:
BROOKS/COLE
511 Forest Lodge Road
Pacific Grove, CA 93950 USA
www.brookscole.com

Printed in the United States of America

10 9 8 7 6 5 4 3 2

ISBN: 0-534-37327-5

CONTENTS

Foreword to the Student

This Student Solutions Manual is a supplement to *Beginning and Intermediate Algebra: A Graphical Approach* by Debbie J. Cochener, Bonnie M. Hodge and R. David Gustafson, and is written to be your personal mathematical assistant to help you to be more successful in beginning and intermediate algebra. Complete step-by-step solutions are presented to all odd-numbered exercises as well as all Chapter Summaries, Chapter Tests, Cumulative Reviews and the Appendices.

This supplement is not intended to replace your text or be a substitute for regular class attendance or working your homework assignments. To derive maximum benefit from the material presented here, you should thoroughly read and study your text and lecture notes, and make a serious effort to work assigned problems, referring to these solutions only when you get stuck or want to confirm a solution. Remember that there is often more than one way to work a math problem, so don't be alarmed if your method is different from mine.

Please send any suggestions or corrections by e-mail to cochenerdj@apsu.edu . I hope that your study of algebra is enjoyable and successful.

David Cochener

CHAPTER 1 BASIC ARITHMETIC AND GRAPHING CALCULATOR CONCEPTS

Section 1.1 Review of Fractions

Vocabulary

1. natural

3. rational

5. identity; identity

Concepts

7. a. and c. are in lowest terms

9. b. and c. are correct

11. Factor each denominator into prime factors. Multiply together the primes that occur, using each one the greatest number of times it occurs as a factor in any one factorization.

13. Reversing the order produces different answers: 8/4 = 2 while 4/8 = 0.5

15. Do it as $4+\frac{17}{19}$ ▸Frac.

Practice

17. $\frac{24}{18}=\frac{2\cdot 2\cdot 2\cdot 3}{2\cdot 3\cdot 3}=\frac{4}{3}$

19. $\frac{72}{64}=\frac{2\cdot 2\cdot 2\cdot 3\cdot 3}{2\cdot 2\cdot 2\cdot 2\cdot 2\cdot 2}=\frac{9}{8}$

21. $\frac{5}{12}\cdot\frac{18}{5}=\frac{5\cdot 2\cdot 3\cdot 3}{2\cdot 2\cdot 3\cdot 5}=\frac{3}{2}$

23. $12\cdot\frac{5}{6}=\frac{2\cdot 2\cdot 3\cdot 5}{2\cdot 3}=10$

25. $2\frac{2}{3}\cdot 1\frac{5}{8}=\frac{8}{3}\cdot\frac{13}{8}=\frac{13}{3}$

27. $\frac{2}{12}\div\frac{8}{13}=\frac{2}{12}\cdot\frac{13}{8}=\frac{2\cdot 13}{2\cdot 2\cdot 3\cdot 2\cdot 2\cdot 2}=\frac{13}{48}$

29. $6\div\frac{3}{14}=6\cdot\frac{14}{3}=\frac{2\cdot 3\cdot 2\cdot 7}{3}=28$

31. $1\frac{12}{30}\div 7=\frac{42}{30}\cdot\frac{1}{7}=\frac{2\cdot 3\cdot 7\cdot 1}{2\cdot 3\cdot 5\cdot 7}=\frac{1}{5}$

33. $\frac{3}{5}+\frac{3}{5}=\frac{3+3}{5}=\frac{6}{5}$

35. $\frac{1}{6}+\frac{1}{24}=\frac{1\cdot 4}{6\cdot 4}+\frac{1}{24}=\frac{4}{24}+\frac{1}{24}=\frac{5}{24}$

37. $\frac{5}{14}-\frac{4}{21}=\frac{5\cdot 3}{14\cdot 3}-\frac{4\cdot 2}{21\cdot 2}=\frac{15}{42}-\frac{8}{42}=\frac{7}{42}=\frac{1}{6}$

39. $\frac{17}{3}+4=\frac{17}{3}+\frac{4\cdot 3}{3}=\frac{17}{3}+\frac{12}{3}=\frac{29}{3}$

41. $4\frac{3}{5}+\frac{3}{5}=\frac{23}{5}+\frac{3}{5}+\frac{26}{5}=5\frac{1}{5}$

43. $3\frac{3}{4}-2\frac{1}{2}=3\frac{3}{4}-2\frac{2}{4}=\frac{15}{4}-\frac{10}{4}=\frac{5}{4}=1\frac{1}{4}$

45. The parentheses are needed to override the natural order of operations.

Applications

47. $\frac{1545}{8667}=\frac{3\cdot 5\cdot 103}{3\cdot 3\cdot 3\cdot 3\cdot 107}=\frac{515}{2889}$

49. $\frac{7}{25}\cdot 8667=2426.76\approx 2427$ of the accepted applicants

51. a. $\frac{1}{5}\cdot 750=150$ b. $\frac{4}{5}\cdot 750=600$

53. $\frac{13}{25}+\frac{9}{50}=\frac{26}{50}+\frac{9}{50}=\frac{35}{50}=\frac{7}{10}$

55. $1-\left(\frac{3}{25}+\frac{1}{10}\right)=1-\left(\frac{12}{100}+\frac{10}{100}\right)=1-\frac{22}{100}=$
$\frac{78}{100}=\frac{39}{50}$

Section 1.2 Exponents and Order of Operations

Vocabulary

1. exponent

3. base; exponent

Concepts

5. a. x·x b. 3z·z·z·z c. 5x·5x d. 5·2x·2x·2x

7. "x multiplied by itself n times" means
x · (n factors of x) = x^{n+1}

9. Smaller numbers are produced.

11. 'Evaluate' means to give a value or number as an answer. 'Simplify' means to write in a simpler alternate form.

Practice

13. $5^2 = 5\cdot5 = 25$

15. $7^3 = 7\cdot7\cdot7 = 343$

17. $4x^2 = 4(3^2) = 4\cdot9 = 36$

19. $(5y)^3 = (5\cdot2)^3 = 10^3 = 1000$

21. $2y^4 = 2\cdot3^2 = 2\cdot9 = 18$

23. $(3y)x = (3\cdot2)\cdot3 = 6\cdot3 = 18$

25. $4\cdot6 + 5 = 24 + 5 = 29$

27. $4(6 + 5) = 4\cdot11 = 44$

29. $4^2 - 2^2 = 16 - 4 = 12$

31. $(5 - 2)^3 = 3^3 = 27$

33. $12 + 2\cdot3 + 2 = 12 + 6 + 2 = 20$

35. $16 \div (5 + 3) = 16 \div 8 = 2$

37. $(7 + 9) \div 2\cdot4 = 16 \div 2\cdot4 = 8\cdot4 = 32$

39. $36 \div 9\cdot4 - 2 = 4\cdot4 - 2 = 16 - 2 = 14$

41. $100 \div [10\cdot(10 \div 100)] = 100 \div [10\cdot(1/10)]$
$= 100 \div 1 = 100$

43. $4\cdot3 + 2(5 - 2) - 2^3 = 12 + 2(3) - 8 =$
$12 + 6 - 8 = 18 - 8 = 10$

45. $3^3 - (3 - 1)^3 = 3^3 - 2^3 = 27 - 8 = 19$

47. $(3\cdot5 - 2\cdot6)^2 = (15 - 12)^2 = 3^2 = 9$

49. $\frac{25-(2\cdot3-1)}{2\cdot9-8}=\frac{25-(6-1)}{18-8}=\frac{25-5}{10}=\frac{20}{10}$
$=2$

51. $\frac{\left(4^2-2\right)\div7}{5(2+4)-3^2}=\frac{(16-2)\div7}{5\cdot6-9}=\frac{14\div7}{21}=\frac{2}{21}$

53. $2z + y = 2\cdot4 + 2 = 8 + 2 = 10$

55. $(4 + z)y = (4 + 4)2 = 8\cdot2 = 16$

57. $zx + y^2 - 2z = 4\cdot3 + 2^2 - 2\cdot4 = 12 + 4 - 8 = 8$

59. $3x^2 + (2y)^2 = 3(3^2) + (2\cdot2)^2 = 3\cdot9 + 4^2 =$
$27 + 16 = 43$

61. $\frac{2z^2-y}{2x-y^2}=\frac{2\cdot4^2-2}{2\cdot3-2^2}=\frac{32-2}{6-4}=\frac{30}{2}=15$

Applications

63. Area A = 13^2 = 169 square units

65. Volume V = 6^3 = 216 cubic units

67. Total enrollment = (8791 + 2385 + 5468 + 6101 + 1282 + 7181) = 31,208

69. Average cost per child = $\dfrac{45+30+21}{12} = \dfrac{96}{12} =$ \$8.

Review

71. $\dfrac{7}{8}+\dfrac{7}{12}=\dfrac{7\cdot 3}{8\cdot 3}+\dfrac{7\cdot 2}{12\cdot 2}=\dfrac{21}{24}+\dfrac{14}{24}=\dfrac{35}{24}$

73. $5\dfrac{3}{8}-2\dfrac{5}{6}=\dfrac{43}{8}-\dfrac{17}{6}=\dfrac{43\cdot 3}{8\cdot 3}-\dfrac{17\cdot 4}{6\cdot 4}=\dfrac{129}{24}-\dfrac{68}{24}$
$=\dfrac{61}{24}$

Section 1.3 Adding and Subtracting Real Numbers

Vocabulary

1. absolute value

3. associative

Concepts

5. a. |- 4| = 4 b. - |-6| = -(6) = -6 c. - |5| = -5

7. a. 8 – 4 = 4 b. -8 – 4 = -12
c. 8 – (- 4) = 8 + 4 = 12
d. -8 – (- 4) = -8 + 4 = - 4

9. Change the sign of the number to be subtracted and add.

Practice

11. |9 + (-11)| = |-2| = 2

13. (-5) + (-7) = -(5 + 7) = -12

15. (-0.4) + 0.9 = 0.9 – 0.4 = 0.5

17. $\dfrac{1}{5}+\dfrac{1}{7}=\dfrac{7}{35}+\dfrac{5}{35}=\dfrac{12}{35}$

19. $\dfrac{-3}{4}+\dfrac{2}{3}=-\left(\dfrac{3}{4}-\dfrac{2}{3}\right)=-\left(\dfrac{9}{12}-\dfrac{8}{12}\right)=-\dfrac{1}{12}$

21. 5 + [4 + (-2)] = 5 + 2 = 7

23. -2 + (- 4 + 5) = -2 + 1 = -1

25. [- 4 + (-3)] + [2 + (-2)] = -7 + 0 = -7

27. - 4 + (-3 + 2) + (-3) = - 4 + (-1) + (-3) = -8

29. - |-9 + (-3)| + (-6) = - |-12| + (-6) = - (12) + (-6) = -12 + (-6) = -18

31. -5.2 + |-2.5 + (- 4)| = -5.2 + |-6.5| = -5.2 + 6.5 = 1.3

33. $\dfrac{5}{3}-\dfrac{7}{6}=\dfrac{10}{6}-\dfrac{7}{6}=\dfrac{3}{6}=\dfrac{1}{2}$

35. $-5-\left(-\dfrac{3}{5}\right)=-5+\dfrac{3}{5}=\dfrac{-25}{5}+\dfrac{3}{5}=\dfrac{-22}{5}$

37. |-6.7 – (-2.5)| = |-6.7 + 2.5| = |- 4.2| = 4.2

39. 3 – [(-4) – 3] = 3 – (-7) = 3 + 7 = 10

41. (5 – 3) + (3 – 5) = 2 + -2 = 0

43. 5 – [4 + (-2) – 5] = 5 – [2 – 5] = 5 – (-3) = 8

45. [5 – (-34)] – [-2 + (-23)] = 39 – (-25) = 64

47. $\left(\dfrac{5}{2}-3\right)-\left(\dfrac{3}{2}-5\right)=\left(\dfrac{5}{2}-\dfrac{6}{2}\right)-\left(\dfrac{3}{2}-\dfrac{10}{2}\right)=$
$-\dfrac{1}{2}-\left(-\dfrac{7}{2}\right)=-\dfrac{1}{2}+\dfrac{7}{2}=\dfrac{6}{2}=3$

49. (x + y) + 3 = (- 4 + 5) + 3 = 1 + 3 = 4

51. $|2x + y| = |2(-4) + 5| = |-8 + 5| = |-3| = 3$

53. $x - y - z = -4 - 5 - (-6) = -4 - 5 + 6 =$
$-9 + 6 = -3$

55. $z - (x - y) + 10 = -6 - (-4 - 5) + 10 =$
$-6 - (-9) + 10 = -6 + 9 + 10 = 13$

57. $\dfrac{y-x}{3-z} = \dfrac{5-(-4)}{3-(-6)} = \dfrac{5+4}{3+6} = \dfrac{9}{9} = 1$

59. $\dfrac{|y+x|}{y-x} = \dfrac{|5+(-4)|}{5-(-4)} = \dfrac{|1|}{5+4} = \dfrac{1}{9}$

61.

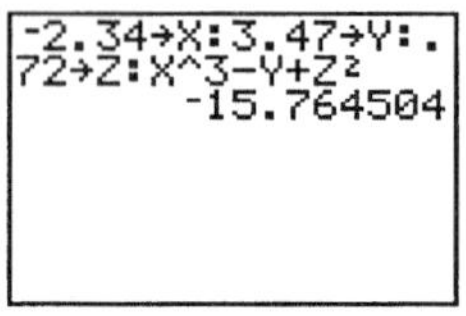

≈ -15.8

63.

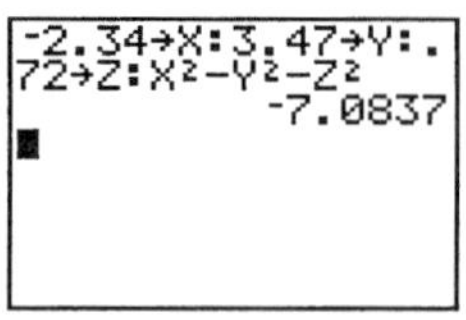

≈ -7.1

Applications

65. The temperature at 5 p.m. was
$0 + (-14°) + 10° = -4°$.

67. They lived 1700 – (-300) = 2000 years apart.

69. 5 + 7 – 5 + 1 – 2 – 6 = 0 yards gained/lost

71. The difference is -3° – (-21°) = -3° + 21° = 18°

73. Amount received = 115,000 – (78 + 446 + 216 + 7612.32 + 23,445.11) + ½ (446) = 115,000 – 31,797.43 + 223 = $83,425.57

75. $28\frac{1}{2} - 19\frac{3}{4} - 5\frac{1}{10} = 28\frac{10}{20} - 19\frac{15}{20} - 5\frac{2}{20} =$
$\frac{570}{20} - \frac{395}{20} - \frac{102}{20} = \frac{73}{20} = 3\frac{13}{20}$ tons left.

77. a. $\left|\frac{1}{2} - \left(-\frac{2}{3}\right)\right| = \left|\frac{3}{6} + \frac{4}{6}\right| = \left|\frac{7}{6}\right| = \frac{7}{6}$

b. $\left|\frac{-4}{5} - \left(-\frac{1}{8}\right)\right| = \left|\frac{-32}{40} + \frac{5}{40}\right| = \left|\frac{-27}{40}\right| = \frac{27}{40}$

79. Various possible answers involving two decreases for negative numbers and two increases for positive numbers.

81. Answers agree when the numbers have the same sign.

Review

83. $2 + 5 \cdot 3^2 = 2 + 5 \cdot 9 = 2 + 45 = 47$

85. $6x^2 - 2y^2 = 6 \cdot 3^2 - 2 \cdot 2^2 = 6 \cdot 9 - 2 \cdot 4 = 54 - 8 =$
46

87. average = (90 + 89 + 82 + 76 +71) ÷ 5

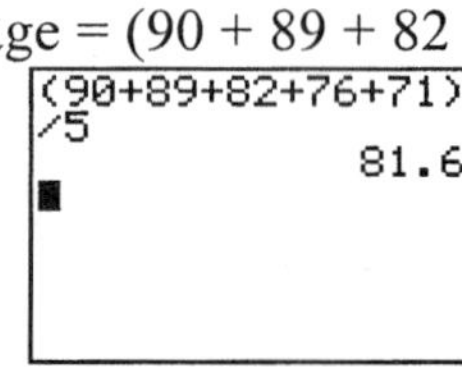

Section 1.4 Multiplying and Dividing Real Numbers

Vocabulary

1. product

Concepts

3. The factors and divisors are the same.

5. Multiply the absolute values. Use a minus sign if only one factor is negative, otherwise a (understood) plus sign.

7. Divide the absolute values. If the number of minus signs is odd, the answer will have a minus sign. Otherwise, it will have a plus sign.

9. If the quotient of two numbers is undefined, the denominator must be 0, and hence the product of the two numbers is 0.

Practice

11. $\left(\frac{1}{2}\right)(32)=\frac{32}{2}=16$

13. $\left(\frac{-3}{4}\right)\left(\frac{-8}{3}\right)=\frac{24}{12}=2$

15. $3(-4)(-6) = 3(24) = 72$

17. $(-2)(3)(4) = -(2\cdot 3\cdot 4) = -24$

19. $\frac{8-12}{-2}=\frac{-4}{-2}=\frac{4}{2}=2$

21. $\frac{20-25}{7-12}=\frac{-5}{-5}=\frac{5}{5}=1$

23. $x^2 = (-1)^2 = 1$

25. $-xz = -(-1)(-3) = -(1\cdot 3) = -3$

27. $z - xy = -3 - (-1)(2) = -3 - (-2) = -3 + 2 = -1$

29. $y(x - z) = 2(-1 - (-3)) = 2(-1 + 3) = 2(2) = 4$

31. $zx - zy = -3(-1) - (-3)(2) = 3 - (-6) = 9$

33. $z^3y = (-3)^3(2) = (-27)(2) = -54$

35. $z(y - x)^2 = -3(2 - (-1))^2 = -3(3^2) = -3\cdot 9 = -27$

37. $\frac{zx}{y}=\frac{(-3)(-1)}{2}=\frac{3}{2}$

39. $\frac{xzy}{y-1}=\frac{(-1)\cdot 2\cdot(-3)}{2-1}=\frac{1\cdot 2\cdot 3}{1}=\frac{6}{1}=6$

41. $\frac{x^2y^3}{yz}=\frac{(-1)^2 2^3}{2(-3)}=\frac{1\cdot 8}{-6}=-\frac{4}{3}$

43. $\frac{y^2+z^2}{y+z}=\frac{2^2+(-3)^2}{2+(-3)}=\frac{4+9}{-1}=\frac{13}{-1}=-13$

45. $x+y=\frac{1}{2}+\frac{-2}{3}=\frac{3}{6}+\frac{-4}{6}=-\frac{1}{6}$

47. $x+y+z=\frac{1}{2}+\frac{-2}{3}+\frac{-3}{4}=\frac{6}{12}+\frac{-8}{12}+\frac{-9}{12}=$
$\frac{6}{12}+\frac{-17}{12}=\frac{-11}{12}$

49. $(x + y)(x - y) =$
$\left(\frac{1}{2}+\frac{-2}{3}\right)\left(\frac{1}{2}-\frac{-2}{3}\right)=\left(\frac{3}{6}+\frac{-4}{6}\right)\left(\frac{3}{6}-\frac{-4}{6}\right)=$
$\frac{-1}{6}\cdot\frac{7}{6}=\frac{-7}{36}$

51. $(x+y+z)(xyz)=\underbrace{\frac{-11}{12}}_{\text{exercise \# 47}}\left(\frac{1}{2}\right)\left(\frac{-2}{3}\right)\frac{-3}{4}=$
$\frac{-11}{12}\cdot\frac{6}{24}=-\frac{66}{288}=-\frac{11}{48}$

53. An odd number of factors could be negative.

Applications

55. He began with $94 - 8\cdot 8 = 94 - 64 = \$30$

57. $\frac{-37.5\text{ lbs.}}{-2.5\text{ lbs/week}}=15\text{ weeks}$

59. $\frac{18+21+33+(-5)+(-71)}{5}=-0.8$ The average loss was 8/10 of one million or \$800,000 per year.

61. $2.40+\frac{3}{4}+2-1\frac{1}{2}-\frac{7}{8}+\frac{1}{4}-\frac{3}{8}=2.40+\frac{6}{8}+\frac{16}{8}$
$-\frac{12}{8}-\frac{7}{8}+\frac{2}{8}-\frac{3}{8}=2.40+\frac{24}{8}-\frac{2}{8}=2.40+\frac{2}{8}=$
$2.40+.25=2.65$
The stock is worth \$2.65 after 6 months.

Review

63. $\frac{-2}{3}+\frac{4}{5}=\frac{-10}{15}+\frac{12}{15}=\frac{2}{15}$

65. Let y = -2 and x = - ½ .
a. x + y = - ½ + (-2) = -5/2
b. x – y = - ½ – (-2) = - ½ + 2 = 3/2

Section 1.5 Basic Applications

Vocabulary

1. irrational

3. area

5. volume

Concepts

7. a. 3 b. 0, 3 c. -3 d. all but $\sqrt{2}$ e. $\sqrt{2}$

9. $\frac{2}{3}=0.\bar{6}$ The value 0.67 is a two-digit decimal approximation.

11. a. 5% = .05 b. 4 ½ % = .045
c. 8 ¾ % = 8.75% = .0875

Practice

13. P = 2(10) + 2(3) cm. = 26 cm.

15. P = 4· 4 in. = 16 in.

17. A = ½ (8)(4) cm^2 = 16 cm^2

19. A = ½ (12) (16 + 22) = 6(38) $cm.^2$ = 228 $cm.^2$

21. C = π (21) m. = 65.97344573 cm.

23. A = π (7^2) $m.^2 \approx 153.938$ $m.^2$

25. V = 6 · 3 · 2 ft^3 = 36 $ft.^3$

27. $V=\frac{1}{3}\pi\left(\frac{12}{2}\right)\cdot 14\text{ in.}^3 \approx 527.788\text{ in.}^3$

29. V = 6 · 3· 4 $in.^3$ + ⅓· 6 ·3 · 4$in.^3$ =
(72 + 24)$in.^3$ = 96 $in.^3$

Applications

31. A = 52^2 $in.^2$ = 2704 $in.^2$

33. V = 36(24)(10¼) $in.^3$ = 8856 $in.^3$

35. V = π $(1.5)^2$ (4.5) $in.^3 \approx 31.81$ $in.^3$

37. In square feet, the area of the 4 sides of the coop is A = 2(6·12) + 2(6·10) = 2(72) + 2(60) = 144 + 120 = 264 $ft.^2$ The netting will cover an area of 2·50 = 100 $ft.^2$, so one roll is not enough.

39. Area to be papered = 2(8.5)(17.3) $ft.^2$ = 294.1 $ft.^2$. Total rolls needed = 294.1/33 ≈8.9, so 9 rolls must be bought. Total cost is 9(27.50) = \$247.50.

41. Area to be carpeted = 23(17.5) $ft.^2$ + 17.5(14) $ft.^2$ = 647.5 $ft.^2$
Square yards of carpet needed = 647.5/9 ≈71.94 so 72 square yards must be bought.
Total cost = 72(29.79) = \$2144.88.

43. a. 2° C b. 50° F
c. 72° C is not displayed exactly in the table.

45. a. I = 2300(.0375)2 = \$172.50
b. I = 2300(.0375)(6/12) = \$43.13

47. Done by phone: .24(52,000) = 12,480

49. Don't snack: .06(871) ≈ 52 kids

51. Gallons of gas used = $\dfrac{15{,}675.2 \text{ miles}}{25.5 \text{ miles/gallon}}$ = 614.71 gal.
Total cost = (614.71)(1.27) = \$780.68

53. Holstein: 7600(.035) = 266 gal. of butterfat
Guernsey: 6500(.05) = 325 gal. of butterfat
The Guernsey produces more butterfat.

55. Cost for second contractor = 150(27.50) + 4500 = \$8625. Thus, he has the lower bid.

Review

57. a. $-3^2 = -3\cdot3 = -9$ b. $(-3)^2 = (-3)(-3) = 9$
c. $3^2 = 3\cdot3 = 9$ d. $4\cdot3^2 = 4\cdot3\cdot3 = 36$

59. $4 - 2\frac{7}{9} = 3\frac{9}{9} - 2\frac{7}{9} = 1\frac{2}{9}$ gallons left over

Chapter 1 Summary

1. a. $\frac{1}{3}+\frac{1}{7}=\frac{1\cdot7}{3\cdot7}+\frac{1\cdot3}{7\cdot3}=\frac{7}{21}+\frac{3}{21}=\frac{10}{21}$

b. $\frac{2}{3}-\frac{1}{7}=\frac{2\cdot7}{3\cdot7}-\frac{1\cdot3}{7\cdot3}=\frac{14}{21}-\frac{3}{21}=\frac{11}{21}$

c. $\frac{31}{15}\cdot\frac{10}{62}=\frac{31\cdot2\cdot5}{3\cdot5\cdot2\cdot31}=\frac{1}{3}$

d. $\frac{18}{21}\div\frac{6}{7}=\frac{18}{21}\cdot\frac{7}{6}=\frac{3\cdot6\cdot7}{3\cdot7\cdot6}=1$

e. $3\frac{2}{3}+4\frac{1}{2}=\frac{11}{3}+\frac{9}{2}=\frac{11\cdot2}{3\cdot2}+\frac{9\cdot3}{2\cdot3}=\frac{22}{6}+\frac{27}{6}$
$=\frac{49}{6}=8\frac{1}{6}$

f. $6\frac{4}{5}-2\frac{1}{3}=\frac{34}{5}-\frac{7}{3}=\frac{34\cdot3}{5\cdot3}-\frac{7\cdot5}{3\cdot5}=\frac{102}{15}-\frac{35}{15}$
$=\frac{67}{15}=4\frac{7}{15}$

g. $3\frac{1}{2}\cdot4\frac{1}{3}=\frac{7}{2}\cdot\frac{13}{3}=\frac{7\cdot13}{2\cdot3}=\frac{91}{6}=15\frac{1}{6}$

h. $3\frac{3}{4}\div1\frac{5}{8}=\frac{15}{4}\div\frac{13}{8}=\frac{15}{4}\cdot\frac{8}{13}=\frac{15\cdot2\cdot4}{4\cdot13}=\frac{30}{13}$

2. a. $1-\left(\frac{1}{5}+\frac{2}{3}\right)=1-\left(\frac{3}{15}+\frac{10}{15}\right)=1-\frac{13}{15}=\frac{2}{15}$

b. First day: $\frac{1}{5}(600)=120$

Second day: $\frac{2}{3}(600)=400$

Third day: $\frac{2}{15}(600)=80$

3. a. $3^4 = 3\cdot3\cdot3\cdot3 = 81$ b. $\left(\frac{2}{3}\right)^2=\frac{2}{3}\cdot\frac{2}{3}=\frac{4}{9}$
c. $-5^2 = -(5\cdot5) = -25$ d. $(-5)^2 = (-5)(-5) = 25$

4. a. $4 + (8 \div 4) = 4 + 2 = 6$
b. $(4 + 8) \div 4 = 12 \div 4 = 3$
c. $(5-2)^2 + 5^2 + 2^2 = 3^2 + 5^2 + 2^2 =$
$9 + 25 + 4 = 38$

d. $\frac{4\cdot3+3^4}{31}=\frac{12+81}{31}=\frac{93}{31}=3$

e. $\left(\frac{3}{2}\right)^2-\frac{1}{3}=\frac{9}{4}-\frac{1}{3}=\frac{27}{12}-\frac{4}{12}=\frac{23}{12}$

f. $\frac{4}{3}\cdot\frac{9}{2}+\frac{1}{2}\cdot18=\frac{36}{6}+\frac{18}{2}=6+9=15$

5. a. $\left(\frac{-1}{4}\right)^3$ b. $\frac{4-8}{-12+1}$ c. $4-\frac{8}{-12}+1$

6. Average time spent = $\frac{(5.2+4.7+9.5+8)}{4}$ = 27.4/4 = 6.85 hours

7. a. $1-[5-(-3)] = 1-8 = -7$

b. $\left|\frac{3}{7}-\frac{-4}{7}\right|=\left|\frac{3}{7}+\frac{4}{7}\right|=\left|\frac{7}{7}\right|=1$

c. $\frac{3}{7}-\left|-\frac{4}{7}\right|=\frac{3}{7}-\frac{4}{7}=-\frac{1}{7}$

d. $\frac{2}{3}-\left(\frac{1}{3}-\frac{2}{3}\right)=\frac{2}{3}-\left(-\frac{1}{3}\right)=\frac{2}{3}+\frac{1}{3}=\frac{3}{3}=1$

e. $-\frac{3}{14}\cdot\left(-\frac{7}{6}\right)=\frac{3\cdot7}{14\cdot6}=\frac{3\cdot7}{2\cdot7\cdot2\cdot3}=\frac{1}{4}$

f. $-\frac{1}{2}\cdot\frac{4}{3}=-\frac{1\cdot4}{2\cdot3}=-\frac{4}{6}=-\frac{2}{3}$

8. a. $\left(\frac{-3+(-3)}{3}\right)\left(-\frac{15}{5}\right)=\left(\frac{-6}{3}\right)(-3)=$ (-2)(-3) = 6.

b. $\frac{-2-(-8)}{5+(-1)}=\frac{-2+8}{4}=\frac{6}{4}=\frac{3}{2}$

c. $\frac{[-3+(-4)]^2}{10+(-3)}=\frac{(-7)^2}{7}=\frac{49}{7}=7$

d. $\left(\frac{-10}{2}\right)^2-(-1)^3=(-5)^2-(-1)=25+1=$ 26

9. y + z = -3 + (-1) = -(3 +1) = - 4

10. x + y = 2 + (-3) = -(3 – 2) = -1

11. x + (y + z) = 2 + (-3 + (-1)) = 2 + (- 4) = -2

12. x – y = 2 – (-3) = 2 + 3 = 5

13. x – (y – z) = 2 – [-3 – (-1)] = 2 – [-3 + 1] = 2 – (-2) = 2 + 2 = 4

14. (x – y) – z = [2 – (-3)] – (-1) = [2 + 3] + 1 = 5 + 1 = 6

15. xy = 2(-3) = -(2·3) = -6

16. yz = (-3)(-1) = 3·1 = 3

17. x(x + z) = 2[2 + (-1)] = 2(1) = 2

18. xyz = 2(-3)(-1) = 2·3·1 = 6

19. $y^2z + x = (-3)^2(-1) + 2 = 9(-1) + 2 = -9 + 2 = -(9-2) = -7$

20. $yz^3 + (xy)^2 = -3(-1)^3 + [2(-3)]^2 = -3(-1) + (-6)^2 = 3 + 36 = 39$

21. $\frac{xy}{z}=\frac{2(-3)}{-1}=\frac{-6}{-1}=6$

22. $\frac{|xy|}{3z}=\frac{|2(-3)|}{3(-1)}=\frac{|-6|}{-3}=\frac{6}{-3}=-2$

23. $\frac{3y^2-x^2+1}{y|z|}=\frac{3(-3)^2-2^2+1}{(-3)|-1|}=\frac{3\cdot9-4+1}{(-3)\cdot1}=$ $\frac{27-4+1}{-3}=\frac{24}{-3}=-8$

24. $\frac{2y^2-xyz}{x^2|yz|}=\frac{2(-3)^2-2(-3)(-1)}{2^2|(-3)(-1)|}=\frac{2\cdot9-6}{4|3|}=$ $\frac{18-6}{12}=\frac{12}{12}=1$

25. Strapping needed = [2(4.2) + 2(2.7)] + [2(4.2) + 2(1.2)] + [2(2.7) + 2(1.2)]·2 = 13.8 + 10.8 + (7.8)·2 = 40.2 feet

26. Volume =
$\pi\left(\frac{32.1}{2}\right)^2(18.7)=\pi(16.05)^2(18.7)\approx$
15,133.576 ft.^{3}

27. $C=\frac{5}{9}(F-32)=\frac{5}{9}(95-32)=\frac{5}{9}(63)=35°$

28. Sale price = 0.80(42.95) = \$34.36
Tax = 0.06(34.36) + \$2.06
Total cost = \$34.36 + \$2.06 = \$36.42

Chapter 1 Test

1. $3(4-2)-2(2-6)=3\cdot 2-2(-4)=6-(-8)=$ $6+8=14$

2. $6+4\cdot 3-2(-7)=6+12-(-14)=6+12+14$ $=32$

3. $-13.7-|-13.7|=-13.7-13.7=-27.4$

4. $\frac{7}{8}\cdot\frac{24}{21}=\frac{7\cdot 3\cdot 8}{8\cdot 3\cdot 7}=1$

5. $\frac{18}{35}\div\frac{9}{14}=\frac{18}{35}\cdot\frac{14}{9}=\frac{2\cdot 9\cdot 2\cdot 7}{5\cdot 7\cdot 9}=\frac{4}{5}$

6. $\frac{24}{16}+3=\frac{24}{16}+\frac{48}{16}=\frac{72}{16}=\frac{9\cdot 8}{2\cdot 8}=\frac{9}{2}$

7. $\frac{17-5}{36}-\frac{2(13-5)}{12}=\frac{12}{36}-\frac{2\cdot 8}{12}=\frac{12}{36}-\frac{16\cdot 3}{12\cdot 3}=$
$\frac{12}{36}-\frac{48}{36}=\frac{-36}{36}=-1$

8. $\frac{|-7-(-6)|}{-7-|-6|}=\frac{|-7+6|}{-7-6}=\frac{|-1|}{-13}=\frac{1}{-13}=-\frac{1}{13}$

9. Area = (12.8)(23.56)ft.^{2} = 301.57 ft.^{2}

10. Area = ½ (16)(8) cm.^{2} = 64 cm.^{2}

11. Volume =
$\pi\left(\frac{14}{2}\right)^2\cdot 10=\pi\cdot 7^2\cdot 10=490\pi\approx 1539.380\text{ in}^3$

12. Total interest = 0.055(5000) + 0.03(1500) = 275 + 45 = \$320

13. Heated by gas: 0.517(150,000) = 77,550 homes

14. Volume = (14.5)(14.5)(18.5) in.^{3} = 3889.625 in.^{3}

15. $xy+z=(-2)3+4=-6+4=-2$

16. $x(y+z)=-2(3+4)=-2\cdot 7=-14$

17. $\frac{z+4y}{2x}=\frac{4+4\cdot 3}{2(-2)}=\frac{4+12}{-4}=\frac{16}{-4}=-4$

18. $|x^y-z|=|(-2)^3-4|=|-8-4|=|-12|=12$

19. $x^3+y^2+z=(-2)^3+3^2+4=-8+9+4=5$

20. $|x|-3|y|-4|z|=|-2|-3|3|-4|4|=$
$2-3\cdot 3-4\cdot 4=2-9-16=-23$

Chapter 2 Exponents and Polynomials

Section 2.1 Whole Number Exponents

Vocabulary

1. base; exponent

Concepts

3. base: 4 exponent: 3

5. base: -x exponent: 2

7. base: x exponent: 1

9. base: x exponent: 3

11. $2\cdot2\cdot2 = 2^3$

13. $x\cdot x\cdot x\cdot x\cdot x = x^5$

15. $(2x)(2x)(2x) = (2x)^3$

Practice

17. $3x^3 = 3(-2)^3 = 3(-8) = -24$

19. $(-2y)^4 = (-2 \cdot -3)^4 = 6^4 = 1296$

21. $x^4x^3 = x^{4+3} = x^7$

23. $x^5x^5 = x^{5+5} = x^{10}$

25. $y^3(y^2y^4) = y^3(y^{2+4}) = y^3y^6 = y^9$

27. $4x^2(3x^5) = 4\cdot3\ x^2x^5 = 4\cdot3\ x^{2+5} = 12\ x^7$

29. $(-y^2)(4y^3) = (-1)4\ y^2y^3 = -4\ y^{2+3} = -4y^5$

31. $(3^2)^4 = 3^{2\cdot4} = 3^8$

33. $(y^5)^3 = y^{5\cdot3} = y^{15}$

35. $(a^3)^7 = a^{3\cdot7} = a^{21}$

37. $(x^2x^3)^5 = (x^{2+3})^5 = (x^5)^5 = x^{5\cdot5} = x^{25}$

39. $(2x^5)^2(x^7)^3 = 2^2(x^5)^2(x^7)^3 = 4x^{5\cdot2}x^{7\cdot3} = 4x^{10}x^{21}$
$= 4x^{10+21} = 4x^{31}$

41. $(xy)^3 = x^3y^3$

43. $(r^3\ s^2)^2 = (r^3)^2\ (s^2)^2 = r^6s^4$

45. $(4ab^2)^2 = 4^2a^2(b^2)^2 = 16a^2b^{2\cdot2} = 16a^2b^4$

47. $(-2r^3s^3t)^3 = (-2)^3(r^3)^3(s^3)^3t^3 = -8r^{3\cdot3}s^{3\cdot3}t^3 =$
$-8r^9s^9t^3$

49. $\left(\frac{a}{b}\right)^3 = \frac{a^3}{b^3}$

51. $\left(\frac{x^2}{y^3}\right)^5 = \frac{(x^2)^5}{(y^3)^5} = \frac{x^{10}}{y^{15}}$

53. $\left(\frac{-2a}{b}\right)^5 = \frac{(-2a)^5}{b^5} = \frac{(-2)^5a^5}{b^5} = \frac{-32a^5}{b^5}$

55. $\left(\frac{b^2}{3a}\right)^3 = \frac{(b^2)^3}{(3a)^3} = \frac{b^6}{3^3a^3} = \frac{b^6}{27a^3}$

57. $\frac{x^5}{x^3} = x^{5-3} = x^2$

59. $\frac{x^3y^4}{y\cdot y^2} = \frac{x^3y^4}{y^3} = x^3y^{4-3} = x^3y$

61. $\frac{(ab^2)^3}{(ab)^2} = \frac{a^3(b^2)^3}{a^2b^2} = \frac{a^3b^6}{a^2b^2} = a^{3-2}b^{6-2} = ab^4$

63. $\frac{20(r^4s^3)^4}{6(rs^3)^3} = \frac{20(r^4)^4(s^3)^4}{6r^3(s^3)^3} = \frac{20r^{16}s^{12}}{6r^3s^9} =$

$$\frac{10r^{16-3}s^{12-9}}{3} = \frac{10r^{13}s^{3}}{3}$$

65. $\left(\frac{y^3y}{2yy^2}\right)^3 = \left(\frac{y^4}{2y^3}\right)^3 = \left(\frac{y^{4-3}}{2}\right)^3 = \left(\frac{y}{2}\right)^3 = \frac{y^3}{8}$

67. $2x^0 = 2\cdot1 = 2$

69. $\frac{x^0 - 5x^0}{2x^0} = \frac{1-5\cdot1}{2\cdot1} = \frac{1-5}{2} = -2$

71. No, $2^3 \neq 3^2$, for example.

Applications

73. Area = $(5x)(3x^2) = 15x^3$

75. Area = $\frac{1}{2}(6x)(5x + 10x) = 3x(15x) = 45x^2$

Review

77. a. -9 b. 30 c. -3 d. -3 e. -3 f. 3

79. a. 45% = .45 b. 3% = .03 c. 2.5% = .025

Section 2.2 Algebraic Expressions and Polynomials

Vocabulary

1. algebraic expressions

3. algebraic terms

5. polynomial

7. degree; degree; degree

Concepts

9. b and d are non-equivalent pairs

11. a. trinomial b. degree is 2 c. -5

13. Circle the line with x = -2.

Practice

15. xy

17. 2x + 2y

19. y/x

21. y − x

23. y decreased by 2

25. the product of x and z

27. three times x divided by the sum of y and z

29. binomial

31. monomial

33. neither

35. monomial

37. 5 (the terms have degree 5 and 2)

39. 5 (the terms have degree 5,2 and 1)

41. 6 (the terms have degree 5 = 2 + 3 and 6 = 3 + 2 + 1)

43. 10 (the terms have degree 8 = 2 + 3 + 3 , 10 = 2 + 8 and 0)

45. 1 ($17^2x = 289x^1$)

47. 0 ($-25 = -25x^0$)

49. P(0) = 5(0) – 3 = 0 – 3 = -3

51. P(-2) = 5(-2) – 3 = -10 – 3 = -13

53. P(t) = 5t – 3

55. P(2t) = 5(2t) – 3 = 10t – 3

57. $P(1/10) = 5(1/10) - 2 = ½ - 2 = -3/2$

59. $P(-v^4) = 5(-v^4) - 2 = -5v^4 - 2$

61. $P(10x^7) = 5(10x^7) - 2 = 50x^7 - 2$

63. $P(x - h) = 5(x - h) - 2 = 5x - 5h - 2$

65. $P(x) - P(h) = (5x - 2) - (5h - 2) =$
$5x - 2 - 5h + 2 = 5x - 5h$

67. $P(-3r + 2s) = 5(-3r + 2s) - 2 = -15r + 10s - 2$

Applications

69. m + 25000 miles

71. 35000 a dollars

73. length of each piece = $\frac{12}{x}$

75. Wendy's sister has 2x – 8 shares. Wendy can't own x = 3 shares, since this would mean her sister would own -2 shares.

77. $h(2) = -16(2^2) + 120(2) + 16 =$
$- 64 + 240 + 16 = 192$ ft.

79. $h(7) = -16(7^2) + 120(7) + 16 =$
$- 16(49) + 840 + 16 = 72$ ft.

81. a. base value = 9 b. base value is 16

Section 2.3 Adding and Subtracting Polynomials

Vocabulary

1. distributive

Concepts

3. $3y + 4y = (3 + 4)y = 7y$

5. not like terms, cannot be combined

7. $3x^3 + 4x^3 + 6x^3 = (3 + 4 + 6)x^3 = 13x^3$

9. $32x^5y^3 + (-21x^5y^3) + (-11x^5y^3) = [32 + (-21) + (-11)]x^5y^3 = 0 \cdot x^5y^3 = 0$

11. $4y + 5y = (4 + 5)y = 9y$

13. $-8t^2 - 4t^2 = (-8 - 4)t^2 = -12t^2$

15. $32u^3 - 16u^3 = (32 - 16)u^3 = 16u^3$

17. $18x^5y^2 - 11x^5y^2 = (18 - 11)x^5y^2 = 7x^5y^2$

Practice

19. $3rst + 4rst + 7rst = (3 + 4 + 7)rst = 14rst$

21. $- 4a^2bc + 5a^2bc - 7a^2bc = (- 4 + 5 - 7)a^2bc =$
$-6a^2bc$

23. $3mn^3 - 5mn^3 + 3m^3n = (3 - 5)mn^3 + 3m^3n =$
$-2mn^3 + 3m^3n$

25. Perimeter $= 2(8x) + 2(2x) = 16x + 4x =$
$(16 + 4)x = 20x$

27. Perimeter $= 5x + 8x + 5y = (5 + 8)x + 5y =$
$13x + 5y$

29. $(3x + 7) + (4x - 3) = 3x + 7 + 4x - 3 =$
$3x + 4x + 7 - 3 = (3 + 4)x + 4 = 7x + 4$

31. $(4a + 3) - (2a - 4) = 4a + 3 - 2a + 4 =$
$4a - 2a + 3 + 4 = (4 - 2)a + 7 = 2a + 7$

33. $(3x^2 - 3x - 2) + (3x^2 + 4x - 3) = 3x^2 - 3x - 2$
$+ 3x^2 + 4x - 3 = 3x^2 + 3x^2 - 3x + 4x - 2 - 3 =$
$6x^2 + x - 5$

35. $(3a^2 - 2a + 4) - (a^2 - 3a + 7) = 3a^2 - 2a + 4 -$
$a^2 + 3a - 7 = 3a^2 - a^2 - 2a + 3a + 4 - 7 =$
$2a^2 + a - 3$

37. $5(x + y) + 7(x + y) = 5x + 5y + 7x + 7y = 5x$
$+ 7x + 5y + 7y = 12x + 12y$

39. $- 4(a - b) - 5(a - b) = - 4a + 4b - 5a + 5b =$
$- 4a - 5a + 4b + 5b = -9a + 9b$

41. $-5(y^2 - 2y - 6) + 6(2y^2 + 2y - 5) = -5y^2 + 10y + 30 + 12y^2 + 12y - 30 = -5y^2 + 12y^2 + 10y + 12y + 30 - 30 = 7y^2 + 22y$

43. $3y(xy + y) - 2y^2(x - 4 + y) + 2(y^3 + y^2) = 3xy^2 + 3y^2 - 2xy^2 + 8y^2 - 2y^3 + 2y^3 + 2y^2 = 3xy^2 - 2xy^2 + 3y^2 + 8y^2 + 2y^2 = xy^2 + 13y^2$

45. Perimeter $= 5(2x - 5) + 2x + 2(3x^2 - 4x + 1) = 10x - 25 + 2x + 6x^2 - 8x + 2 = 6x^2 + 10x + 2x - 8x - 25 + 2 = 6x^2 + 4x - 23$

47. Perimeter $= (3x + 1) + (x + 1) + (2x^2 - 3x - 8) + (x - 2) = 3x + 1 + x + 1 + 2x^2 - 3x - 8 + x - 2 = 2x^2 + 3x + x - 3x + x + 1 + 1 - 8 - 2 = 2x^2 + 2x - 8$

49.

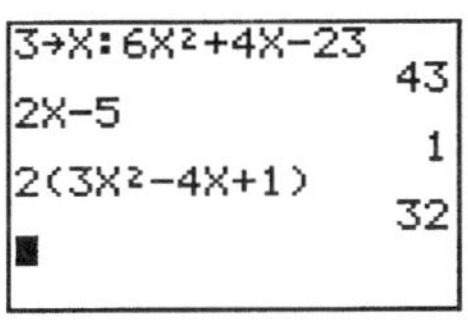

a. perimeter = 43
b. sides: 1, 3, 32
c. It is not a valid value since the horizontal distance across the top is 9, but across the bottom is 32.

51.

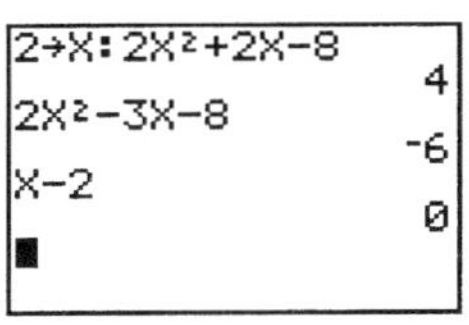

a. perimeter = 4
b. sides are 7, 3, 0, -6 (only two are shown)
c. It is not a valid value since the length of a side cannot be negative or zero

53. $[(2x^2 - 3x + 4) + (3x^2 - 2)] + (x^2 + x - 3) = (2x^2 - 3x + 4 + 3x^2 - 2) + (x^2 + x - 3) = 5x^2 - 3x + 2 + x^2 + x - 3 = 6x^2 - 2x - 1$

55. $[(3t^3 + t^2) + (-t^3 + 6t - 3)] - (t^3 - 2t^2 + 2) = (3t^3 + t^2 - t^3 + 6t - 3) - t^3 + 2t^2 - 2 = 2t^3 + t^2 + 6t - 3 - t^3 + 2t^2 - 2 = t^3 + 3t^2 + 6t - 5$

57. $[(-2x^2 - 7x + 1) + (-4x^2 + 8x - 1)] + (3x^2 + 4x - 7) = (-2x^2 - 7x + 1 - 4x^2 + 8x - 1) + (3x^2 + 4x - 7) = -6x^2 + x + 3x^2 + 4x - 7 = -3x^2 + 5x - 7$

Applications

59. Cost = 75 + 10 x dollars

61. Total interest $= 0.06x + 0.035(5000 - x) = 0.06x + 175 - 0.035x = 0.025x + 175$ dollars

63. Total distance $= 45x + 72(x + 2) = 45x + 72x + 144 = 117x + 144$ miles

65. Total pages $= x + (x - 2) + 40x + (x + 4) = x + x - 2 + 40x + x + 4 = 43x + 2$ pages

Review

67. $4^3 = 4 \cdot 4 \cdot 4$

69. a. 8 1/4% = 8.25% = 0.0825
b. Tax = 0.0825(299.99) = $24.75
c. Since it is money, round to the nearest penny.

Section 2.4 Multiplying Polynomials

Vocabulary

1. conjugate

3. FOIL

Concepts

5. a. $(2x)(3x) = 2{\cdot}3{\cdot}x{\cdot}x = 6x^2$
b. $(-4x^2)(-x) = 4{\cdot}x^2{\cdot}x = 4x^3$
c. $\frac{1}{2}x \cdot \frac{4}{5}x = \frac{2}{5}x \cdot x = \frac{2}{5}x^2$

Practice

7. $(3x^2)(4x^3) = 3{\cdot}4x^{2+3} = 12x^5$

9. $(3b^2)(-2b)(4b^3) = 3(-2)4\ b^{2+1+3} = -24b^6$

11. $(x^2y^3)^5 = (x^2)^5\ (y^3)^5 = x^{10}y^{15}$

13. $(a^3b^2c)(abc^3)^2 = a^3b^2c\ a^2b^2(c^3)^2 = a^{3+2}b^{2+2}\ c{\cdot}c^6 = a^5b^4c^{1+6} = a^5b^4c^7$

15. $3x(x-2) = 3x{\cdot}x - 3x{\cdot}2 = 3x^2 - 6x$

17. $-2x^2(3x^2 - x) = (-2x^2)(3x^2) - (-2x^2)x = -6x^4 + 2x^3$

19. $2x^2(3x^2 + 4x - 7) = 2x^2(3x^2) + 2x^2(4x) - 2x^2{\cdot}7 = 6x^4 + 8x^3 - 14x^2$

21. $\frac{1}{4}x^2(8x^5 - 4) = \frac{1}{4}x^2(8x^5) - \frac{1}{4}x^2{\cdot}4 = 2x^7 - x^2$

23. $(a+4)(a+5) = a{\cdot}a + 4a + 5a + 4{\cdot}5 = a^2 + 9a + 20$

25. $(3x-2)(x+4) = 3x{\cdot}x - 2x + 12x - 8 = 3x^2 + 10x - 8$

27. $(2a+4)(3a-5) = 2a{\cdot}3a + 12a - 10a - 20 = 6a^2 + 2a - 20$

29. $(3x-5)(2x+1) = 3x{\cdot}2x - 10x + 3x - 5 = 6x^2 - 7x - 5$

31. $(2t+3s)(3t-s) = 2t{\cdot}3t + 3s{\cdot}3t - s{\cdot}2t - 3s{\cdot}s = 6t^2 + 9st - 2st - 3s^2 = 6t^2 + 7st - 3s^2$

33. $(x+4)(x+4) = x^2 + 2{\cdot}4{\cdot}x + 4^2 = x^2 + 8x + 16$

35. $(x+5)^2 = x^2 + 2{\cdot}5{\cdot}x + 5^2 = x^2 + 10x + 25$

37. $(r+4)(r-4) = r^2 - 4^2 = r^2 - 16$

39. $(4x+5y)(4x-5y) = (4x)^2 - (5y)^2 = 16x^2 - 25y^2$

41. $(2a-3b)^2 = (2a)^2 - 2(2a)(3b) + (3b)^2 = 4a^2 - 12ab + 9b^2$

43. Volume $= 2(x+1)(x-4) = 2(x^2 + x - 4x - 4) = 2(x^2 - 3x - 4) = 2x^2 - 6x - 8$ cubic units

45. Area $= (4x+3)(x^2+2x+3) = (4x+3)x^2 + (4x+3)(2x) + (4x+3){\cdot}3 = 4x^3 + 3x^2 + 8x^2 + 6x + 12x + 9 = 4x^3 + 11x^2 + 18x + 9$

47. Area $= (x-2y)(x^2+2xy+4y^2) = (x-2y)x^2 + (x-2y)2xy + (x-2y)4y^2 = x^3 - 2x^2y + 2x^2y - 4xy^2 + 4xy^2 - 8y^3 = x^3 - 8y^3$

49. $2t(t+2) + 3t(t-5) = 2t^2 + 4t + 3t^2 - 15t = 5t^2 - 11t$

51. $3xy(x+y) - 2x(xy-x) = 3x^2y + 3xy^2 - 2x^2y + 2x^2 = x^2y + 3xy^2 + 2x^2$

53. $(x+y)(x-y) + x(x+y) = x^2 - y^2 + x^2 + xy = 2x^2 + xy - y^2$

55. $(x+2)^2 - (x-2)^2 = x^2 + 2{\cdot}2{\cdot}x + 2^2 - (x^2 - 2{\cdot}2{\cdot}x + 2^2) = x^2 + 4x + 4 - x^2 + 4x - 4 = 8x$

57. $(2s-3)(s+2) + (3s+1)(s-3) = (2s^2 - 3s + 4s - 6) + (3s^2 + s - 9s - 3) = 2s^2 + s - 6 + 3s^2 - 8s - 3 = 5s^2 - 7s - 9$

Applications

59. Area $= x(5+x) = 5x + x^2$ square units

61. Volume $= x(x+2)(x-2) = x(x^2 - 2^2) = x(x^2 - 4) = x^3 - 4x^2$ cubic units

63. Product $= x(x+2)(x+4) = x(x^2 + 2x + 4x + 8) = x(x^2 + 6x + 8) = x^3 + 6x^2 + 8x$

65. $x^2 + (x+2)^2 = x^2 + (x^2 + 2{\cdot}2{\cdot}x + 2^2) = x^2 + x^2 + 4x + 4 = 2x^2 + 4x + 4$

67. Area of desktop $= (2x^2 - 5x)(3x^2 + 2x) = 6x^4 - 15x^3 + 4x^3 - 10x^2 = 6x^4 - 11x^3 - 10x^2$. Area not covered by the calendar $= 6x^4 - 11x^3 - 10x^2 - (12x^2 + 17x - 5) = 6x^4 - 11x^3 - 10x^2 - 12x^2 - 17x + 5 = 6x^4 - 11x^3 - 22x^2 - 17x + 5$ square units

Review

69. Let Y1 = 65x. Set TblStart = .5, ΔTbl = .5

X	Y_1
.5	32.5
1	65
1.5	97.5
2	130
2.5	162.5
3	195
3.5	227.5

X=2.5

71. Total distance = 70t + 60t = 130t miles

Section 2.5 Dividing Polynomials by Monomials

Concepts

1. $\frac{8x}{2}=\frac{2\cdot 4x}{2}=4x$

3. $\frac{xy}{yx}=\frac{xy}{xy}=1$

5. $\frac{4x+12y}{4}=\frac{1}{4}(4x+12y)=\frac{1}{4}(4x)+\frac{1}{4}(12y)=$
$x+3y$

7. $\frac{7x+9xy}{x}=\frac{1}{x}(7x+9xy)=\frac{1}{x}(7x)+\frac{1}{x}(9xy)$
$=7+9y$

Practice

9. $\frac{r^3s^2}{rs^3}=\frac{r^3}{r}\cdot\frac{s^2}{s^3}=r^{3-1}s^{2-3}=r^2s^{-1}=\frac{r^2}{s}$

11. $\frac{8x^3y^2}{4xy^3}=\frac{8}{4}\cdot\frac{x^3}{x}\cdot\frac{y^2}{y^3}=2x^{3-1}y^{2-3}=2x^2y^{-1}=$
$\frac{2x^2}{y}$

13. $\frac{12u^5v}{-4u^2v^2}=\frac{12}{-4}\cdot\frac{u^5}{u^2}\cdot\frac{v}{v^2}=-3u^{5-2}v^{1-2}=-3u^3v^{-1}$
$=\frac{-3u^3}{v}$

15. $\frac{-16r^3y^2}{-4r^2y^3}=\frac{-16}{-4}\cdot\frac{r^3}{r^2}\cdot\frac{y^2}{y^3}=4r^{3-2}y^{2-3}=4ry^{-1}=$
$\frac{4r}{y}$

17. $\frac{-65rs^2t}{15r^2s^3t}=\frac{-65}{15}\cdot\frac{r}{r^2}\cdot\frac{s^2}{s^3}=\frac{-13}{3}\cdot r^{1-2}s^{2-3}=$
$\frac{-13}{3}r^{-1}s^{-1}=\frac{-13}{3rs}$

19. $\frac{x^2x^3}{xy}=\frac{x^5}{xy}=\frac{x^{5-1}}{y}=\frac{x^4}{y}$

21. $\frac{(a^3b^4)^3}{ab^4}=\frac{(a^3)^3(b^4)^3}{ab^4}=\frac{a^9b^{12}}{ab^4}=a^{9-1}b^{12-4}=$
a^8b^8

23. $\frac{15(r^2s^3)^2}{-5(rs^2)^3}=\frac{15}{-5}\cdot\frac{(r^2)^2(s^3)^2}{r^3(s^2)^3}=-3\frac{r^4s^6}{r^3s^6}=-3r$

25.
$$\frac{-32(x^3y)^3}{128(x^2y)^3}=\frac{-32}{128}\cdot\frac{(x^3)^3y^3}{(x^2)^3y^3}=\frac{-1}{4}\cdot\frac{x^9y^3}{x^6y^3}$$
$$=\frac{-1}{4}x^{9-6}y^{3-3}=\frac{-1}{4}x^3y^0=\frac{-x^3}{4}$$

27. $\frac{(5a^2b)^3}{(2a^2b)^3}=\frac{5^3(a^2)^3b^3}{2^3(a^2)^3b^3}=\frac{125a^6b^3}{8a^6b^3}=\frac{125}{8}$

29. $\frac{-(3x^3y^4)^3}{-(9x^4y^5)^2}=\frac{-3^3(x^3)^3(y^4)^3}{-9^2(x^4)^2(y^5)^2}=\frac{-27x^9y^{12}}{-81x^8y^{10}}=$
$\frac{x^{9-8}y^{12-10}}{3}=\frac{xy^2}{3}$

31. $\frac{(a^2a^3)^4}{(a^4)^3} = \frac{(a^5)^4}{a^{12}} = \frac{a^{20}}{a^{12}} = a^{20-12} = a^8$

33. $\frac{(z^3z^4)^3}{(z^3)^2} = \frac{(z^7)^3}{z^6} = \frac{z^{21}}{z^6} = z^{21-6} = z^{15}$

35. $\frac{6x+9y}{3xy} = \frac{1}{3xy}(6x+9y) = \frac{6x}{3xy} + \frac{9y}{3xy} =$
$\frac{2}{y} + \frac{3}{x}$

37. $\frac{5x-10y}{25xy} = \frac{1}{25xy}(5x-10y) = \frac{5x}{25xy} - \frac{10y}{25xy} =$
$\frac{1}{5y} - \frac{2}{5x}$

39. $\frac{3x^2+6y^3}{3x^2y^2} = \frac{1}{3x^2y^2}(3x^2+6y^3) = \frac{3x^2}{3x^2y^2} +$
$\frac{6y^3}{3x^2y^2} = \frac{1}{y^2} + \frac{2y}{x^2}$

41. $\frac{15a^3b^2-10a^2b^3}{5a^2b^2} = \frac{1}{5a^2b^2}(15a^3b^2-10a^2b^3) =$
$\frac{15a^3b^2}{5a^2b^2} - \frac{10a^2b^3}{5a^2b^2} = 3a-2b$

43. $\frac{4x-2y+8z}{4xy} = \frac{1}{4xy}(4x-2y+8z) = \frac{4x}{4xy} -$
$\frac{2y}{4xy} + \frac{8z}{4xy} = \frac{1}{y} - \frac{1}{2x} + \frac{2z}{xy}$

45. $\frac{12x^3y^2-8x^2y-4x}{4xy} = \frac{1}{4xy}(12x^3y^2-8x^2y-4x) =$
$\frac{12x^3y^2}{4xy} - \frac{8x^2y}{4xy} - \frac{4x}{4xy} = 3x^2y - 2x - \frac{1}{y}$

47. $\frac{-25x^2y+30xy^2-5xy}{-5xy} = \frac{1}{-5xy}(-25x^2y +$
$30xy^2 - 5xy) = \frac{-25x^2y}{-5xy} + \frac{30xy^2}{-5xy} - \frac{5xy}{-5xy} =$
$5x - 6y + 1$

49. $\frac{5x(4x-2y)}{2y} = \frac{20x^2-10xy}{2y} = \frac{20x^2}{2y} - \frac{10xy}{2y} =$
$\frac{10x^2}{y} - 5x$

51. $\frac{(-2x)^2+(3x^2)^2}{6x^2} = \frac{4x^2+9x^4}{6x^2} = \frac{1}{6x^2}(4x^2 +$
$9x^4) = \frac{4x^2}{6x^2} + \frac{9x^4}{6x^2} = \frac{2}{3} + \frac{3x^2}{2}$

53. $\frac{4x^2y^2-2(x^2y^2+xy)}{2xy} = \frac{4x^2y^2-2x^2y^2-2xy}{2xy}$
$= \frac{1}{2xy}(2x^2y^2-2xy) = \frac{2x^2y^2}{2xy} - \frac{2xy}{2xy} = xy-1$

55. $\frac{(3x-y)(2x-3y)}{6xy} = \frac{6x^2-2xy-9xy+3y^2}{6xy} =$
$\frac{6x^2-11xy+3y^2}{6xy} = \frac{1}{6xy}(6x^2-11xy+3y^2) =$
$\frac{6x^2}{6xy} - \frac{11xy}{6xy} + \frac{3y^2}{6xy} = \frac{x}{y} - \frac{11}{6} + \frac{y}{2x}$

57. $\frac{(a+b)^2-(a-b)^2}{2ab} = \frac{a^2+2ab+b^2}{2ab} -$
$\frac{a^2-2ab+b^2}{2ab} = \frac{a^2+2ab+b^2-a^2+2ab-b^2}{2ab}$
$= \frac{4ab}{2ab} = 2$

Applications

59. length of a side = $\dfrac{8x+12}{4} = \dfrac{8x}{4} + \dfrac{12}{4} = 2x + 3$

61. width =

$$\frac{8x^3+2x^2-12}{4x^2} = \frac{8x^3}{4x^2} + \frac{2x^2}{4x^2} - \frac{12}{4x^2} = 2x + \frac{1}{2} - \frac{3}{x^2}$$

63. depth = $\dfrac{18x^4-12x^2+8}{(2x)(3x^2)} = \dfrac{18x^4-12x^2+8}{6x^3} =$

$$\frac{18x^4}{6x^3} - \frac{12x^2}{6x^3} + \frac{8}{6x^3} = 3x - \frac{2}{x} + \frac{4}{3x^3}$$

65.

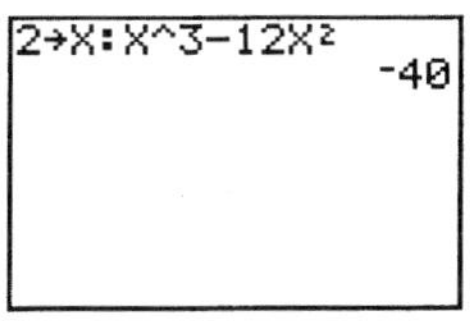

x = 2 is not a valid value since it gives a negative value for the perimeter.

67.

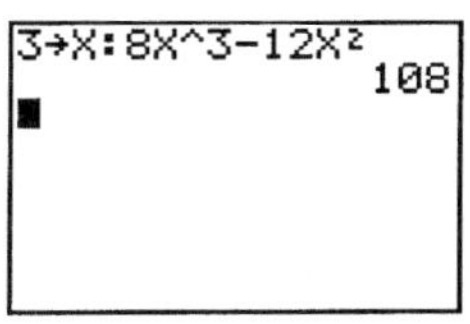

For x = 3, the perimeter = 108 so that each side would have length ¼ (108) = 27. Thus, x = 3 is a valid value.

Review

69. a. $P(x) = 2x - 5$ and b. $P(3) = 1$

Section 2.6 Dividing Polynomials by Polynomials

Concepts

1. a. $\dfrac{6x^3}{2x} = 3x^2$ b. $\dfrac{8x^2}{2x} = 4x$ c. $\dfrac{-4x}{2x} = -2$

d. $\dfrac{6x^3+8x^2-4x}{2x} = \dfrac{6x^3}{2x} + \dfrac{8x^2}{2x} - \dfrac{4x}{2x} = 3x^2 + 4x - 2$

3. a. $x^2 + 0 \cdot x - 9$ b. $x^3 + 0 \cdot x^2 + 0 \cdot x - 8$

Practice

5.
$$\begin{array}{r} x-3 \\ x-2\overline{)x^2-5x+6} \\ \underline{x^2-2x} \\ -3x+6 \\ \underline{-3x+6} \\ 0 \end{array}$$

quotient: $x - 3$ remainder: 0

7.
$$\begin{array}{r} z-4 \\ z-3\overline{)z^2-7z+12} \\ \underline{z^2-3z} \\ -4z+12 \\ \underline{-4z+12} \\ 0 \end{array}$$

quotient: $z - 4$ remainder: 0

9.
$$\begin{array}{r} a-b \\ a-b\overline{)a^2-2ab+b^2} \\ \underline{a^2-ab} \\ -ab+b^2 \\ \underline{-ab+b^2} \\ 0 \end{array}$$

quotient: $a - b$ remainder: 0

11.
$$\begin{array}{r} 4a+3 \\ 2a-1\overline{)8a^2+2a-3} \\ \underline{8a^2-4a} \\ 6a-3 \\ \underline{6a-3} \\ 0 \end{array}$$

quotient: $4a + 3$ remainder: 0

13. $$\begin{array}{r} b-1 \\ 3b-2\overline{)3b^2-5b+2} \\ \underline{3b^2-2b} \\ -3b+2 \\ \underline{-3b+2} \\ 0 \end{array}$$

quotient: $b-1$ remainder: 0

15. $$\begin{array}{r} 3x-y \\ x+2y\overline{)3x^2+5xy-2y^2} \\ \underline{3x^2+6xy} \\ -xy-2y^2 \\ \underline{-xy-2y^2} \\ 0 \end{array}$$

quotient: $3x-y$ remainder: 0

17. $$\begin{array}{r} x+3 \\ 2x-7\overline{)2x^2-x-21} \\ \underline{2x^2-7x} \\ 6x-21 \\ \underline{6x-21} \\ 0 \end{array}$$

quotient: $x+3$ remainder: 0

19. $$\begin{array}{r} 3x+1 \\ 3x+1\overline{)9x^2+6x+1} \\ \underline{9x^2+3x} \\ 3x+1 \\ \underline{3x+1} \\ 0 \end{array}$$

quotient: $3x+1$ remainder: 0

21. $$\begin{array}{r} 2x-y \\ x+2y\overline{)2x^2+3xy-2y^2} \\ \underline{2x^2+4xy} \\ -xy-2y^2 \\ \underline{-xy-2y^2} \\ 0 \end{array}$$

quotient: $2x-y$ remainder: 0

23. $$\begin{array}{r} 5x-4y \\ 2x+3y\overline{)10x^2+7xy-12y^2} \\ \underline{10x^2+15xy} \\ -8xy-12y^2 \\ \underline{-8xy-12y^2} \\ 0 \end{array}$$

quotient: $5x-4y$ remainder: 0

25. $$\begin{array}{r} x^2+2x-1 \\ 2x+3\overline{)2x^3+7x^2+4x-3} \\ \underline{2x^3+3x^2} \\ 4x^2+4x \\ \underline{4x^2+6x} \\ -2x-3 \\ \underline{-2x-3} \\ 0 \end{array}$$

quotient: x^2+2x-1 remainder: 0

27. $$\begin{array}{r} 2x^2+2x+1 \\ 3x+2\overline{)6x^3+10x^2+7x+2} \\ \underline{6x^3+4x^2} \\ 6x^2+7x \\ \underline{6x^2+4x} \\ 3x+2 \\ \underline{3x+2} \\ 0 \end{array}$$

quotient: $2x^2+2x+1$ remainder: 0

29. $$\begin{array}{r} x^2+xy+y^2 \\ 2x+y\overline{)2x^3+3x^2y+3xy^2+y^3} \\ \underline{2x^3+x^2y} \\ 2x^2y+3xy^2 \\ \underline{2x^2y+xy^2} \\ 2xy^2+y^3 \\ \underline{2xy^2+y^3} \\ 0 \end{array}$$

quotient: $x^2 + xy + y^2$ remainder: 0

31. $$\begin{array}{r} x+1 \\ 2x+3\overline{)2x^2+5x+2} \\ \underline{2x^2+3x} \\ 2x+2 \\ \underline{2x+3} \\ -1 \end{array}$$

answer: $x+1+\dfrac{-1}{2x+3}$

33. $$\begin{array}{r} 2x+2 \\ 2x+1\overline{)4x^2+6x-1} \\ \underline{4x^2+2x} \\ 4x-1 \\ \underline{4x+2} \\ -3 \end{array}$$

answer: $2x + 2 + \dfrac{-3}{2x+1}$

35. $$\begin{array}{r} x^2+2x+1 \\ x+1\overline{)x^3+3x^2+3x+1} \\ \underline{x^3+x^2} \\ 2x^2+3x \\ \underline{2x^2+2x} \\ x+1 \\ \underline{x+1} \\ 0 \end{array}$$

answer: x^2+2x+1

37. $$\begin{array}{r} x^2+2x-1 \\ 2x+3\overline{)2x^3+7x^2+4x+3} \\ \underline{2x^3+3x^2} \\ 4x^2+4x \\ \underline{4x^2+6x} \\ -2x+3 \\ \underline{-2x-3} \\ 6 \end{array}$$

answer: $x^2+2x-1+\dfrac{6}{2x+3}$

39. $$\begin{array}{r} 2x^2+8x+14 \\ x-2\overline{)2x^3+4x^2-2x+3} \\ \underline{2x^3-4x^2} \\ 8x^2-2x \\ \underline{8x^2-16x} \\ 14x+3 \\ \underline{14x-28} \\ 31 \end{array}$$

answer : $2x^2+8x+14+\dfrac{31}{x-2}$

41. $$\begin{array}{r} x+1 \\ x-1\overline{)x^2+0\cdot x-1} \\ \underline{x^2-x} \\ x-1 \\ \underline{x-1} \\ 0 \end{array}$$

quotient: $x + 1$ remainder: 0

43. $$\begin{array}{r} 2x-3 \\ 2x+3\overline{)4x^2+0\cdot x-9} \\ \underline{4x^2+6x} \\ -6x-9 \\ \underline{-6x-9} \\ 0 \end{array}$$

quotient: $2x - 3$ remainder: 0

45. $x+1\overline{\big)x^3+0\cdot x^2+0\cdot x+1}$ with quotient x^2-x+1

$$\begin{array}{r} x^3+x^2 \\ \hline -x^2+0\cdot x \\ -x^2-x \\ \hline x+1 \\ x+1 \\ \hline 0 \end{array}$$

quotient: x^2-x+1

47. $a+3\overline{\big)a^3+0\cdot a^2+a+0}$ with quotient $a^2-3a+10$

$$\begin{array}{r} a^3+3a^2 \\ \hline -3a^2+a \\ -3a^2-9a \\ \hline 10a+0 \\ 10a+30 \\ \hline -30 \end{array}$$

$$\text{answer}: a^2-3a+10+\frac{-30}{a+3}$$

49. $3x-4\overline{\big)15x^3-23x^2+16x+0}$ with quotient $5x^2-x+4$

$$\begin{array}{r} 15x^3-20x^2 \\ \hline -3x^2+16x \\ -3x^2+4x \\ \hline 12x+0 \\ 12x-16 \\ \hline 16 \end{array}$$

$$\text{answer}: 5x^2-x+4+\frac{16}{3x-4}$$

Applications

51. length of a side $= \dfrac{4x^2-20x+16}{4} =$

$$\frac{4x^2}{4}-\frac{20x}{4}+\frac{16}{4}=x^2-5x+4$$

53. length of a side $= \dfrac{27x^3-6x^2+9}{3} =$

$$\frac{27x^3}{3}-\frac{6x^2}{3}+\frac{9}{3}=9x^3-2x^2+3$$

55. $x+5\overline{\big)x^2+12x+35}$ with quotient $x+7$

$$\begin{array}{r} x^2+5x \\ \hline 7x+35 \\ 7x+35 \\ \hline 0 \end{array}$$

a. The width is $x + 7$

b. The perimeter is $2(x + 7) + 2(x + 5) =$
$2x + 14 + 2x + 10 = 4x + 24$

57. $V = lwh$, so $h = V/lw$. $lw = (6x^4)(2x + 5) = 12x^5 + 30x^4$, and V/lw is given by

$12x^5+30x^4\overline{\big)12x^6+54x^5+60x^4}$ with quotient $x+2$

$$\begin{array}{r} 12x^6+30x^5 \\ \hline 24x^5+60x^4 \\ 24x^5+60x^4 \\ \hline 0 \end{array}$$

The depth (height) is $x + 2$ feet.

Review

59. Volume $= (2x - 3)(x + 2)3x = (2x^2 - 3x + 4x - 6)3x = (2x^2 + x - 6)3x = 6x^3 + 3x^2 - 18x$

61. a. $(4x^2)^3 = 4^3(x^2)^3 = 64x^6$

b. $\dfrac{4x^3y}{8x^2y^2}=\dfrac{1}{2}x^{3-2}y^{1-2}=\dfrac{1}{2}xy^{-1}=\dfrac{x}{2y}$

c. $\dfrac{(2x)^2}{(6x)^3}=\dfrac{4x^2}{216x^3}=\dfrac{1}{54}x^{2-3}=\dfrac{1}{54}x^{-1}=\dfrac{1}{54x}$

63. x + 4, 2x – 5 and 4x. The sum is (x + 4) + (2x – 5) + 4x = x + 4 + 2x – 5 + 4x = 7x – 1

Chapter 2 Summary

1. a. $5^3 = 5\cdot5\cdot5 = 125$
   b. $3^5 = 3\cdot3\cdot3\cdot3\cdot3 = 243$
   c. $(-8)^2 = (-8)(-8) = 64$
   d. $-8^2 = -8\cdot8 = -64$
   e. $3^2 + 2^2 = 9 + 4 = 13$
   f. $(3 + 2)^2 = 5^2 = 25$
   g. $3(3^3 + 3^3) = 3(27 + 27) = 3(54) = 162$
   h. $1^{17} + 17^1 = 1 + 17 = 18$

2. a. $x^3 \cdot x^2 = x^{3+2} = x^5$
   b. $x(x^2y) = x \cdot x^2 \cdot y = x^{1+2}y = x^3y$
   c. $y^7y^3 = y^{7+3} = y^{10}$
   d. $x^0y^5 = 1 \cdot y^5 = y^5$
   e. $2b^3b^4b^5 = 2b^{3+4+5} = 2b^{12}$
   f. $(-z^2)(z^3y^2) = -z^2z^3y^2 = -z^{2+3}y^2 = -z^5y^2$
   g. $(4^4x)s^2 = 256xs^2$
   h. $-3y(y^5) = -3y^{1+5} = -3y^6$

3. a. $(x^2x^3)^3 = (x^5)^3 = x^{5\cdot3} = x^{15}$
   b. $(2x^2y)^2 = 2^2(x^2)^2y^2 = 4x^{2\cdot2}y^2 = 4x^4y^2$
   c. $(3x^0)^2 = (3 \cdot 1)^2 = 3^2 = 9 \quad (x \neq 0)$
   d. $(3x^2y^2)^0 = 1 \quad (x \neq 0, y \neq 0)$
   e. $\dfrac{x^7}{x^3} = x^{7-3} = x^4$
   f. $\left(\dfrac{x^2y^3}{xy^2}\right)^2 = \left(x^{2-1}y^{3-2}\right)^2 = (xy)^2 = x^2y^2$
   g. $\dfrac{8\left(y^2x\right)^2}{2^3(yx)^2} = \dfrac{8\left(y^2\right)^2x^2}{8y^2x^2} = \dfrac{y^4x^2}{y^2x^2} = y^{4-2}x^{2-2}$
   $= y^2x^0 = y^2 \cdot 1 = y^2$
   h. $\dfrac{\left(5x^0y^2z^3\right)^3}{25(yz)^5} = \dfrac{5^3 \cdot 1^3\left(y^2\right)^3\left(z^3\right)^3}{25y^5z^5} = \dfrac{5y^6z^9}{y^5z^5} =$
   $5y^{6-5}z^{9-5} = 5yz^4$

4. a. degree = 7; monomial
   b. degree = 2; binomial
   c. degree = 5; trinomial
   d. degree = 3; binomial

5. $P(x) = 5x^4 - x$
   a. $P(3) = 5(3^4) - 3 = 5\cdot81 - 3 = 405 - 3 = 402$
   b. $P(0) = 5(0^4) - 0 = 5\cdot0 - 0 = 0$
   c. $P(-2) = 5(-2)^4 - (-2) = 5\cdot16 + 2 = 82$
   d. $P(2t) = 5(2t)^4 - (2t) = 5(16t^4) - 2t =$
   $80t^4 - 2t$

6. a. a – c
   2. b + c
   3. ab

7. He has 10000 – b dollars to invest in stocks.

8. 55·2 miles in 2 hours; 55·5 miles in 5 hours; 55b miles in b hours

9. a. $3x + 5x - x = (3 + 5 - 1)x = 7x$
   b. $3x + 2y$ cannot be simplified further
   c. $(xy)^2 + 3x^2y^2 = x^2y^2 + 3x^2y^2 = (1 + 3)x^2y^2 =$
   $4x^2y^2$
   d. $-2x^2yz + 3yx^2z = (-2 + 3)x^2yz = x^2yz$
   e. $3x^2y^0 + 2x^2 = 3x^2\cdot1 + 2x^2 = (3 + 2)x^2 = 5x^2$
   f. $2(x + 7) + 3(x + 7) = (2 + 3)(x + 7) =$
   $5(x + 7) = 5x + 35$
   g. $(7a^2 + 2a - 5) - (3a^2 - 2a + 1) = 7a^2 + 2a -$
   $5 - 3a^2 + 2a - 1 = 7a^2 - 3a^2 + 2a + 2a - 5$
   $-1 = 4a^2 + 4a - 6$
   h. $4(4x^3 + 2x^2 - 3x - 8) - 5(2x^3 - 3x + 8) =$
   $16x^3 + 8x^2 - 12x - 32 - 10x^3 + 15x - 40$
   $=16x^3 - 10x^3 + 8x^2 - 12x + 15x - 32 - 40$
   $=6x^3 + 8x^2 + 3x - 72$

10. Perimeter $= (x^2 + 7) + (2x^2 + 3x - 5) + 2x +$
   $(3x^2 - 4x + 6) = x^2 + 7 + 2x^2 + 3x - 5 + 2x +$
   $3x^2 - 4x + 6 = 6x^2 + x + 8$

11. $[(3x^2 + 2x - 9) + (5x^2 + 6x + 1)] - (4x^2 - 5x)$
   $= [3x^2 + 2x - 9 + 5x^2 + 6x + 1] - 4x^2 + 5x =$
   $8x^2 + 8x - 8 - 4x^2 + 5x = 4x^2 + 13x - 8$

12. a. $(2x^2y^3)(5xy^2) = 2\cdot5x^2xy^3y^2 = 10x^3y^5$
   b. $5(x + 3) = 5x + 5\cdot3 = 5x + 15$
   c. $x^2(3x^2 - 5) = x^2(3x^2) - x^2(5) = 3x^4 - 5x^2$
   d. $-3xy(xy - x) = -3xy(xy) - (-3xy)x =$
   $-3x^2y^2 + 3x^2y$
   e. $(x + 3)(x + 2) = x^2 + 3x + 2x + 2\cdot3 =$
   $x^2 + 5x + 6$

13. a. $(3a-3)(2a+2) = 3(a-1)\cdot 2(a+1) =$
$6(a+1)(a-1) = 6(a^2-1^2) = 6a^2-6$
b. $(a-1)(a+1) = a^2-1^2 = a^2-1$
c. $6(y-2)(y+2) = 6(y^2-2^2) = 6y^2-24$
d. $(x+4)^2 = x^2+2\cdot 4\cdot x+4^2 = x^2+8x+16$
e. $(x-3)^2 = x^2-2\cdot 3\cdot x+3^2 = x^2-6x+9$

14. Area $= (x^2+2x+1)(3x+1) = x^2(3x)+x^2\cdot 1 +$
$(2x)(3x)+(2x)1+1(3x)+1\cdot 1 =$
$3x^3+7x^2+5x+1$

15. $(y^2+1)(y^2-1) = (y^2)^2-1^2 = y^4-1$

16. a. $\dfrac{3x+6y}{2xy} = \dfrac{1}{2xy}(3x+6y) = \dfrac{3x}{2xy}+\dfrac{6y}{2xy} =$
$\dfrac{3}{2y}+\dfrac{3}{x}$

b. $\dfrac{14xy-21x}{7xy} = \dfrac{1}{7xy}(14xy-21x) =$
$\dfrac{14xy}{7xy}-\dfrac{21x}{7xy} = 2-\dfrac{3}{y}$

c. $\dfrac{15a^2bc+20ab^2c-25abc^2}{-5abc} = \dfrac{15a^2bc}{-5abc}+$
$\dfrac{20ab^2c}{-5abc}-\dfrac{25abc^2}{-5abc} = -3a-4b+5c$

d. $\dfrac{(x+y)^2+(x-y)^2}{-2xy} =$
$\dfrac{x^2+2xy+y^2+x^2-2xy+y^2}{-2xy} = \dfrac{2x^2+2y^2}{-2xy}$
$= \dfrac{2x^2}{-2xy}+\dfrac{2y^2}{-2xy} = \dfrac{-x}{y}+\dfrac{-y}{x}$

17. a.
$$x-1\overline{)x^2-6x+5} \quad \text{quotient line: } x-5$$
$$\underline{x^2-x}$$
$$-5x+5$$
$$\underline{-5x+5}$$
$$0$$
quotient: $x-5$ remainder: 0

b.
$$x+2\overline{)x^2+3x+5} \quad \text{quotient line: } x+1$$
$$\underline{x^2+2x}$$
$$x+5$$
$$\underline{x+2}$$
$$3$$
answer: $x+1+\dfrac{3}{x+2}$

c.
$$3x-1\overline{)3x^2+14x-2} \quad \text{quotient line: } x+5$$
$$\underline{3x^2-x}$$
$$15x-2$$
$$\underline{15x-5}$$
$$3$$
answer: $x+5+\dfrac{3}{3x-1}$

d.
$$3x+1\overline{)9x^3+0\cdot x^2-13x-4} \quad \text{quotient line: } 3x^2-x-4$$
$$\underline{9x^3+3x^2}$$
$$-3x^2-13x$$
$$\underline{-3x^2-x}$$
$$-12x-4$$
$$\underline{-12x-4}$$
$$0$$
quotient: $3x^2-x-4$ remainder: 0

Chapter 2 Test

1. $2\cdot x\cdot x\cdot x\cdot y\cdot y\cdot y\cdot y = 2x^3y^4$

2. $y^2(yy^3) = y^2(y^{1+3}) = y^2y^4 = y^{2+4} = y^6$

3. $(2x^3)^5(x^2)^3 = 2^5(x^3)^5(x^2)^3 = 32x^{3\cdot 5}x^{2\cdot 3} = 32x^{15}x^6$
$= 32x^{15+6} = 32x^{21}$

4. $(2rr^2r^3)^3 = (2r^{1+2+3})^3 = (2r^6)^3 = 2^3(r^6)^3 = 8r^{6\cdot 3} =$
$8r^{18}$

5. $3x^0 = 3 \cdot 1 = 3$

6. $3x^2 + 2$ is a **binomial.**

7. The degree of $3x^2y^3z^4 + 2x^3y^2z - 5x^2y^3z^5$ is 10, the degree of the third term, the term of largest degree.

8. $P(x) = x^2 + x - 2$. $P(-2) = (-2)^2 + (-2) - 2 = 4 - 2 - 2 = 0$

9. $(xy)^2 + 5x^2y^2 - (3x)^2y^2 = x^2y^2 + 5x^2y^2 - 9x^2y^2 = (1 + 5 - 9)x^2y^2 = -3x^2y^2$

10. $(3x^3 + 4x^2 - x - 7) + (2x^3 - 2x^2 + 3x + 2) =$
$3x^3 + 4x^2 - x - 7 + 2x^3 - 2x^2 + 3x + 2 =$
$3x^3 + 2x^3 + 4x^2 - 2x^2 - x + 3x - 7 + 2 =$
$5x^3 + 2x^2 + 2x - 5$

11. $(2x^2 - 7x + 3) - (3x^2 - 2x - 1) = 2x^2 - 7x + 3 - 3x^2 + 2x + 1 = 2x^2 - 3x^2 - 7x + 2x + 3 + 1 = -x^2 - 5x + 4$

12. $-6(x - y) + 2(x + y) - 3(x + 2y) = -6x + 6y + 2x + 2y - 3x - 6y = (-6 + 2 - 3)x + (6 + 2 - 6)y = -7x + 2y$

13. $-2(x^2 + 3x - 1) - 3(x^2 - x + 2) + 5(x^2 + 2) =$
$-2x^2 - 6x + 2 - 3x^2 + 3x - 6 + 5x^2 + 10 =$
$-2x^2 - 3x^2 + 5x^2 - 6x + 3x + 2 - 6 + 10 =$
$-3x + 6$

14. $(-2x^3)(2x^2y) = (-2)2x^3x^2y = -4x^{3+2}y = -4x^5y$

15. $3y^2(y^2 - 2y + 3) = 3y^2y^2 - 3y^2(2y) + 3y^2 \cdot 3 = 3y^{2+2} - 6y^{2+1} + 9y^2 = 3y^4 - 6y^3 + 9y^2$

16. $(2x - 5)(3x + 4) = 2x \cdot 3x - 5 \cdot 3x + 4 \cdot 2x - 5 \cdot 4 = 6x^2 - 15x + 8x - 20 = 6x^2 - 7x - 20$

17. $(2x - 3)(x^2 - 2x + 4) = 2x \cdot x^2 - 2x(2x) + 2x \cdot 4 - 3 \cdot x^2 + 3(2x) - 3 \cdot 4 = 2x^3 - 4x^2 + 8x - 3x^2 + 6x - 12 = 2x^3 - 7x^2 + 14x - 12$

18. $\dfrac{8x^2y^3z^4}{16x^3y^2z^4} = \dfrac{8}{16}x^{2-3}y^{3-2}z^{4-4} = \dfrac{1}{2}x^{-1}y \cdot 1 = \dfrac{y}{2x}$

19. $\dfrac{6a^2 - 12b^2}{24ab} = \dfrac{1}{24ab}\left(6a^2 - 12b^2\right) =$
$\dfrac{6a^2}{24ab} - \dfrac{12b^2}{24ab} = \dfrac{a}{4b} - \dfrac{b}{2a}$

20. $$\begin{array}{r} x - 2 \\ 2x + 3 \overline{\smash{)}\,2x^2 - x - 6} \\ \underline{2x^2 + 3x} \\ -4x - 6 \\ \underline{-4x - 6} \\ 0 \end{array}$$

quotient: $x - 2$ remainder: 0

Cumulative Review: Chapters 1 - 2

Vocabulary/Concepts

1. prime

2. a. Factor each denominator into prime factors
b. Write down each prime that appears, putting it down the greatest number of times it appears in any one factorization.
c. The product of these is the LCD.

3. $4x = x + x + x + x$ while $x^4 = x \cdot x \cdot x \cdot x$

4. Evaluate $3x + 5x$ means to find its numerical value when given a value of x. Simplify $3x + 5x$ means to write it in a simpler alternative form (8x in this case).

5. a. $|-3| = 3$ b. $-|-3| = -3$ c. $-|3| = -(3) = -3$

6. The double negative rule.

7. If $x \neq 0$, $0 \cdot x = 0$ so $\dfrac{0}{x} = 0$, but if $\dfrac{x}{0} = k$, this means $x = k \cdot 0 = 0$, a contradiction. Thus, $\dfrac{x}{0}$ is undefined.

8. Letters b. and c. represent products.

9. a. 0, 6
b. -5, -1, 0, 6

c. all but $\sqrt{2}$
d. $\sqrt{2}$
e. 0, 6

10. a. Perimeter b. Volume c. Volume d Area

11. a. x^{m+n} b. $x^{m\cdot n}$ c. $x^n y^n$ d. $\frac{x^n}{y^n}$ e. x^{m-n} f. 1

12. In $3x^5$, x is the base for the exponent 5.

13. Numerical coefficient (or constant factor)

14. Binomials

15. Kevin lives 15 + x blocks from the building.

16. distributive property

17. $(x + 2)^3$ is the same as b.

18. conjugate; $(4 + x)(4 - x) = 4^2 - x^2 = 16 - x^2$

19. $\frac{4x+12}{3} = \frac{1}{3}(4x+12) = \frac{4x}{3} + \frac{12}{3} = \frac{4x}{3} + 4$

20. In descending powers: $-x^2 - 3x + 4$

Practice

21. $\frac{5}{6}\cdot\frac{3}{20} = \frac{15}{120} = \frac{15\cdot 1}{15\cdot 8} = \frac{1}{8}$

22. $\frac{2}{3}\left(\frac{4}{5} \div \frac{28}{35}\right) = \frac{2}{3}\left(\frac{4}{5}\cdot\frac{35}{28}\right) = \frac{2}{3}\cdot\frac{4\cdot 5\cdot 7}{5\cdot 4\cdot 7} = \frac{2}{3}$

23. $5^3 - (5 - 1)^3 = 5^3 - 4^3 = 125 - 64 = 61$

24. $4 + 2\cdot 3 - [6 - (5 - 1)] = 4 + 2\cdot 3 - [6 - 4] = 4 + 2\cdot 3 - 2 = 4 + 6 - 2 = 8$

25. $-|8 - 12| - 5 = -|-4| - 5 = -(4) - 5 = -9$

26. $\frac{-7}{9} - \frac{7}{3} = \frac{-7}{9} - \frac{7\cdot 3}{3\cdot 3} = \frac{-7}{9} - \frac{21}{9} = \frac{-28}{9}$

27. $\frac{-4-11}{4-7} = \frac{-15}{-3} = 5$

28. $\frac{3(2^2)-3}{-2^3-1} = \frac{3(4)-3}{-8-1} = \frac{12-3}{-9} = \frac{9}{-9} = -1$

29. $zx + x^2 - z = 2\cdot 3 + 3^2 - 2 = 6 + 9 - 2 = 13$

30. $\frac{x^2-y}{x+y} = \frac{(-2)^2-(-3)}{-2+-3} = \frac{4+3}{-5} = -\frac{7}{5}$

31. $-3x(x^4) = -3x^{1+4} = -3x^5$

32. $\left(\frac{x^8}{x^2}\right)^3 = \left(x^{8-2}\right)^3 = \left(x^6\right)^3 = x^{6\cdot 3} = x^{18}$

33. a. xy (or yx) b. 2x + 4 (or 4 + 2x)

34. $P(x) = 3x^2 - 4x + 2$
a. degree = 2
b. $P(-2) = 3(-2)^2 - 4(-2) + 2 = 12 + 8 + 2 = 22$

35. $(4x^2 - 2x + 6) - 2(x^2 - 2x - 3) = 4x^2 - 2x + 6 - 2x^2 + 4x + 6 = 4x^2 - 2x^2 - 2x + 4x + 6 + 6 = 2x^2 + 2x + 12$

36. $(3x + 6y) + (x - 4y) - (2x + 4y) = 3x + 6y + x - 4y - 2x - 4y = (3 + 1 - 2)x + (6 - 4 - 4)y = 2x - 2y$

37. $(3x + 5)(x^2 - 3x + 1) = 3x\cdot x^2 - 3x(3x) + 3x\cdot 1 + 5\cdot x^2 - 5(3x) + 5\cdot 1 = 3x^3 - 9x^2 + 3x + 5x^2 - 15x + 5 = 3x^3 - 4x^2 - 12x + 5$

38. $(x + 2)^2 - (x + 3)^2 = x^2 + 2\cdot 2\cdot x + 2^2 - (x^2 + 2\cdot 3\cdot x + 3^2) = x^2 + 4x + 4 - x^2 - 6x - 9 = -2x - 5$

39. $\frac{-30x^4 + 25xy^2 - 5xy}{-5xy} = \frac{1}{-5xy}(-30x^4 + 25xy^2 - 5xy) = \frac{-30x^4}{-5xy} + \frac{25xy^2}{-5xy} - \frac{5xy}{-5xy} = \frac{6x^3}{y} - 5y + 1$

40.
$$\begin{array}{r} 4x^2 - 5x + 4 \\ 2x+1 \overline{)\, 8x^3 - 6x^2 + 3x + 2} \\ \underline{8x^3 + 4x^2} \quad\quad\quad\quad \\ -10x^2 + 3x \quad\quad \\ \underline{-10x^2 - 5x} \quad\quad \\ 8x - 2 \\ \underline{8x + 4} \\ -6 \end{array}$$

answer: $4x^2 - 5x + 4 + \frac{-6}{2x+1}$

41.
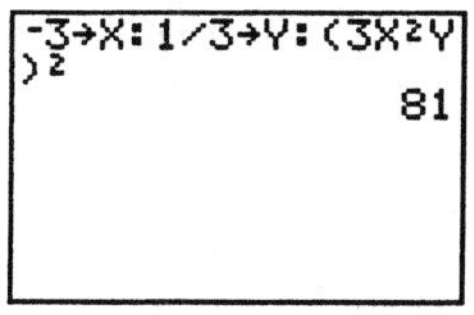

42. Line 3, where x = -1 should be circled.

43.
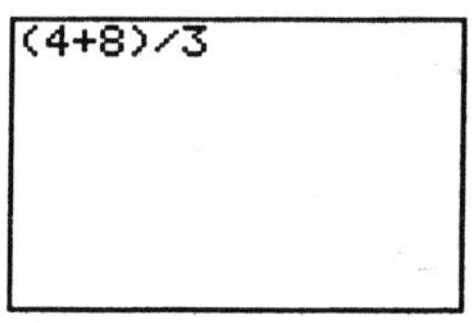

44.
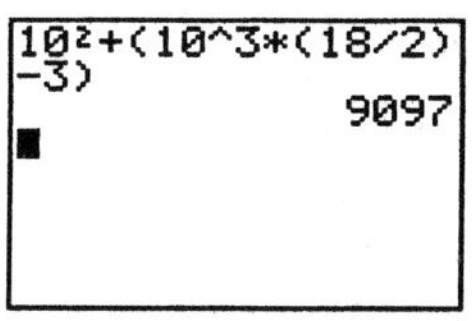

45.
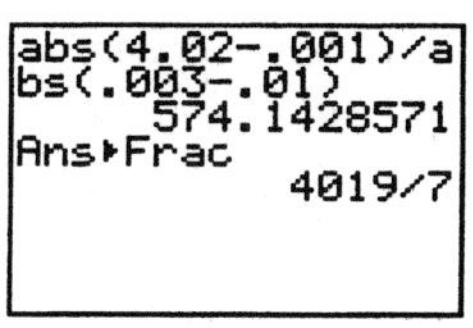

46. $\frac{1}{2(5+7)}$

Applications

47. a. $\frac{5}{6}(24) = \frac{120}{6} = 20$ students

b. $\frac{1}{4}(20) = 5$ students

48. Earnings = $2.50 + 20(.10) = \$4.50$

49. Length = $\frac{\text{area}}{\text{width}} = \frac{x^3 + 5x^2 + 6x + 8}{x + 4}$

$$\begin{array}{r} x^2 + x + 2 \\ x+4 \overline{)\, x^3 + 5x^2 + 6x + 8} \\ \underline{x^3 + 4x^2} \quad\quad\quad\quad \\ x^2 + 6x \quad\quad \\ \underline{x^2 + 4x} \quad\quad \\ 2x + 8 \\ \underline{2x + 8} \\ 0 \end{array}$$

The length is $x^2 + x + 2$.

50. Total students = $x + 2x + 3x + (x + 4) + (x + 1) + (x - 2) = 9x + 3$

Chapter 3 Modeling Linear Equations in One and Two Variables

Section 3.1 Introduction to Linear Equations in One Variable

Vocabulary

1. linear

3. equivalent

5. symmetric

7. no

Concepts

9. ϕ or { }

11. a. 6(A – 2) = 6.6, {3.1}
    b. 2(5A + 4) = 2, {-0.6}
    c. 2(T – 5) = 3(2T + 5) – 6, {-19/4}
    d. 6x + 4 = 3x – 14, {-6}
    e. 2x – 5 = x + 3, {8}
    f. 4(T + 1) – 3 = 3T + 2(T – 5), {11}

13. x + 2, x + 4

15. a. x + 4 b. x – 7

Practice

17. Solution: 1 + 2 = 3

19. Solution: 2·2 = 4

21. Solution: 0/5 = 0

23. Solution: 3(3) + 5 = 14 = 5(3) – 1

25. Solution: 2(-9 + 5) – 3(-9) = 19 = -2(-9) + 1

27. Solution: $\frac{5+0}{10} - 0 = \frac{5}{10} = \frac{1}{2}$

29. $\frac{b}{3} = \frac{1}{3}$ { 1 }

$$3 \cdot \frac{b}{3} = 3 \cdot \frac{1}{3}$$
$$b = 1$$

31. $2V - \frac{7}{3} = \frac{-5}{6}$ { 3/4 }

$$2V - \frac{7}{3} + \frac{7}{3} = \frac{-5}{6} + \frac{7}{3} = \frac{9}{6} = \frac{3}{2}$$
$$\frac{1}{2} \cdot 2V = \frac{1}{2} \cdot \frac{3}{2}$$
$$V = \frac{3}{4}$$

33. 5(b + 7) = 6b – 2 { 37 }
    5b + 35 = 6b – 2
    5b + 35 – 5b = 6b – 2 – 5b
    35 = b – 2
    35 + 2 = b – 2 + 2
    37 = b

35. 2 + 3(x – 5) = 4(x – 1) {-9}
    2 + 3x – 15 = 4x – 4
    3x – 13 = 4x – 4
    3x – 13 – 3x = 4x – 4 – 3x
    -13 = x – 4
    -13 + 4 = x – 4 + 4
    -9 = x

37. 11x + 6(3 – x) = 3 + x {-15/4}
    11x + 18 – 6x = 3 + x
    18 + 5x = 3 + x
    18 + 5x – x = 3 + x – x
    18 + 4x = 3
    18 + 4x – 18 = 3 – 18
    4x = - 15
    4x/4 = -15/4
    x = -15/4

39. 9(x + 11) + 5(13 – x) = 2x {-82}
    9x + 99 + 65 – 5x = 2x
    4x + 164 = 2x

$4x + 164 - 2x = 2x - 2x = 0$
$2x + 164 = 0$
$2x + 164 - 164 = 0 - 164 = -164$
$2x = -164$
$2x/2 = -164/2 = -82$
$x = -82$

41. $\frac{3(t-7)}{2} = t - 6$ {9}
$2 \cdot \frac{3(t-7)}{2} = 2(t-6)$
$3(t - 7) = 2t - 12$
$3t - 21 = 2t - 12$
$3t - 21 - 2t = 2t - 12 - 2t$
$t - 21 = -12$
$t - 21 + 21 = -12 + 21 = 9$
$t = 9$

43. $8x + 3(2 - x) = 5(x + 2) - 4$ $\mathbb{R}$
$8x + 6 - 3x = 5x + 10 - 4$
$5x + 6 = 5x + 6$
x = any real number

45. $3(b - 1) + 18 = 4(b + 5)$ {-5}
$3b - 3 + 18 = 4b + 20$
$3b + 15 = 4b + 20$
$3b + 15 - 3b = 4b + 20 - 3b$
$15 = b + 20$
$15 - 20 = b + 20 - 20$
$-5 = b$

47. $\frac{2t-1}{6} - 2 = \frac{t+2}{6}$ {16}
$6 \cdot \frac{2(t-1)}{6} - 6 \cdot 2 = 6 \cdot \frac{t+2}{6}$
$2(t - 1) - 12 = t + 2$
$2t - 2 - 12 = t + 2$
$2t - 14 = t + 2$
$2t - 14 - t = t + 2 - t$
$t - 14 = 2$
$t - 14 + 14 = 2 + 14 = 16$
$t = 16$

49. $4.8x + 6 = 0.8x - 2$ {-2}
$10(4.8x + 6) = 10(0.8x - 2)$
$48x + 60 = 8x - 20$
$48x - 8x + 60 = 8x - 20 - 8x$
$40x + 60 = -20$
$40x + 60 - 60 = -20 - 60 = -80$
$40x = -80$
$40x/40 = -80/40 = -2$

51. $3.1(x - 2) = 1.3x + 2.8$ {5}
$10(3.1)(x - 2) = 10(1.3x + 2.8)$
$31(x - 2) = 13x + 28$
$31x - 62 = 13x + 28$
$31x - 62 - 13x = 13x + 28 - 13x$
$18x - 62 = 28$
$18x - 62 + 62 = 28 + 62 = 90$
$18x = 90$
$18x/18 = 90/18$
$x = 5$

53. $3.2x + 5.64 = 5.6x + x - 1.16$ {2}
$100(3.2x + 5.64) = 100(5.6x + x - 1.16)$
$320x + 564 = 560x + 100x - 116$
$320x + 564 = 660x - 116$
$320x + 564 - 320x = 660x - 320x - 116$
$564 = 340x - 116$
$564 + 116 = 340x - 116 + 116$
$680 = 340x$
$680/340 = 340x/340$
$2 = x$

Applications

55. a. $4(2x + 3) = 28$
b. $8x + 12 = 28$
c. $8x + 12 - 12 = 28 - 12$
$8x = 16$
$x = 2$
d. Each side is of length $2 \cdot 2 + 3 = 7$ inches

57. a. $2(x - 4) + 2x = 20$
b. $2x - 8 + 2x = 20$
$4x - 8 = 20$
c. $4x - 8 + 8 = 20 + 8$
$4x = 28$
$x = 7$
d. The width is 7 feet and the length is $7 - 4 = 3$ feet.

59. a. $x + (x + 2) = 54$

b. $2x + 2 = 54$
c. $2x + 2 - 2 = 54 - 2$
$2x = 52$
$2x/2 = 52/2 = 26$
$x = 26$
d. The integers are 26 and 28.

61. a. $2x + 5 = 15$
b. already simplified
c. $2x + 5 - 5 = 15 - 5$
$2x = 10$
$2x/2 = 10/2$
$x = 5$
d. The number is 5.

63. a. $3.75x + 29.95 = 52.45$
b. already simplified
c. $3.75x + 29.95 - 29.95 = 52.45 - 29.95$
$3.75x = 22.50$
$3.75x/3.75 = 22.50/3.75 = 6$
$x = 6$
d. The subscriber watched 6 movies.

65. a. $12x + 16 = 100$
b. already simplified
c. $12x + 16 - 16 = 100 - 16 = 84$
$12x = 84$
$12x/12 = 84/12 = 7$
d. He was gone 7 days.

Review

67. a. $\frac{4}{5} - \frac{1}{3} = \frac{4 \cdot 3}{5 \cdot 3} - \frac{1 \cdot 5}{3 \cdot 5} = \frac{12}{15} - \frac{5}{15} = \frac{7}{15}$

b. $\left(3\frac{1}{2}\right) \cdot \frac{-3}{14} = \frac{7}{2} \cdot \frac{-3}{14} = \frac{-21}{28} = \frac{-3 \cdot 7}{4 \cdot 7} = \frac{-3}{4}$

c. $\frac{-4}{5} \div \left(1\frac{1}{2}\right) = \frac{-4}{5} \div \frac{3}{2} = \frac{-4}{5} \cdot \frac{2}{3} = \frac{-8}{15}$

69. $(x\,x^7\,y^3)^2 = (x^8\,y^3)^2 = (x^8)^2(y^3)^2 = x^{16}\,y^6$

71. $2x^2(x^3 - 4xy) = 2x^2(x^3) - 2x^2(4xy) =$
$2x^5 - 8x^3y$

Section 3.2 Literal Equations and Formulas

Vocabulary

1. literal equations

Concepts

3. $P = a + b + c$
$P - a - b = c$ Subtraction property

5. $d = rt$
$d/r = t$ Division property

7. $i = prt$
$\frac{i}{pt} = r$ Division property

9. $C = 2\pi r$
$\frac{C}{2\pi} = r$ Division property

Practice

11. $P = 2l + 2w$
$P - 2l = 2w$
$\frac{P - 2l}{2} = w$ (or $P/2 - l = w$)

13. $a = ½\,(B + b)h$
$\frac{2a}{B + b} = h$

15. $I = \frac{E}{R}$

$IR = E$ so $R = \frac{E}{I}$

17. $½\,x + 3y = 10$
$2(½\,x + 3y) = 2 \cdot 10$
$x + 6y = 20$
$6y = 20 - x$
$y = \frac{20 - x}{6} = \frac{10}{3} - \frac{1}{6}x$

19. $\frac{1}{2}x + \frac{2}{3}y = 5$

$6\left(\frac{1}{2}x + \frac{2}{3}y\right) = 6 \cdot 5$

$3x + 4y = 30$

$4y = 30 - 3x$

$y = \frac{30 - 3x}{4} = \frac{15}{2} - \frac{3}{4}x$

21. $\frac{2}{3}(x + 4) + y = -1$

$y = -1 - \frac{2}{3}(x + 4) = -\frac{3}{3} - \frac{2}{3}x - \frac{8}{3} = \frac{-2x - 11}{3}$

(or $-\frac{2}{3}x - \frac{11}{3}$)

23. $2(x - 4) + 3(y + 2) = 6$

$2x - 8 + 3y + 6 = 6$

$2x + 3y = 8$

$3y = 8 - 2x$

$y = \frac{8 - 2x}{3} = \frac{8}{3} - \frac{2}{3}x$

25. $d = rt$

$t = d/r$ $\quad t = 135/45 = 3$

27. $i = prt$

$t = \frac{i}{pr}$ $\quad t = \frac{12}{100(.06)} = \frac{12}{6} = 2$

29. $P = a + b + c$

$P - a - b = c$ $\quad c = 37 - 15 - 19 = 3$

31. $K = \frac{1}{2} h(a + b)$

$2K = h(a + b)$

$h = \frac{2K}{a + b}$ $\quad h = \frac{2 \cdot 48}{7 + 5} = \frac{96}{12} = 8$

Applications

33. $E = IR$

$\frac{E}{R} = I$ $\quad I = \frac{48}{12} = 4$ amperes

35. $C = 2\pi r$

$r = \frac{C}{2\pi}$ $\quad r = \frac{14.32}{2\pi} \approx 2.28$ feet

37. $P = I^2 R$

$R = \frac{P}{I^2}$ $\quad R = \frac{2700}{14^2} \approx 13.78$ ohms

39. $F = \frac{GmM}{d^2}$

$Fd^2 = GmM$

$m = \frac{Fd^2}{GM}$

41. $L = 2D + 3.25(r + R)$

$L - 3.25(r + R) = 2D$

$D = \frac{L - 3.25(r + R)}{2}$

$D = \frac{25 - 3.25(1 + 3)}{2} = \frac{12}{2} = 6$

Review

43. Let x = -3. Then $2(x - 1) + 2 = 2(-3 - 1) + 2$
$= 2(-4) + 2 = -8 + 2 = -6$

45. $2(x - 1) + 2 = 2x - 2 + 2 = 2x$

Section 3.3 Introduction to Equations in Two Variables

Vocabulary

1. two

1. incremented

5. function

7. range

Concepts

9. The pair (2,2) is not a solution to $3x - 4y = 2$ since $3\cdot2 - 4\cdot2 = -2 \neq 2$.

11. Since in general (a,b) ≠ (b,a), having one pair as a solution of a linear equation in two variables does not guarantee that the other pair is a solution also.

13. The table represents a function since each value of x produces a single value of Y1.

15. Range = {-14,-13,-12,-11,-10,-9,-8}

Practice

17. $x - y = 3$

$-1 - y = 3$ $\qquad$ $5 - y = 3$
$-1 - 3 = y$ $\qquad$ $5 - 3 = y$
$-4 = y$ $\qquad$ $2 = y$

The pairs are (-1,- 4) and (5,2)

19. $x + 3y = -4$

$-5 + 3y = -4$ $\qquad$ $-8 + 3y = -4$
$3y = 1$ $\qquad$ $3y = 4$
$y = 1/3$ $\qquad$ $y = 4/3$

The pairs are (-5,1/3) and (-8,4/3)

21. $y = -4x + 1$

$y = -4(1/4) + 1$ $\qquad$ $y = -4(-3/4) + 1$
$= -1 + 1$ $\qquad$ $= 3 + 1$
$= 0$ $\qquad$ $= 4$

The pairs are (1/4,0) and (-3/4,4).

23. $x^2 - 20 = y$

$10^2 - 20 = y$ $\qquad$ $(-5)^2 - 20 = y$
$100 - 20 = y$ $\qquad$ $25 - 20 = y$
$80 = y$ $\qquad$ $5 = y$

The pairs are (10,80) and (-5,5).

25. $y = 3(x + 2)^2$

$y = 3(2 + 2)^2$ $\qquad$ $y = 3(-6 + 2)^2$
$y = 3 \cdot 4^2$ $\qquad$ $y = 3(-4)^2$
$y = 48$ $\qquad$ $y = 48$

The pairs are (2,48) and (-6,48)

27. $y = \frac{2}{3}(x + 3)^2 - 2$

$y = \frac{2}{3}(-1 + 3)^2 - 2,$ $\qquad$ $y = \frac{2}{3}(2 + 3)^2 - 2$
$y = \frac{2}{3}(2^2) - 2$ $\qquad$ $y = \frac{2}{3}(5^2) - 2$
$y = 8/3 - 2$ $\qquad$ $y = 50/3 - 2$
$y = \frac{2}{3}$ $\qquad$ $y = 44/3$

The pairs are (-1,⅔) and (2,44/3).

29. $y = -x^3$

$y = -(-2)^3$ $\qquad$ $y = -(10^3)$
$y = -(-8)$ $\qquad$ $y = -(1000)$
$y = 8$ $\qquad$ $y = -1000$

The pairs are (-2,8) and (10, -1000)

31. $y = \frac{4}{5}x^3 - 1$

$y = \frac{4}{5}(1) - 1 = \frac{-1}{5}, \quad y = \frac{4}{5}(-1) - 1 = \frac{-9}{5}$

The pairs are (1,-1/5) and (-1,-9/5).

33. $5x + y = 7$

$5x + -3 = 7$
$5x = 10$
$x = 2$

$5x + 0 = 7$
$5x = 7$
$x = 7/5$

$5x + 3 = 7$
$5x = 4$
$x = 4/5$

x	y
2	- 3
7/5	0
4/5	3

35. $\frac{1}{2}x + \frac{2}{3}y = 7$

x	y
62/3	-5
14	0
10	3

$\frac{1}{2}x + \frac{2}{3}(-5) = 7$
$\frac{1}{2}x = 7 + \frac{10}{3} = \frac{31}{3}$
$x = \frac{62}{3}$

$\frac{1}{2}x + \frac{2}{3} \cdot 0 = 7$
$\frac{1}{2}x = 7$
$x = 14$

$\frac{1}{2}x + \frac{2}{3} \cdot 3 = 7$
$\frac{1}{2}x + 2 = 7$
$\frac{1}{2}x = 5$ or $x = 10$

37. Function
Domain = {-4,5,3} Range = {5,- 4}

39. Function
Domain = {-1,0,1,2,3,4,5}
Range = {-8,-5,-2,1,4,7,10}

41. $2x + y = 1$ function
Domain = $\mathbb{R}$ Range = $\mathbb{R}$

43. $y = x^3 - 1$ function
Domain = $\mathbb{R}$ Range = $\mathbb{R}$

Applications

45. $y = 0.75x + 5$

x	0	5	10	15
y	**5**	**8.75**	**12.50**	**16.25**

a. The cost of 5 rides is \$8.75 = 0.75(5) + 5.
b. The cost of 15 rides is \$16.25 = 0.75(15)+5.
c. $y = 0.50x + 5$ is the new equation.
d. $y = 0.50(15) + 5 =$ \$12.50 is the new cost.

47. $y = 1250x + 82000$
a. $y = 1250(12) + 82000 =$ \$97,000 after 12 years
b. $1250x + 82000 = 92000$
$1250x = 10000$
$x = 8$
In 8 years it will be worth \$92,000.

49. Let x = hours, y = total cost of repairs
The model is $y = 37.50x + 20$

x (hours)	1	1.5	5	5.75
y (cost)	**57.50**	**76.25**	**207.50**	**235.63**

$y = 37.50(1) + 20 =$ \$57.50
$y = 37.50(1.5) + 20 =$ \$76.25
$y = 37.50(5) + 20 =$ \$207.50
$y = 37.50(5.75) + 20 =$ \$235.63

51. $y = -800x + 4000$ where y = value, x = years
After $x = 4$ years, $y = -800 \cdot 4 + 4000 =$\$800.

53. $y = 5.50x$, where x = hours and y = cost.
The cost for 3.5 hours is $y = 5.50(3.5) =$ \$19.25.

Review

55. $P(x) = 3x^2 - 5x + 2$
a. degree = 2
b. $P(-3) = 3(-3)^2 - 5(-3) + 2 = 27 + 15 + 2 = 44$.

57. Volume $= 3x(x + 1)(3x + 2) = 3x \cdot (3x^2 + 5x + 2) = 9x^3 + 15x^2 + 6x$ cubic units.

59. ⅔(36) = 24 students did not make an A.

Section 3.4 The Rectangular Coordinate System

Vocabulary

1. rectangular coordinate system (or Cartesian coordinate system)

3. origin

5. abscissa

7. y

9. standard

11. scale

Concepts

13. E: quadrant I F: quadrant II G: quadrant III H: quadrant IV

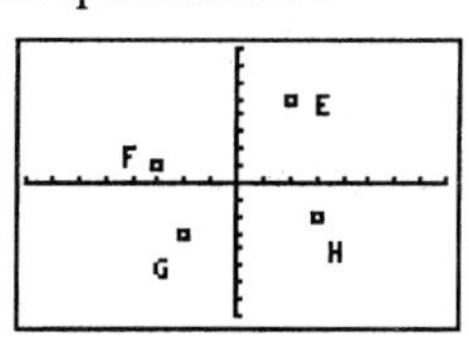

15. a. H(2,-2) b. G(-5,-5) c. F(-4,0) d. E(0,0)
e. B(-3,5) f. C(-2,-3) g. D(4,-5) h. A(2,3)

17. (92, 12.6) or (1992, 12.6)

19. a. [-5,5] by [-5,5] b. [-5,10] by [-10,5]

c. [-2,3] by [-4,4] d. [-6,2] by [-4,8

21. 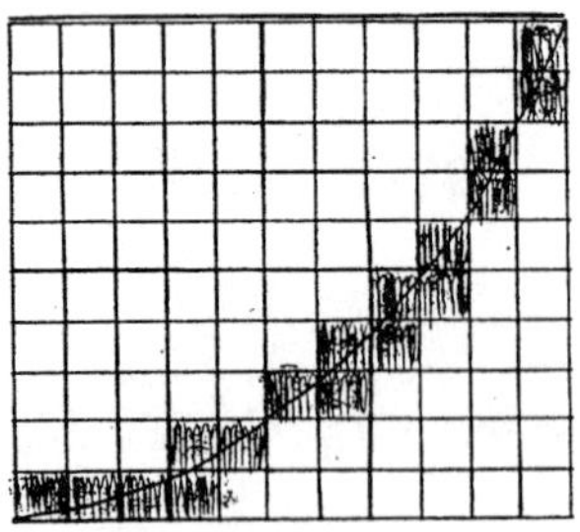

23. Disadvantages of ZInteger: only coordinates with integer values will be displayed; details of the graph near the origin may be hard to see.

25. x-intercepts: (- 4,0) and (4,0)
y-intercept: (0,- 4)

27. You would want to graph 3 points as a check.

29. equation of the vertical line: x = -2. It is not a function since all pairs have -2 as first coordinate.

31. Horizontal line through (3,-2): y = -2

33. Equation of the y-axis: x = 0

Practice

35. y = x – 4

x	y
5	**1**
4	**0**
-1	**-5**

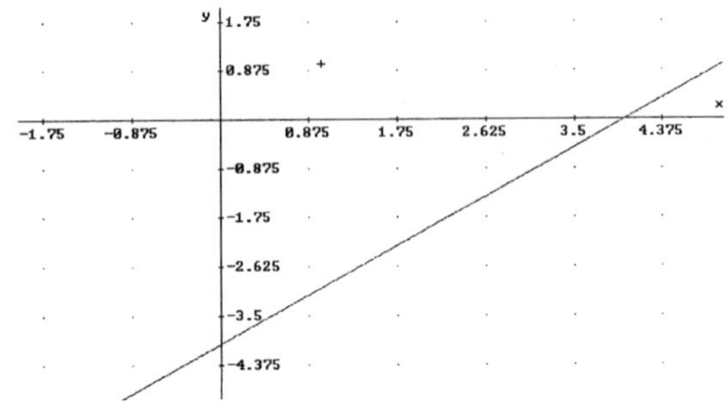

37. y = x/2

x	y
1	**.5**
-1	**-.5**
-4	**-2**

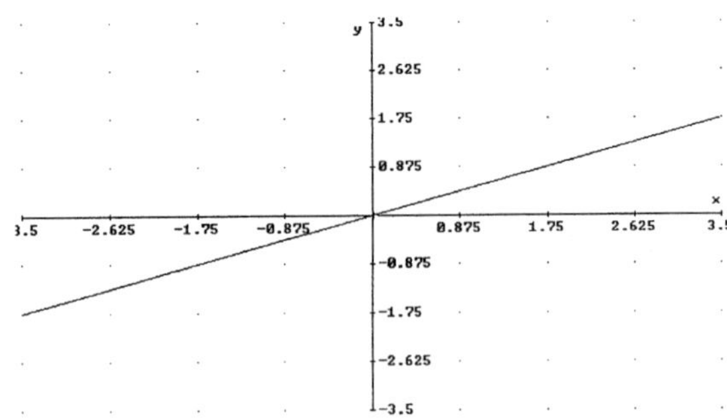

39. y = 3x + 1

x	y
-2	**-5**
0	**1**
1	**4**

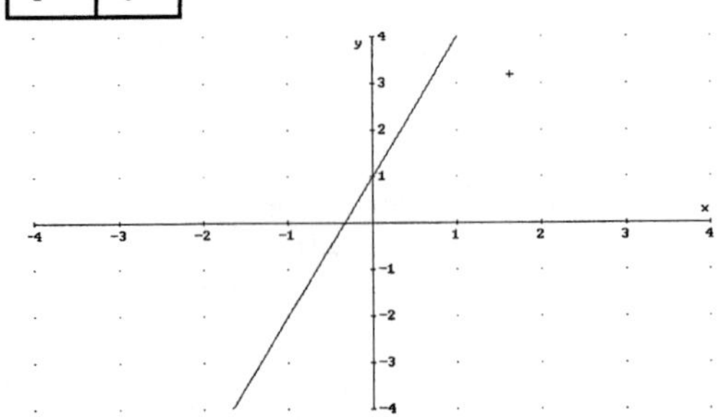

41. $y = x/3 - 3$

x	y
6	**-1**
0	**-3**
-3	**-4**

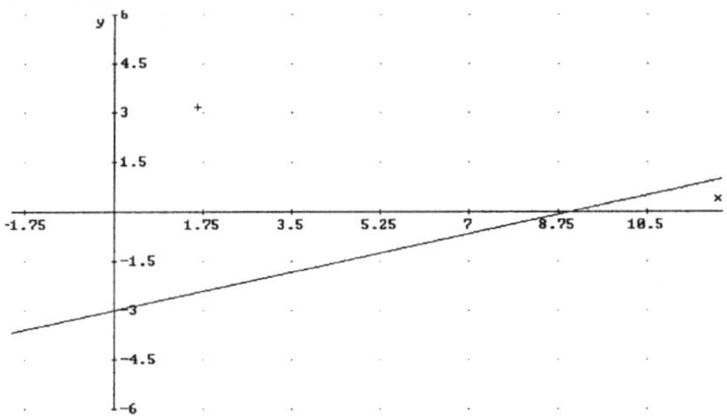

43. $y = x^2 - 3$

x	y
-2	**1**
-1	**-2**
0	**-3**
1	**-2**
2	**1**

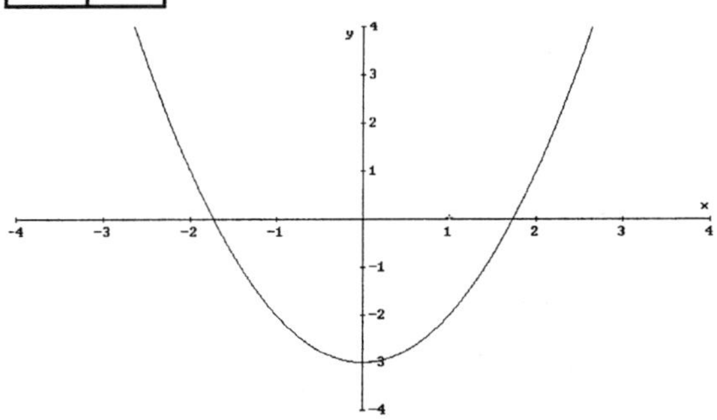

45. $y = (x + 9)^2 - 7$

x	y
-11	**-3**
-10	**-6**
-9	**-7**
-8	**-6**
-7	**-3**

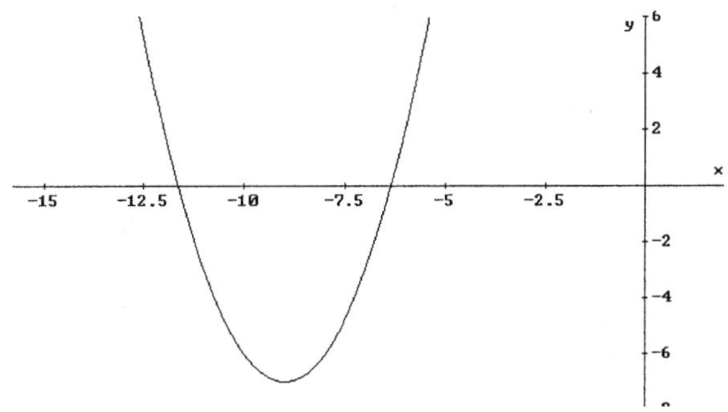

47. $y = x^3 - 4$

x	y
-2	**-12**
-1	**-5**
0	**-4**
1	**-3**
2	**4**

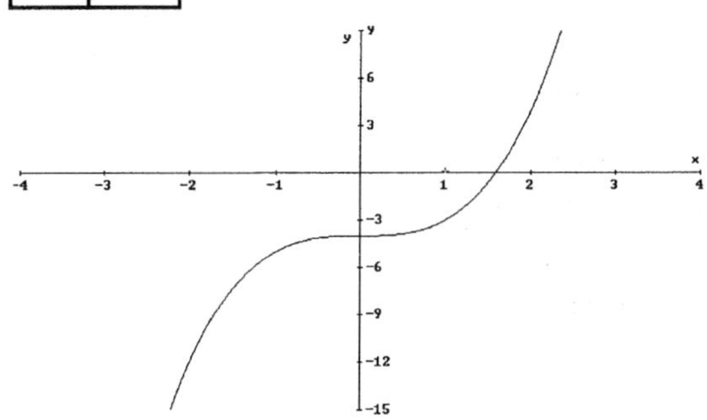

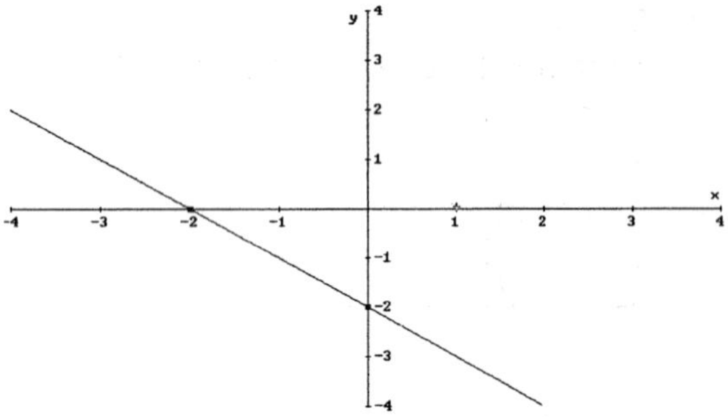

51. $x - y = -2$ Intercepts: (0,2), (-2,0)

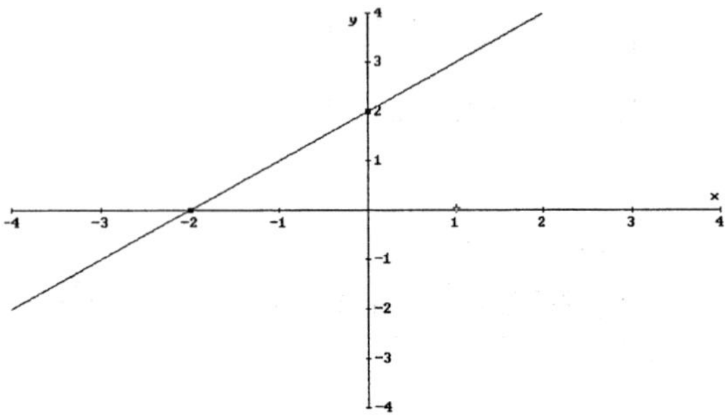

53. $3x + y = -1$ Intercepts: (0,-1), ($-\frac{1}{3}$,0)

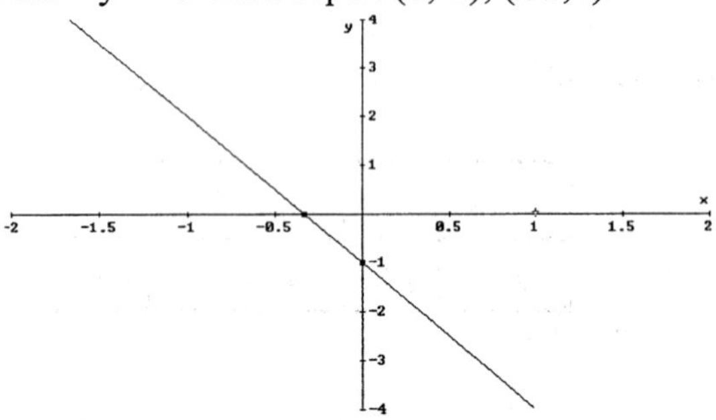

55. $3x - 2y = 6$ Intercepts: (0,-3), (2,0)

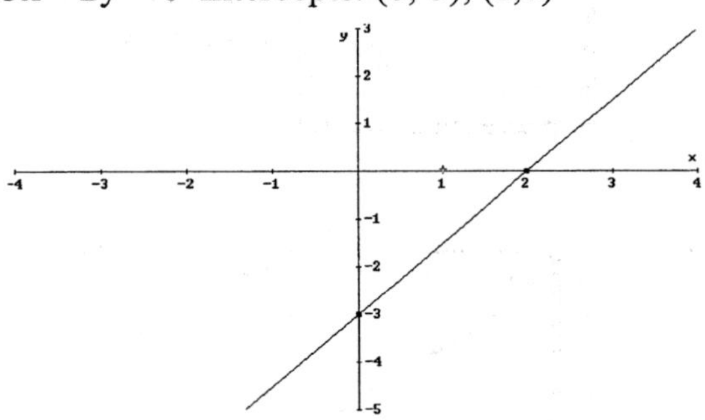

57. $2x + 12 = 9y$ Intercepts: (0,4/3), (-6,0)

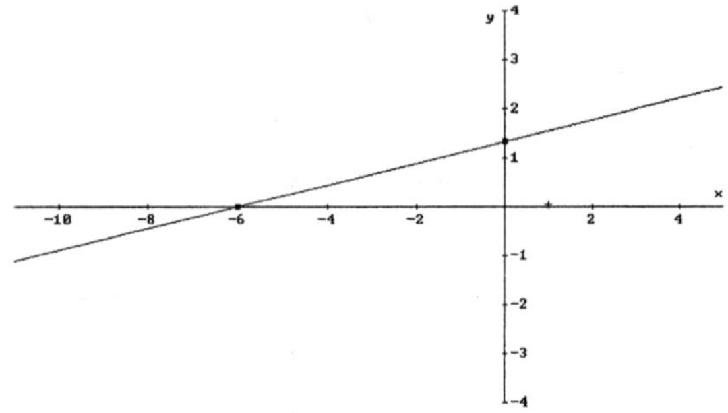

59. $3(y + 1) - x = 4$ Intercepts: (0,$\frac{1}{3}$), (-1,0)

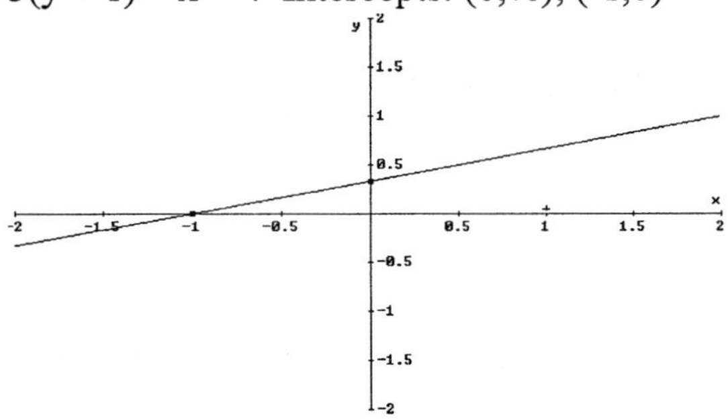

61. $x = 4$

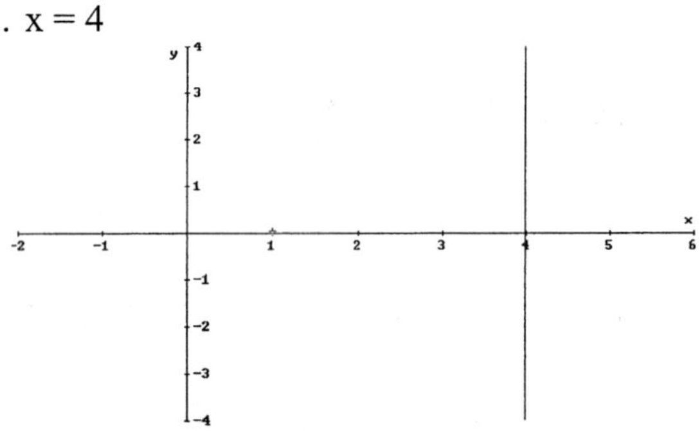

63. $y = 4$

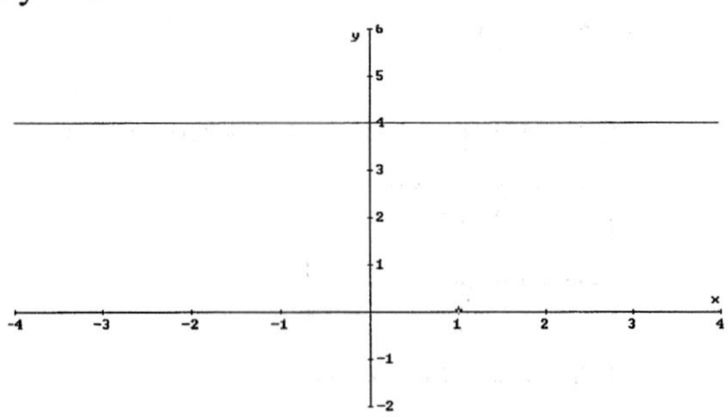

65. $x = 0$ This graph is the y-axis.

67. $3y = 7$

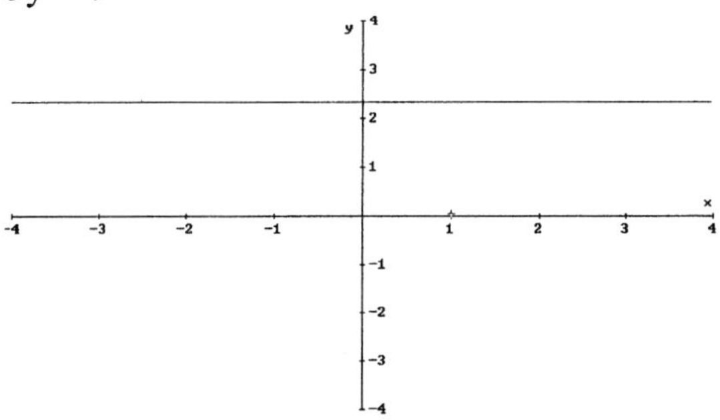

69. $2(y + 3) - y = 5$

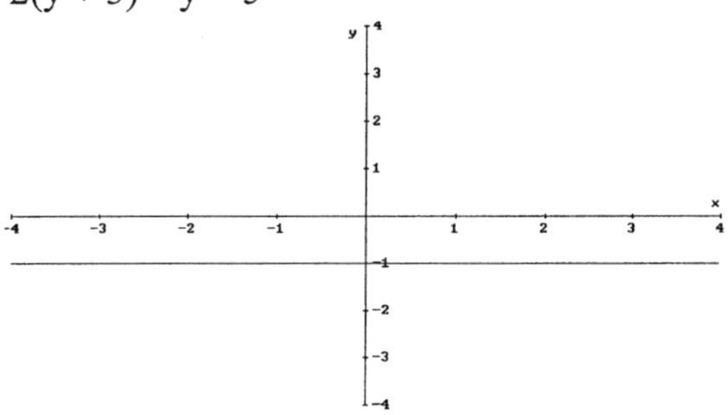

71. $4(x+2) - 2(x - 1) = 6$

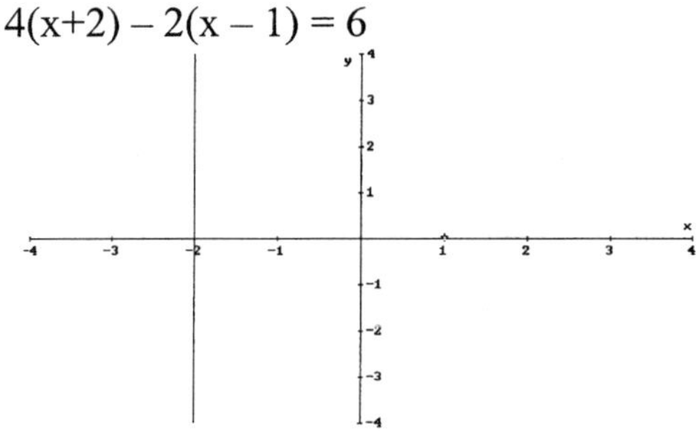

73. $y = -3x - 12$

x-intercept: (- 4, 0), y-intercept: (0,-12)

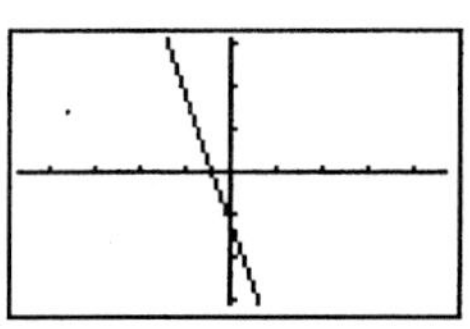

75. $4x - y = 8$
x-intercept: (2,0), y-intercept: (0,-8)

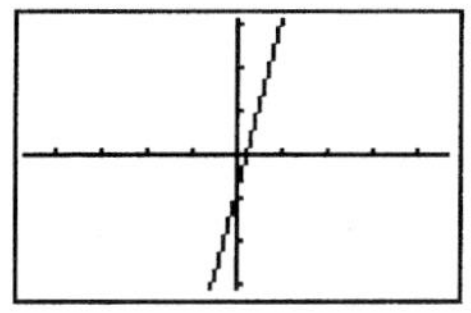

77. $y = -x^2 + 2x - 15$
x-intercepts: none y-intercept: (0,-15)

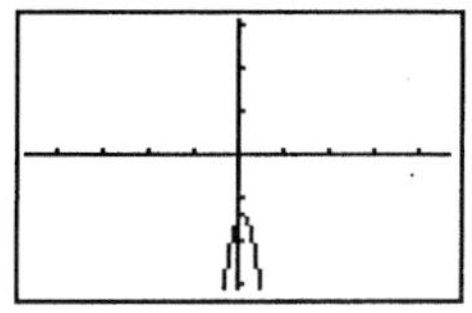

79. $y = (x - 3)^3$
x-intercept: (3,0), y-intercept: (0,-27)

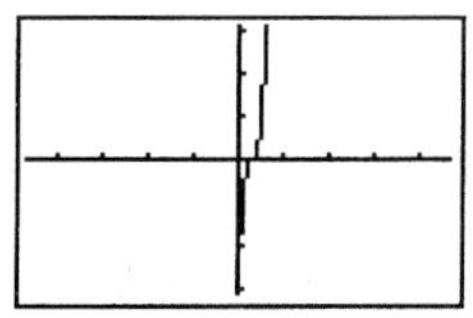

Applications

81.

hours	1	2	3	5
fee	12	24	36	60

83.

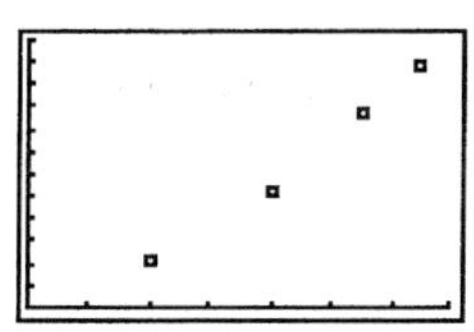

Stopping distance y is a linear function of speed x.

85.

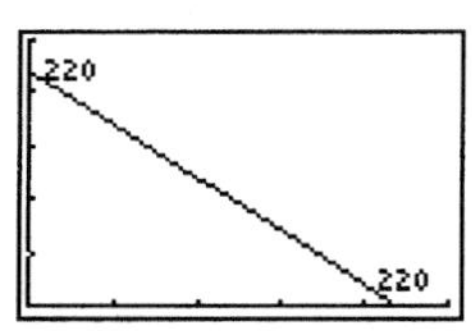

$y = 220 - x$

87.

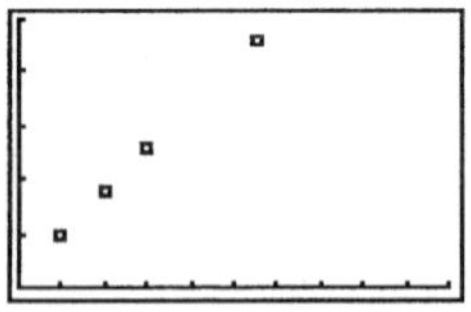

Review

89. Area = $\pi(2x^2)^2 = \pi(4x^4) = 4\pi\ x^4$

91.

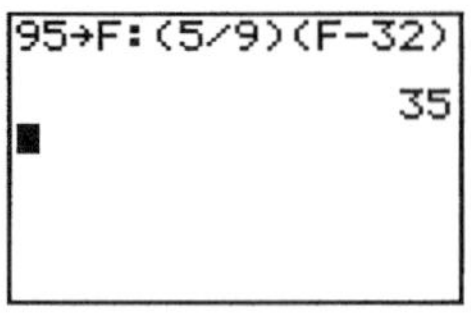

Section 3.5 Linear Applications and Models

Vocabulary

1. best fit

Concepts

3. Sherah makes $35 for sitting a total of 6 hrs.

5. Quick-Copy is better for more than 250 copies.

Practice

7. $y = 5x + 18$ in [-10,10] by [-10,30]; ySc1 = 2

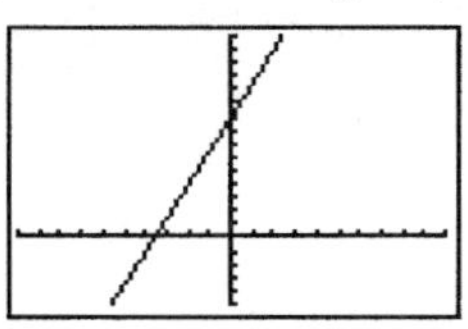

9. $y = -\frac{3}{5}x - 9$ in [-20,10] by [-15,10]
xScl = 2, yScl = 5

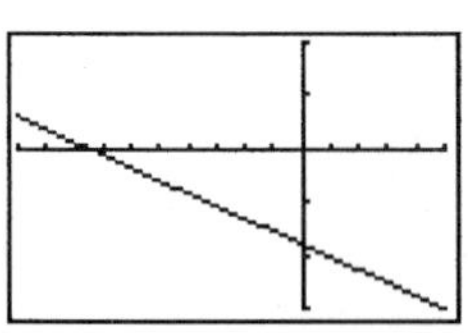

11. $y = |x| + 16$ in [-10,10] by [-5,25]; yScl = 5

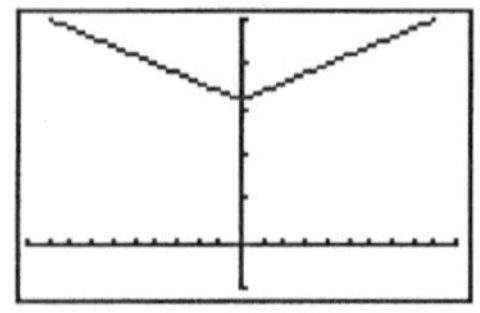

13. $y = x^2 - 15$ in [-5,5] by [-20,10]; yScl = 5

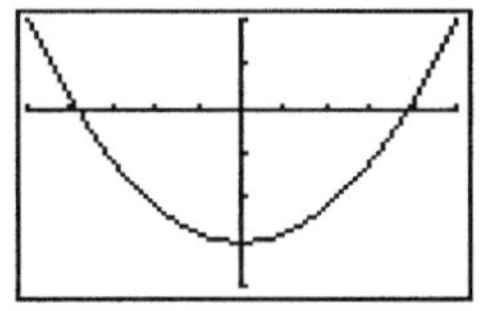

15. $y = x^2 + 14$ in [-5,5] by [-5,25]; yScl = 5

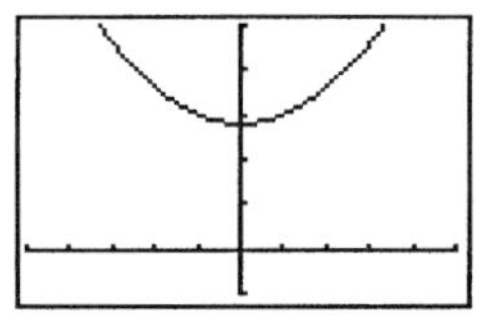

Applications

17. $y = 25x + 50$ in [-10,10] by [-10,75]; yScl = 5

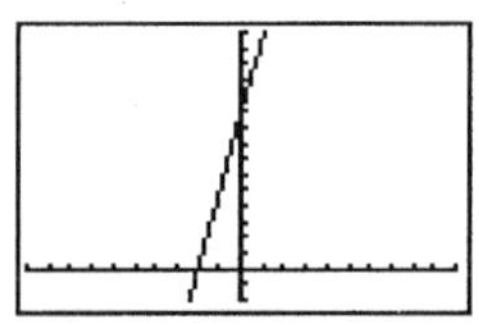

19. a. $y = 0.0875x$
b.

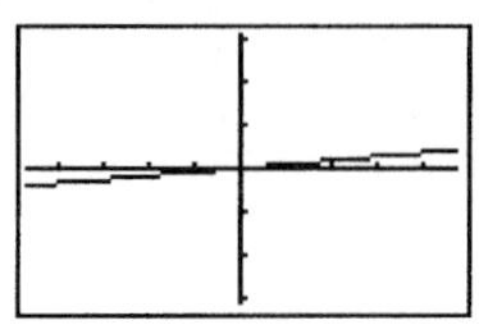

c. $29

21. a. $y = 0.50x + 2.50(25 - x)$
b.

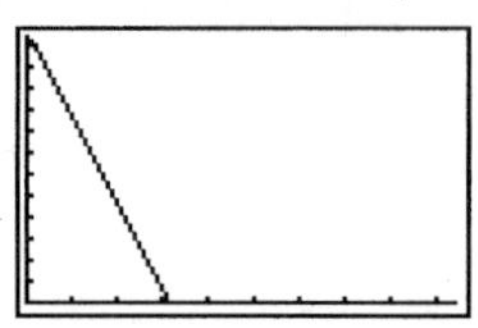

c. 13 d. 4

e. No, since the maximum income occurs when no student/faculty tickets are sold and

$25(2.50) = 62.50 < \$66.50$

23. a. $y = x + 0.40x$ $(= 1.40x)$

b.

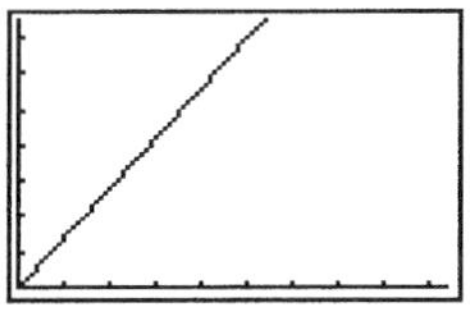

c. \$33

d. \$67.20

25.

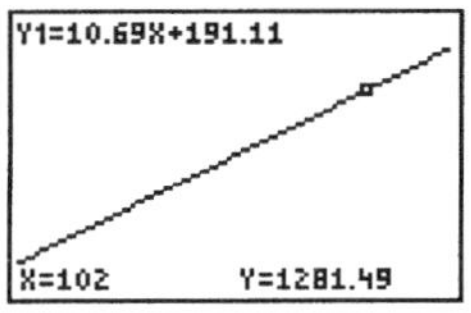

The predicted height of the Empire State Building is 1281.49 ft.

Review

27. a. [-16,10] by [-21,4] for example
b. [-19,26] by [-5,31] for example
c. [-76,51] by [-81,16] for example

29. a. x-intercepts: (5,0) and (9,0)
b. x-intercepts: (-7,0) and (8,0)
c. x-intercepts: (-8,0) and (-3,0)

3.6 Algebraic and Graphical Solutions to One Variable Linear Equations

Vocabulary

1. $ax + c = 0$

Concepts

3. The solution is 2.

In 5 – 13 various answers are possible.

5. [-10,12] by [-10,17]

7. [-10,10] by [- 40,10]

9. [-10,10] by [-20,10]

11. [-10,10] by [-20,10]

13. [-10,10] by [-10,15]

Practice

15. $$\frac{5x+10}{7} = 0$$
$$5x + 10 = 0 \cdot 7 = 0$$
$$5x = -10$$
$$x = -2$$

17. $$\frac{17k-28}{21} + \frac{4}{3} = 0$$
$$17k - 28 + 21\left(\frac{4}{3}\right) = 0 \cdot 21$$
$$17k - 28 + 28 = 0$$
$$17k = 0$$
$$k = 0$$

19. $$\frac{11(x-12)}{2} = 9 - 2x$$
$$11(x-12) = 2(9-2x)$$
$$11x - 132 = 18 - 4x$$
$$15x - 132 = 18$$
$$15x = 150$$
$$x = 10$$

21. $$\frac{5z}{3} + 3 = -2$$
$$3\left(\frac{5z}{3}\right) + 3(3) = -2(3)$$
$$5z + 9 = -6$$
$$5z = -15$$
$$z = -3$$

23. $19.1x - 4(x+0.3) = -46.5$

$$19.1x - 4x - 1.2 = -46.5$$
$$15.1x - 1.2 = -46.5$$
$$15.1x = -45.3$$
$$x = -3$$

25. $\frac{2}{3}(x+6) + \frac{5}{6}(x-6) = x+3$

$$6\left(\frac{2}{3}\right)(x+6) + 6\left(\frac{5}{6}\right)(x-6) = 6(x+3)$$
$$4(x+6) + 5(x-6) = 6x+18$$
$$4x + 24 + 5x - 30 = 6x + 18$$
$$9x - 6 = 6x + 18$$
$$3x = 24$$
$$x = 8$$

27. $\frac{2x+3}{5} = \frac{3x-5}{3}$

$$15\left(\frac{2x+3}{5}\right) = 15\left(\frac{3x-5}{3}\right)$$
$$3(2x+3) = 5(3x-5)$$
$$6x + 9 = 15x - 25$$

$$9 = 9x - 25$$
$$34 = 9x$$
$$\frac{34}{9} = x$$

29. $-[2x - (5x-1) + 3] = -4(x+3)$

$$-[2x - 5x + 1] = -4x - 12$$
$$-[-3x + 4] = -4x - 12$$
$$3x - 4 = -4x - 12$$
$$7x - 4 = -12$$
$$7x = -8$$
$$x = -\frac{8}{7}$$

31. $3(x + 4) - 8 = 5(x - 1) + 16$, $x = -3.5$

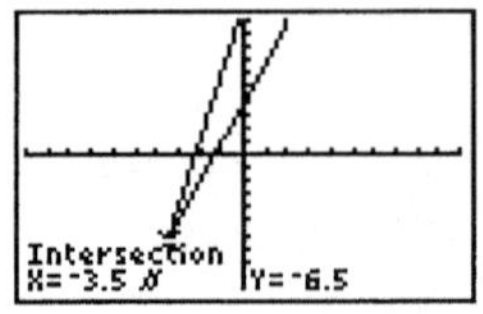

33. $\frac{2x+4}{2} = \frac{5x-2}{8}$, $x = -6$

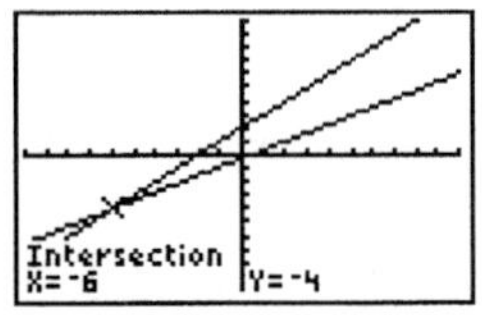

35. $\frac{1}{5}(x-10) - 8 = -11 + 2x$, $x = 5/9$

Using [-10,10] by [-15,10], yScl = 5, gives:

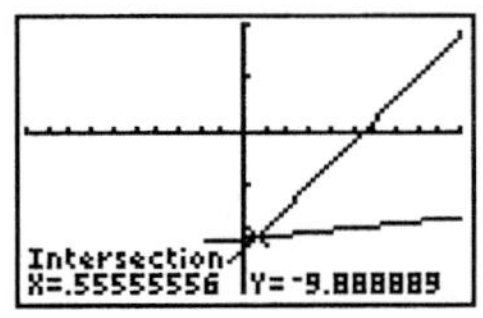

37. $2(x + 6) - 10 = 20$, $x = 9$

Using [-10,10] by [-10,25], yScl = 5, gives:

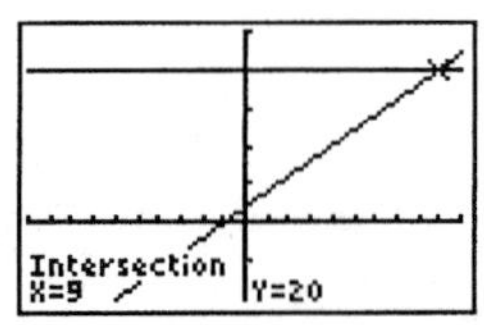

39. $\frac{x}{5} - 3 = 15$, $x = 90$

Using [-10,100] by [-10,20], xScl = 10, yScl = 5 gives;

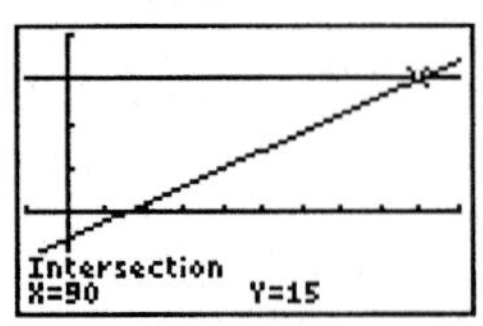

41.

$$3 + 2x = x - 5$$
$$3 + x = -5$$
$$x = -8$$

43. $2(x+3)=5x$

$$2x+6=5x$$
$$6=3x$$
$$2=x$$

45. $x+3(-x)=4$

$$x-3x=4$$
$$-2x=4$$
$$x=-2$$

47. $4(x+6)=x-3$

$$4x+24=x-3$$
$$3x+24=-3$$
$$3x=-27$$
$$x=-9$$

49. $\frac{x}{2}=x+3$

$$x=2(x+3)$$
$$x=2x+6$$
$$-x=6$$
$$x=-6$$

Applications

51. Let x = total number of shares. Then 0.75x = 4.5 million, to that x = 6 million shares.

53. Let x = cost of the brush. Then x + 10.99 + 14.50 = 30.44. Thus x = 30.44 – 25.49 = $4.95.

55. Let x = number of shares owned before the split. Then 1.5x = 555 so $x=\frac{555}{1.5}=370$.

57. Let x = total number of employees. Then 0.90x = 2484 so that $x=\frac{2484}{.90}=2760$.

59. Let x = number of circulars distributed.

$$.05x+5=42.50$$

Then $.05x=37.50$

$$x=\frac{37.50}{.05}=750$$

61. Let x = the integer. Then $\frac{x+7}{2}=5$, so that x + 7 = 10 and x = 3.

63. Let x = number of hours to do the repair. Then the repair cost was 20x + 95 = 155 so that 20x = 60 and thus x = 3 hours.

65. Let x = the amount of her January sales. Then her February sales were 3x – 2000 = 7000 so that 3x = 9000 and x = 3000. Her sales increased by $4000.

67. Let x = length of the shorter piece. Then the total length = x + 2x = 12, so that 3x = 12 and x = 4. The pieces are 4' and 8'.

69. Let x = original price. The discounted price is .80x = 969.20. Thus,

$x=\frac{969.20}{.80}=\$1211.50$, the original price.

71. Let x = number of gallons in the smaller tank. The total the two tanks can hold is x + (2x + 6) = 45 so that 3x + 6 = 45, 3x = 39 and so x = 13 gal. Thus, the smaller tank cannot be emptied into a 10 gal. canister.

73. Let x = number of grams of hydrogen. Then 8x = number of grams of oxygen and so x + 8x = 2700, 9x = 2700 and x = 300 g.

75. Let x = total amount of sales for the team.

Then $x=.45x+.23x+317{,}920$

$$x=.68x+317{,}920$$
$$.32x=317{,}920$$
$$x=\$993{,}900$$

They did not make their sales quota.

77. Let x = number of pounds of cement. Then 3x = number of pounds of gravel and x + 3x = 500, so that 4x = 500 and x = 125. There will be 125 lb. of cement in the mix.

Review

79. The area of the diamond is $90^2 = 8100$ ft.^{2}

81. $\frac{xy}{3}$

83. The total distance = (x +2) + (2x – 1) + (x – 1) = x + 2 + 2x – 1 + x – 1 = 4x miles.

Chapter 3 Summary

1. a. -2 is a solution of 3x + 7 = 1 since 3(-2) + 7 = -6 + 7 = 1
   b. -1 is not a solution of 5 – 2x = 3 since 5 – 2(-1) = 5 – (-2) = 7 ≠ 3.
   c. -3 is not a solution of 2(x + 3) = x since 2(-3 + 3) = 2·0 = 0 ≠ -3.
   d. 13 is a solution of 5(3 – x) = 2 – 4x since 5(3 – 13) = 5(-10) = -50 and 2 – 4x = 2 – 4(13) = 2 – 52 = -50 also.

2. a.
$$\begin{aligned} 3x+7&=1\\ 3x+7-7&=1-7\\ 3x&=-6\\ \frac{3x}{3}&=\frac{-6}{3}\\ x&=-2 \end{aligned}$$

b.
$$\begin{aligned} 7-9x&=8\\ -7+7-9x&=-7+8\\ -9x&=1\\ \frac{-9x}{-9}&=\frac{1}{-9}\\ x&=-\frac{1}{9} \end{aligned}$$

c.
$$\begin{aligned} \frac{x+3}{4}&=2\\ 4\left(\frac{x+3}{4}\right)&=4\cdot 2\\ x+3&=8\\ x+3-3&=8-3\\ x&=5 \end{aligned}$$

d.
$$\begin{aligned} \frac{x-7}{2}&=-2\\ 2\left(\frac{x-7}{2}\right)&=2(-2)\\ x-7&=-4\\ x-7+7&=-4+7\\ x&=3 \end{aligned}$$

e.
$$\begin{aligned} \frac{x}{2}+\frac{7}{2}&=3\\ 2\left(\frac{x}{2}+\frac{7}{2}\right)&=2\cdot 3\\ x+7&=6\\ x+7-7&=6-7\\ x&=-1 \end{aligned}$$

f.
$$\begin{aligned} \frac{x}{3}-3&=2\\ 3\left(\frac{x}{3}-3\right)&=3\cdot 2\\ x-9&=6\\ x-9+9&=6+9\\ x&=15 \end{aligned}$$

3. a.
$$\begin{aligned} 2x-19&=2-x\\ 2x-19+x&=2-x+x\\ 3x-19&=2\\ 3x-19+19&=2+19\\ 3x&=21\\ \frac{3x}{3}&=\frac{21}{3}\\ x&=7 \end{aligned}$$

b. $5x - 19 = 20 - 2x$

$5x - 19 + 2x = 20 - 2x + 2x$

$7x - 19 = 20$

$7x - 19 + 19 = 20 + 19$

$7x = 39$

$\frac{7x}{7} = \frac{39}{7}$

$x = \frac{39}{7}$

c. $10(x - 3) = 3(x + 11)$

$10x - 30 = 3x + 33$

$10x - 30 - 3x = 3x + 33 - 3x$

$7x - 30 = 33$

$7x - 30 + 30 = 33 + 30$

$7x = 63$

$\frac{7x}{7} = \frac{63}{7}$

$x = 9$

d. $2(5x - 7) = 2(x - 16)$

$10x - 14 = 2x - 32$

$10x - 14 - 2x = 2x - 32 - 2x$

$8x - 14 = -32$

$8x - 14 + 14 = -32 + 14$

$8x = -18$

$\frac{8x}{8} = \frac{-18}{8}$

$x = \frac{-18}{8} = \frac{-9}{4}$

e. $\frac{2}{3}(5x - 3) = 38$

$3 \cdot \frac{2}{3}(5x - 3) = 3 \cdot 38$

$2(5x - 3) = 114$

$10x - 6 = 114$

$10x - 6 + 6 = 114 + 6$

$10x = 120$

$\frac{10x}{10} = \frac{120}{10}$

$x = 12$

f. $\frac{3}{2}(3x - 12) = 9$

$2 \cdot \frac{3}{2}(3x - 12) = 2 \cdot 9$

$3(3x - 12) = 2 \cdot 9$

$9x - 36 = 18$

$9x - 36 + 36 = 18 + 36$

$9x = 54$

$\frac{9x}{9} = \frac{54}{9}$

$x = 6$

4. a. Let x = the integer. Then $2x - 7 = 9$, and so $2x = 16$ and $x = 8$. The integer is 8.

b. Let x = the number. Then $3x - 4 = 77$, and so $3x = 81$ and $x = 27$. The number is 27.

c. Let x = number of feet of gutter required. Then the total cost will be $1.50x + 35 = 162.50$. Thus, $1.50x = 127.50$ and $x = 85$. 85 feet of gutter will be needed.

d. Let x = number of kilowatt hours used. Then the bill was $0.18x + 17.50 = 43.96$. Thus, $0.18x = 26.46$ and $x = 147$. The resident used 147 kilowatt hours.

e. Let x = length of the shortest section. Then $45 = x + (3x - 2) + 15$, and so

$$x + 3x - 2 + 15 = 45$$
$$4x + 13 = 45$$
$$4x = 32$$
$$x = 8$$

The shortest piece is 8' long.

5. a. $3x + y = 12$ so $y = 12 - 3x$

b. $4x - 3y = 8$

$$-3y = 8 - 4x$$
$$y = \frac{8 - 4x}{-3} = -\frac{8}{3} + \frac{4}{3}x$$

c. $d = rt$ so $t = \frac{d}{r}$

d. $P = 2l + 2w$

$$P - 2l = 2w$$
$$w = \frac{P - 2l}{2} = \frac{P}{2} - l$$

e. $A = 2\pi rh$ so that $h = \frac{A}{2\pi r}$

f. $A = P + Prt$

$$A - P = Prt$$
$$\frac{A - P}{Pt} = r$$

g. $P = \frac{RT}{V}$

$$PV = RT$$
$$V = \frac{RT}{P}$$

6. a. The pair (2,-1) is a solution of $4x + y = 7$ since $4(2) + (-1) = 8 - 1 = 7$

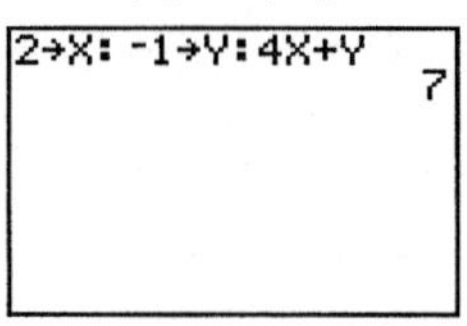

b. The pair (11,2) is a solution of $x - 3y = 5$ since $11 - 3(2) = 11 - 6 = 5$.

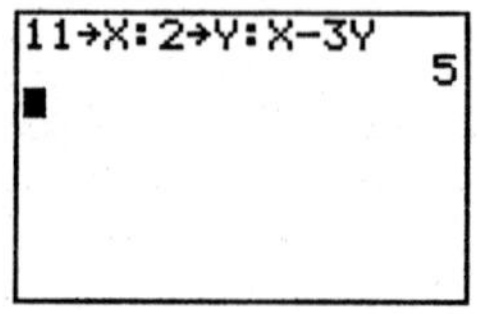

c. The pair (3,- 4) is not a solution of $3x - 2y = -1$ since $3(3) - 2(-4) = 9 - (-8) = 17 \neq -1$.

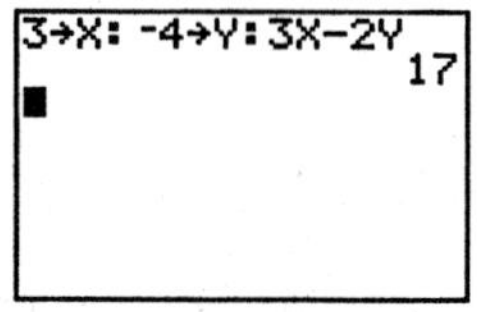

d. The pair (-2,-3) is not a solution of $2x + 3y = 3$ since $2(-2) + 3(-3) = -4 + (-9) = -13 \neq 3$.

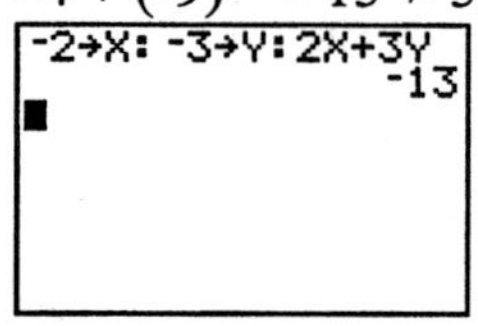

7. a. $x - 2y = 6$; $x = 2$

$$2 - 2y = 6$$
$$-2y = 4$$
$$y = -2$$ The pair is (2,-2).

b. $5x + 2y = 9$; $x = 3$

$$5 \cdot 3 + 2y = 9$$
$$2y = -6$$
$$y = -3$$ The pair is (3,-3).

c. $x^2 + 3x + 2 = y$; $x = -1$

$$(-1)^2 + 3(-1) + 2 = y$$
$$1 - 3 + 2 = y$$
$$0 = y$$ The pair is $(-1,0)$.

d. $x^2 - 9 = y$; $x = 3$

$$3^2 - 9 = y$$
$$9 - 9 = y$$
$$0 = y$$ The pair is (3,0)

8. $3x + y = 6$

x	y
-2	**12**
-1	**9**
0	**6**
1	**3**
2	**0**

$y = 6 - 3x$
$y = 6 - 3(-2) = 6 + 6 = 12$
$y = 6 - 3(-1) = 6 + 3 = 9$
$y = 6 - 3(0) = 6 + 0 = 6$
$y = 6 - 3(1) = 6 - 3 = 3$
$y = 6 - 3(2) = 6 - 6 = 0$

9. a. {(2,3), (4,6), (2,-3)} Not a function.
domain: {2,4} range: {3,6,-3}

b. {(1,5), (2,8), (3,11)} Function.
domain: {1,2,3} range: {5,8,11}

c. $y = 4x + 5$ Function
domain: ℝ range: ℝ

d. $y = x^2 - 1$ Function
domain: ℝ range: all real numbers at least as large as -1

10. a. $y = 3x - 1$

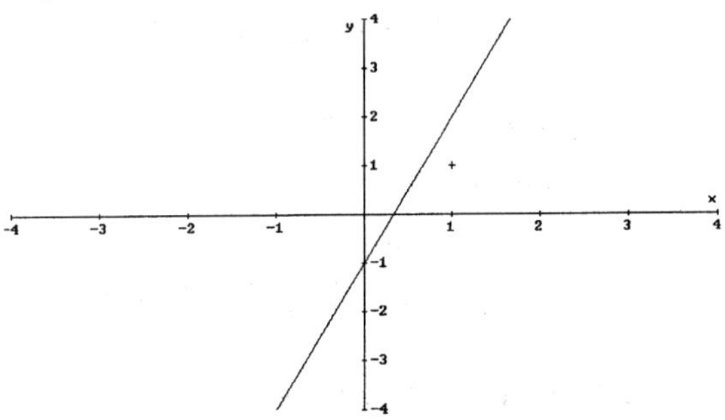

b. $y = -4x + 2$

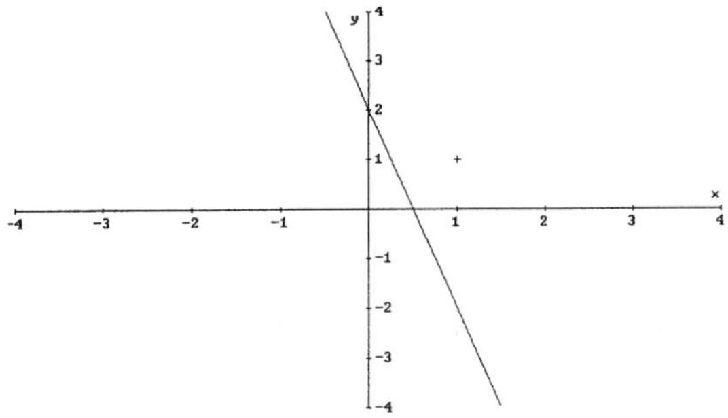

c. $3x + y = 8$

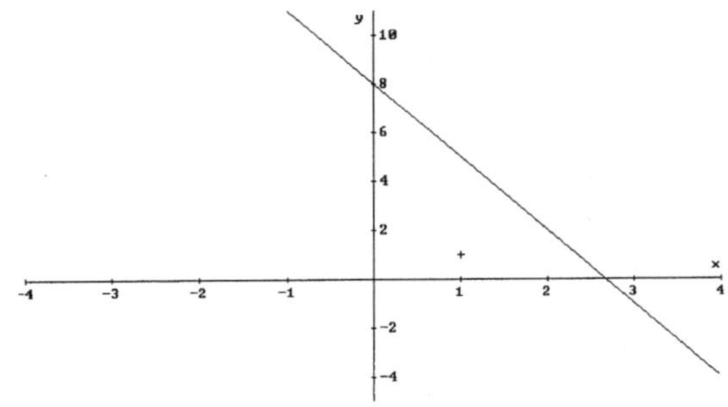

d. $x - 4y = 6$

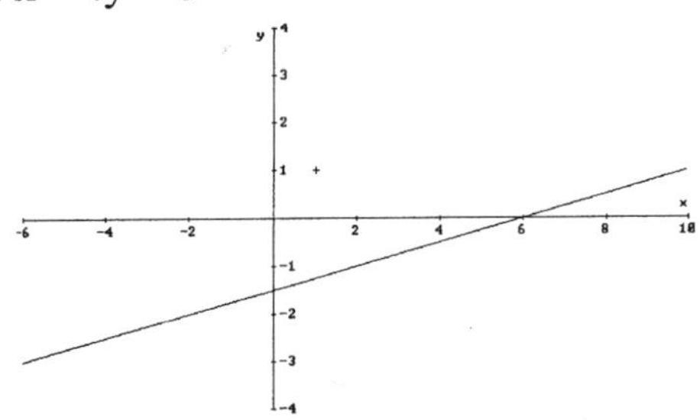

e. $y = 3$

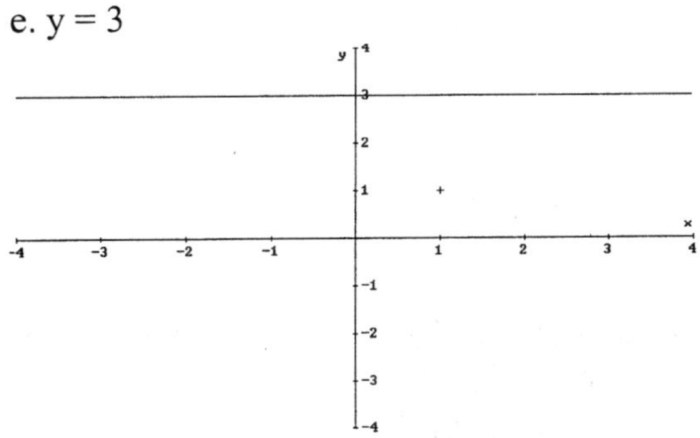

f. $x = -5$

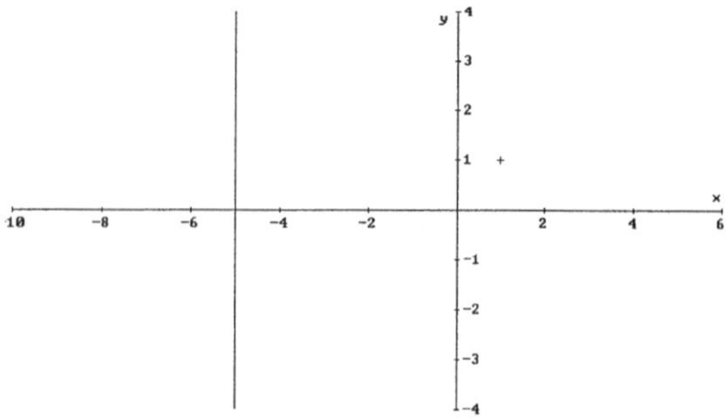

11. a. y-intercept: (0,-1) $y = 3(0) - 1 = -1$
x-intercept: (⅓,0) $0 = 3x - 1,\ x = \frac{1}{3}$

b. y-intercept: (0,2) $y = -4(0) + 2 = 2$
x-intercept: (½,0) $0 = -4x + 2,\ x = \frac{1}{2}$

c. y-intercept: (0,8) $3(0) + y = 8,\ y = 8$
x-intercept: $(\frac{8}{3}, 0)$ $3x + 0 = 8,\ x = \frac{8}{3}$

d. y-intercept: $(0, \frac{-3}{2})$ $0 - 4y = 6,\ y = \frac{-3}{2}$
x-intercept: (6,0) $x - 4(0) = 6,\ x = 6$

e. horizontal line
y-intercept: (0,3) No x-intercept.

f. vertical line
No y-intercept x-intercept: (-5,0)

12. a. $y = 3x - 6$
y-intercept: (0,-6) x-intercept: (2,0)

b. $y = x^2 - 4x - 5$
y-intercept: (0,-5) x-intercepts: (-1,0), (5,0)

c. $y = (x - 4)^3$
y-intercept: (0,-64) x-intercept: (4,0)

d. $4x - y = 12$
y-intercept: (0,-12) x-intercept: (3,0)

13. a. $x^2 - 16x + 48 = y$ [-10,15] by [-16,50]

b. $y = 5x - 100$ [-10,25] by [-110,10]

c. $30x^2 - 155x + 140 = y$
[-10,10] by [-65,150]

14. a. $y = -170x + 1970$ in [0,11.75] by [0,1974]

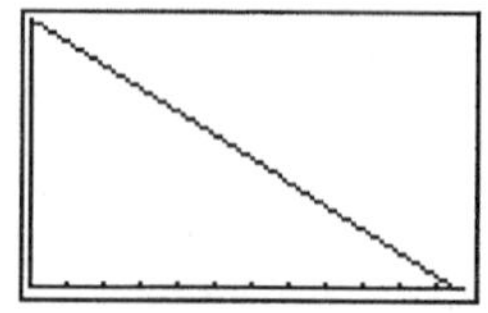

b. The value of the lathe in 8 years is $610.
c. In 6.75 years, it is worth $822.50.

15. a. $y = 125 + 35x$
b. Using [0,9.4] by [-10,500] gives:

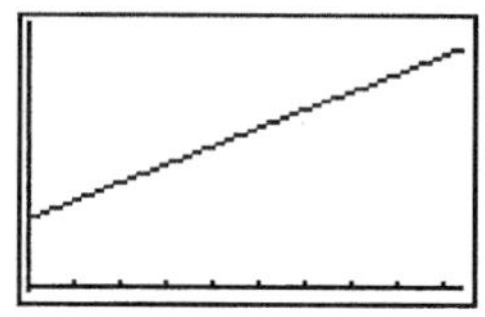

c. The value in 3 years will be $230.
d. It will have a value of $405 in 8 years.

16. a. $12x + 5y = 1069$

b. $5y = 1069 - 12x$ so $y = \frac{1069}{5} - \frac{12}{5}x$

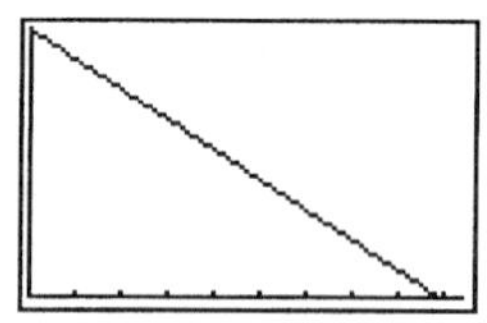

c. If 42 adult tickets are sold, then 113 student tickets are sold.
d. It is not possible for 60 adult tickets to be sold as this would mean that 69.8 student tickets would be sold.
e. It is not possible for 90 adult tickets to be sold as this would mean that -2.2 student tickets would be sold.

17. For answers see Exercises 2 and 3 above in this **Chapter 3 Summary**.

18. For answers see Exercise 4 above in this **Chapter 3 Summary**.

Chapter 3 Test

1. -2 is not a solution of $-3(2 - x) = 0$ since $-3(2-(-2)) = -3(4) = -12 \neq 0$.

2. -1 is not a solution to $x(x + 1) = x^2 + 1$ since $(-1)(-1 + 1) = -1(0) = 0$, while $(-1)^2 = 1 = 2$.

3. $$4x + 3y = 8$$
$$3y = 8 - 4x$$
$$y = \frac{8-4x}{3} = \frac{8}{3} - \frac{4}{3}x$$

4. $$\frac{1}{2}x + 2y = \frac{3}{5}$$
$$2y = \frac{3}{5} - \frac{1}{2}x$$
$$y = \frac{3}{10} - \frac{1}{4}x$$

5. (-1,2) is not a solution of $2x - 5y = 7$ since $2(-1) - 5(2) = -2 - 10 = -12 \neq 7$

6. (1,-1) is a solution of $x + 3y = -2$ since $1 + 3(-1) = 1 - 3 = -2$.

7. $$\frac{x}{7} = 1$$
$$7\left(\frac{x}{7}\right) = 7 \cdot 1$$
$$x = 7$$

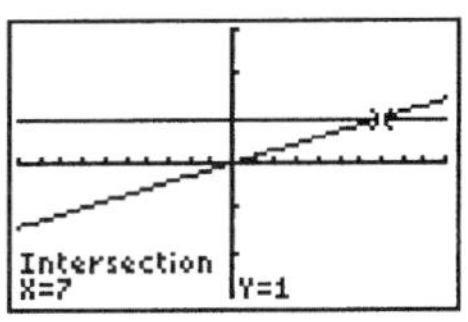

8. $$8x + 2 = -14$$
$$8x + 2 - 2 = -14 - 2$$
$$8x = -16$$
$$x = -2$$

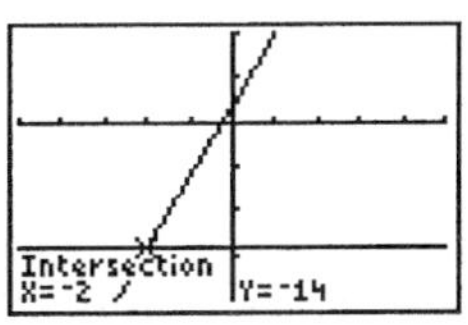

9. $$23 - 5(x + 10) = -12$$
$$23 - 5x - 50 = -12$$
$$-5x - 27 = -12$$
$$-5x = 15$$
$$x = -3$$

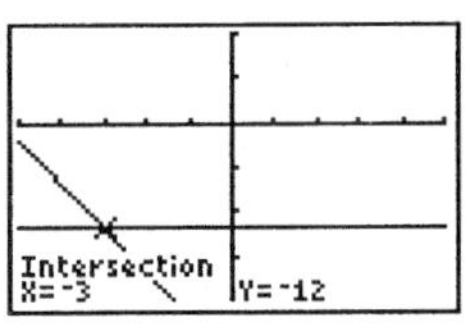

10. $$x(x + 5) = x^2 + 3x - 8$$
$$x^2 + 5x = x^2 + 3x - 8$$
$$5x = 3x - 8$$
$$2x = -8$$
$$x = -4$$

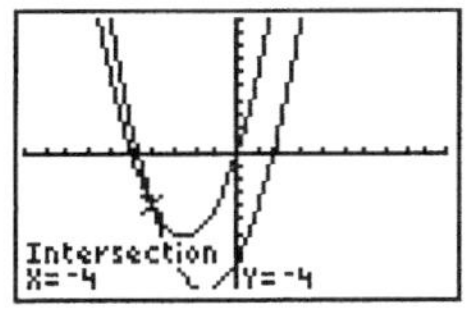

11. $$\frac{3(x-6)}{2} = 6x$$
$$3(x - 6) = 12x$$
$$3x - 18 = 12x$$
$$-18 = 9x$$
$$-2 = x$$

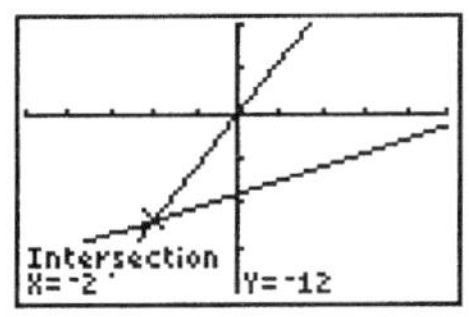

12. $$\frac{5}{3}(x - 7) = 15(x + 1)$$
$$3 \cdot \frac{5}{3}(x - 7) = 3 \cdot 15(x + 1)$$
$$5(x - 7) = 45(x + 1)$$
$$5x - 35 = 45x + 45$$
$$5x = 45x + 80$$
$$-40x = 80$$
$$x = -2$$

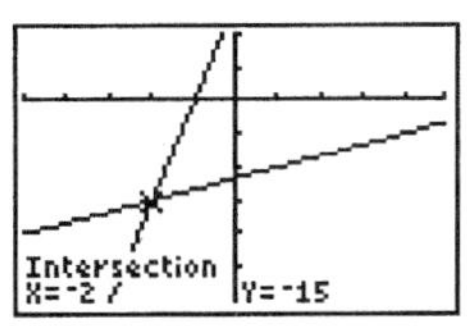

13. $x(x+1) = x^2 + x + 1$

$x^2 + x = x^2 + x + 1$

$0 = 1$, so no solution.

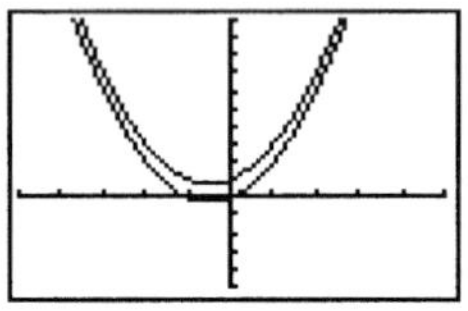

14. $x + 5(x-3) = 2(x-1) + 3(x-5) + 2$

$x + 5x - 15 = 2x - 2 + 3x - 15 + 2$

$6x - 15 = 5x - 15$

$6x = 5x$

$x = 0$

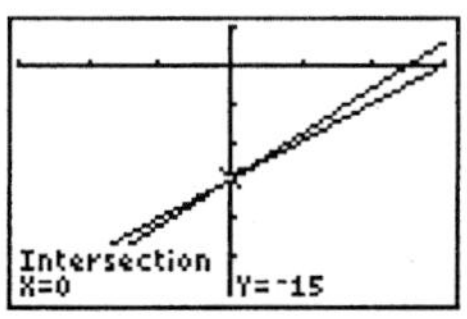

15. $y = 4x - 1$
y-intercept: (0,-1) x-intercept: (¼,0)
Graphing in the standard viewing window,

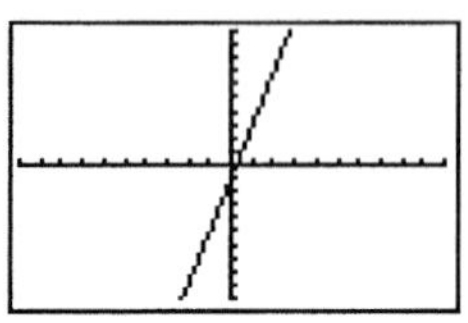

16. $y = -\frac{1}{2}x + 5$

y-intercept: (0,5) x-intercept: (10,0)
Graphing in the window [-10,15] by [-10,10],

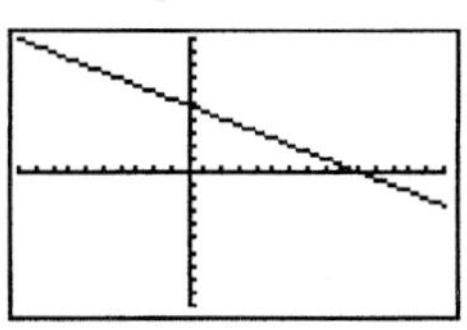

17. $2x + y = 1$
y-intercept: (0,1) x-intercept: (½ ,0)
Graphing in the standard viewing window,

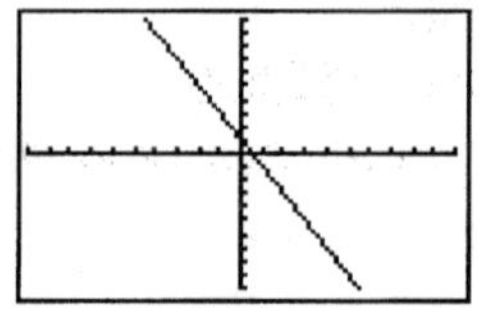

18. $x + 3y = -2$

y-intercept: $(0, \frac{-2}{3})$ x-intercept: (-2,0)

Graphing in the standard viewing window,

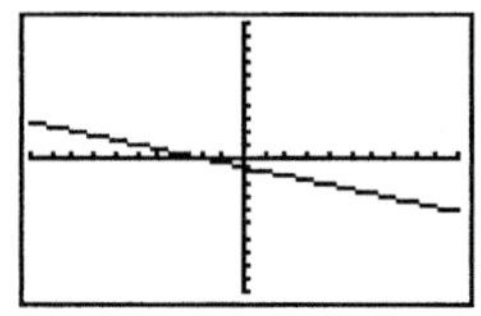

19. a. $y = -\frac{1}{10}x + 170$ in [0,188] by [0,200]

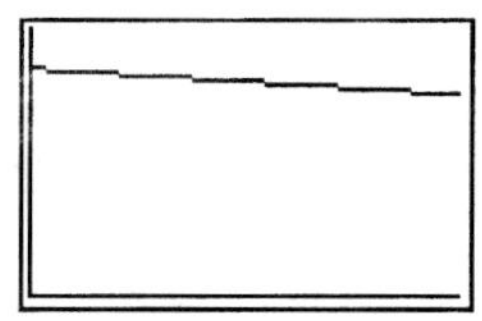

b. 200 sets will be sold at a price of $150.

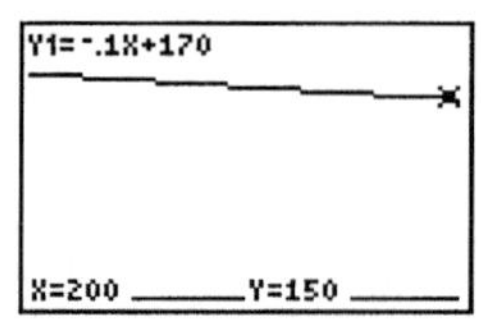

20. Let x and x + 2 represent the integers.
a. $x + (x + 2) = 36$
b. $x + x + 2 = 36$

$2x + 2 = 36$

$2x = 34$

$x = 17$

c. The integers are 17 and 19.

21. Let x = current value of the rug.
a. $x = 560 + 0.96(560)$
b. $x = 560 + 537.60 = 1097.60$
c. The rug is now worth $1097.60.

22. Let x = value of the house.

a. 600 + 0.06x = 5700
b. 0.06x = 5100, so x = 85,000
c. The value of the house was $85,000.

Chapter 4 An Extended Look at Applications and Absolute Value Equations

Section 4.1 Applications - A Numeric Approach

Vocabulary

1. ratio

3. means; extremes

Concepts

5. a. $\frac{5}{7}$ b. $\frac{3}{5}$ c. $\frac{1}{2}$ d. $\frac{1}{2}$ e. $\frac{2}{3}$ f. $\frac{2}{3}$

7. a. similar b. similar c. not similar

Practice

9. $$\frac{x+3}{12}=\frac{-7}{6}$$
$$6(x+3)=12(-7)$$
$$6x+18=-84$$
$$6x=-102$$
$$x=-17$$

11. $$\frac{3x-2}{7}=\frac{x}{28}$$
$$28(3x-2)=7x$$
$$4(3x-2)=x$$
$$12x-8=x$$
$$11x=8$$
$$x=\frac{8}{11}$$

13. $$\frac{7(x+6)}{6}=\frac{6(x+3)}{5}$$
$$35(x+6)=36(x+3)$$
$$35x+210=36x+108$$
$$102=x$$

Applications

15. Let x and x + 2 represent integers. Then
$$x+(x+2)=54$$
$$2x=52$$
$$x=26,\ x+2=28$$
The integers are 26 and 28.

17. Let x, x + 1 and x + 2 represent the integers. Then, $x+(x+1)+(x+2)=120$
$$3x+3=120$$
$$3x=117$$
$$x=39,\ x+1=40,\ x+2=41$$
The integers are 39, 40 and 41.

19. Let x and x + 1 represent the integers. Then,
$$x+2(x+1)=23$$
$$3x+2=23$$
$$3x=21$$
$$x=7$$
The smaller integer is 7.

21. Let x and x + 10 represent the integers. Then
$$x+10=2x-3$$
$$x+13=2x$$
$$13=x$$
The smaller integer is 13.

23. Let x = length of a side. Then, 3x = 57 so that x = 19. Each side has length 19 ft.

25. Let l and $l - 11$ represent the dimensions. Then, $2l + 2(l-11)=94$
$$4l-22=94$$
$$4l=116$$
$$l=29,\ l-11=18$$
The pool is 29 m. by 18 m.

27. Let l and w represent the dimensions. Then, $l = 2w + 5$ and

$$2w + 2(2w+5) = 112$$
$$6w + 10 = 112$$
$$6w = 102$$
$$w = 17, l = 2\cdot 17 + 5 = 39$$

The frame is 39 in. by 17 in.

29. Let x = measure of each angle. Then, 3x = 180 or x = 60. Thus, each angle has a measure of 60°.

31. Let x = cost of 39 packets. Then, $\frac{3}{50} = \frac{39}{x}$

$$3x = 1950$$
$$x = 650$$

Thus, 39 packets cost \$6.50 .

33. Let x = length. Then, $\frac{87}{1} = \frac{x}{\frac{3}{4}}$

$$x \cdot 1 = \frac{3}{4} \cdot 87$$
$$x = 65.25$$

The real engine is 65.25 ft. long.

35. $\frac{6\text{gal.}}{16\text{oz.}} = \frac{6(128)\text{oz.}}{16\text{oz.}} = \frac{768}{16} \neq \frac{50}{1}$ Thus, the instructions are not correct.

37. Let h = height of the flagpole. Then,

$$\frac{h}{5} = \frac{30}{7}$$ The pole is about 21.4 ft.tall.

$$7h = 150$$
$$h = \frac{150}{7}$$
$$h \approx 21.4$$

39. Let x = altitude gained. Then,

$$\frac{100}{1000} = \frac{x}{5280}$$
$$1000x = 528,000$$
$$x = 528$$

The plane gains 528 ft of altitude.

41. Let x = height of the hill. Then,

$$\frac{\frac{1}{2}(5280)}{x} = \frac{300}{100}$$ The hill is 880ft. high.

$$\frac{2640}{x} = \frac{300}{100}$$
$$300x = 264,000$$
$$x = 880$$

Review

43. a. $(4x^2 - 5x + 2) - (6x^2 - 3x - 4) = 4x^2 - 5x + 2 - 6x^2 + 3x + 4 = -2x^2 - 2x + 6$
b. $(2x + 5)^3 = (4x^2 + 2\cdot 2x\cdot 5 + 25)(2x + 5) = (4x^2 + 20x + 25)(2x + 5) = 8x^3 + 40x^2 + 50x + 20x^2 + 100x + 125 = 8x^3 + 60x^2 + 150x + 125$
c.

$$\begin{array}{r} x^2 - 3x + 9 \\ x+3\overline{)x^3 + 0\cdot x^2 + 0\cdot x + 27} \\ \underline{x^3 + 3x^2} \qquad\qquad\quad \\ -3x^2 + 0\cdot x \qquad\quad \\ \underline{-3x^2 - 9x} \qquad\quad \\ 9x + 27 \\ \underline{9x + 27} \\ 0 \end{array}$$

quotient: $x^2 - 3x + 9$ remainder: 0

45. a. The solution is x = 1 or {1}
b. The solution is x = 1 or {1}

Section 4.2 Business Applications

Vocabulary

1. fixed

3. break even

Concepts

5. If C(x) = 4000 + 25x, the unit cost is \$25 and the fixed cost is \$4000.

Applications

7. Let x = number of days beyond deadline.
$$12000(60) = 9400(60) + 10{,}400x$$
$$720000 = 564000 + 10{,}400x$$
$$156000 = 10{,}400x$$
$$15 = x$$
The project can last 15 days beyond deadline.

9. Let x = break-even amount of shoes.
Then $30x = 20x + 9600$
$$10x = 9600$$
$$x = 960$$
960 pairs of shoes is the break-even point.

11. Let x = number of plates
Then $500 + 2x = 800 + x$
$$500 + x = 800$$
$$x = 300$$
300 plates makes the cost the same.

13. Let x = number of spreadsheets sold. Then x database programs and 2x + 15 word processing programs were sold. Thus,
$$150x + 195x + 210(2x + 15) = 72{,}000$$
$$345x + 420x + 3150 = 72000$$
$$765x = 68{,}850$$
$$x = 90$$
Thus 90 each of the spreadsheet and database programs were sold and 195 word processing programs were sold.

15. Let x = number of bolts. Then 2x = number of washers and 2x – 1 = number of locknuts.
Thus, $x + 2x + (2x - 1) = 99$
$$5x - 1 = 99$$
$$5x = 100$$
$$x = 20$$
She has 20 bolts.

17. Let x = number of outstanding stocks. Then 2x = number of average stocks. The total is 3500 so $x + 2x + 200 = 3500$
$$3x + 200 = 3500$$
$$3x = 3300$$
$$x = 1100$$
1100 shares did better than average.

19. Let x = number of wide screen centers in stock. Then x + 40 = number of portables and x + 15 = number of consoles. Thus,
$$1.50(x + 40) + 4(x + 15) + 7.50x = 276$$
$$1.50x + 60 + 4x + 60 + 7.50x = 276$$
$$13x + 120 = 276$$
$$13x = 156$$
$$x = 12$$
There are 12 wide screen centers in stock.

21. Let x = amount invested at 9%.
Then 24000 – x = amount invested at 8%.
Thus, $.09x + .08(24000 - x) = 1965$
$$.09x + 1920 - .01x = 1965$$
$$.01x = 45$$
$$x = 4500$$
\$4500 was invested at 9%, \$19,500 at 8%.

23. Let x = amounts invested at 8% and 11%.
Then $.08x + .11x = 712.50$
$$.19x = 712.50$$
$$x = 3750$$
Thus, \$3750 is invested at each rate.

25. Let x = amount to be invested at 7%. Then
$$.07x + .06(15000) = 1250$$
$$.07x + 900 = 1250$$
$$.07x = 350$$
$$x = 5000$$
She should invest an additional \$5000 at 7%.

27. Let x = lower interest rate. Then x + .01 = higher rate. Then

$$10500x = 21000(x + .01) - 840$$
$$10500x = 21000x + 210 - 840$$
$$-10500x = -630$$
$$x = .06$$

The bond fund rate is 6%, the CD rate is 7%.

Review

29. a. $(4x)^2 = 4^2 x^2 = 16x^2$
 b. $(4x + 3)^2 = (4x)^2 + 2 \cdot 4x \cdot 3 + 3^2 =$
 $16x^2 + 24x + 9$

31. $4\left(\frac{2}{3}x - 1\right) = (2x + 5) - (x - 1)$

$$4\left(\frac{2}{3}x - 1\right) = 2x + 5 - x + 1$$
$$4\left(\frac{2}{3}x - 1\right) = x + 6$$
$$\frac{8}{3}x - 4 = x + 6$$
$$\frac{8}{3}x = x + 10$$
$$\frac{5}{3}x = 10$$
$$x = \frac{3}{5} \cdot 10 = 6$$

a. 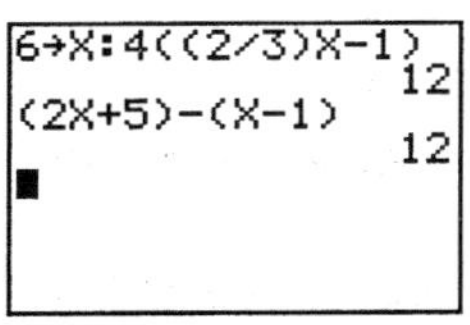

b. 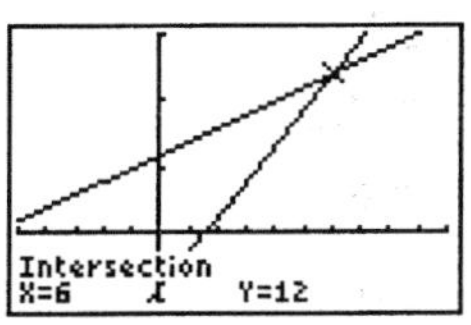

c. 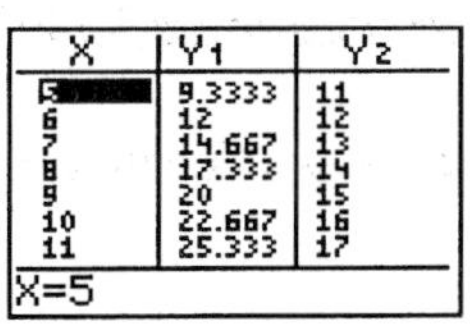

Section 4.3 More Applications of Equations

Concepts

1. a. $3\% = .03 = \frac{3}{100}$
 b. $30\% = .30 = \frac{30}{100} = \frac{3}{10}$
 c. $3\frac{1}{2}\% = 3.5\% = .035 = \frac{35}{1000} = \frac{7}{200}$
 d. $3.5\% = .035 = \frac{35}{1000} = \frac{7}{200}$

3. There was x and 30 – x grams in the bottles.

Applications

5. Let x = number of gallons of \$1.15 per gallon fuel. Then x satisfies the equation

$$1.15x + 20(.85) = (20 + x) \cdot 1$$
$$1.15x + 17 = 20 + x$$
$$.15x = 3$$
$$x = 20$$

You have to mix 20 gal. of the \$1.15 per gallon fuel with 20 gal. of the \$.85 per gallon fuel to get 40 gal. of a mixture worth \$1/gal.

7. Let x = number of gallons of 3% salt solution to be added. Then x satisfies the equation

$$.03x + .07(50) = .05(50 + x)$$
$$.03x + 3.5 = 2.5 + .05x$$
$$1 = .02x$$
$$50 = x$$

50 gal. of a 3% salt solution have to be added to 50 gal. of a 7% salt solution to make a 5% salt solution.

9. Let x = number of ounces of water to be added. Then $.10(30) + 0 \cdot x = .08(30 + x)$

$$3 = 2.4 + .08x$$
$$.6 = .08x$$
$$60 = 8x$$
$$7.5 = x$$

She must add 7.5 ounces of water to dilute the solution from 10% to 8%.

11. Let x = number of pounds of lemon drops to be used. Then x satisfies the equation
$$1.90x + 1.20(100 - x) = 100(1.48)$$
$$1.90x + 120 - 1.20x = 148$$
$$.70x = 28$$
$$x = 40$$
Thus, 40 lbs. of lemon drops and 60 lbs of jelly beans must be used to make a mixture worth \$1.48/lb.

13. Let x = dollar value of a pound of cashews. Then a pound of peanuts is worth x – .30 and x satisfies the equation
$$20x + 20(x - .30) = 40(1.05)$$
$$20x + 20x - 6 = 42$$
$$40x = 48$$
$$x = 1.20$$
Thus, cashews are worth \$1.20/lb.

15. Let x = number of pounds of regular coffee to be used. Then x satisfies the equation
$$40(7) + 4x = (40 + x)5$$
$$280 + 4x = 200 + 5x$$
$$80 = x$$
Thus, 80 lb. of regular coffee should be used.

17. Let x = number of hours for the two cars to meet. Then x satisfies the equation
$$50x + 55x = 315$$
$$105x = 315$$
$$x = 3$$
The two cars will meet in 3 hours.

19. Let x = number of hours for the two cars to be 715 miles apart. Then x satisfies the equation $50x + 60x = 715$
$$110x = 715$$
$$x = 6.5$$
They will be 715 miles apart after 6.5 hours.

21. Let x = number of hours for the cars to be 82.5 miles apart. The distance the faster car travels minus the distance the slower car travels must be 82.5 miles. Thus x satisfies the equation $53x - 42x = 82.5$.
$$11x = 82.5$$
$$x = 7.5$$
The cars will be 82.5 mi. apart after 7.5 hr.

23. Let x = speed of the slower plane in mph. Then x + 200 is the speed of the faster plane. Since they meet in 5 hours x must satisfy
$$5x + 5(x + 200) = 6000$$
$$5x + 5x + 1000 = 6000$$
$$10x = 5000$$
$$x = 500$$
The speed of the slower plane is 500 mph.

Review

25. a. $[4-(2-3)^2+5]-(6-2)^3 = [4-(-1)^2+5] - 4^3 = [4-1+5]-64 = 8-64 = -56$

b. $(3x+5)(2x^2-x+1) = 3x(2x^2) - 3x \cdot x + 3x \cdot 1 + 5(2x^2) - 5x + 5 = 6x^3 - 3x^2 + 3x + 10x^2 - 5x + 5 = 6x^3 + 7x^2 - 2x + 5$

c. $$\frac{4x^2y^3 - 6xy^2 + 8x^3}{4xy} =$$
$$\frac{1}{4xy}\left(4x^2y^3 - 6xy^2 + 8x^3\right) =$$
$$\frac{4x^2y^3}{4xy} - \frac{6xy^2}{4xy} + \frac{8x^3}{4xy} = xy^2 - \frac{3}{2}y + \frac{2x^2}{y}$$

27. 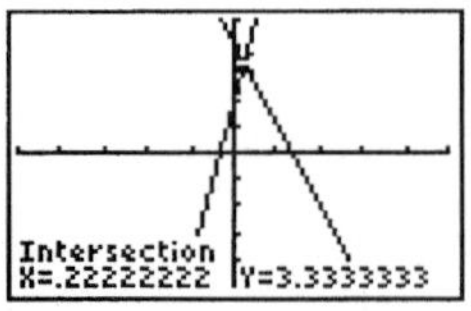

The solution is 2/9.

29. Let x represent the integer. Then x satisfies
$$3(x + 2) + 4 = 13$$
$$3x + 6 + 4 = 13$$
$$3x + 10 = 13$$
$$3x = 3$$
$$x = 1$$
The desired integer is 1.

Section 4.4 Equations Containing Absolute Value

Vocabulary

1. x; -x

Concepts

3. a. $|4| = 4$ b. $|-4| = 4$ c. $-|-4| = -(4) = -4$
   d. $|a| = \pm a$ e. $|3 - \pi| = \pi - 3$

5. Since absolute value is never negative, the statement $|x| = -2$ makes no sense.

Practice

7. $|x| = 8$, $x = \pm 8$

9. $|x - 3| = 6$
   $x - 3 = 6$ or $x - 3 = -6$
   $x = 6 + 3 = 9$ or $x = -6 + 3 = -3$

11. $|2x - 3| = 5$
   $2x - 3 = 5$ or $2x - 3 = -5$
   $2x = 5 + 3 = 8$ or $2x - 3 = -5 + 3 = -2$
   $x = 4$ or $x = -1$

13. $|3x + 2| = 16$
   $3x + 2 = 16$ or $3x + 2 = -16$
   $3x = 14$ or $3x = -18$
   $x = \frac{14}{3}$ or $x = -6$

15. $\left|\frac{7}{2}x + 3\right| = -5$ No solution since absolute value can never be negative.

17. $\left|\frac{x}{2} - 1\right| = 3$
   $\frac{x}{2} - 1 = 3$ or $\frac{x}{2} - 1 = -3$
   $\frac{x}{2} = 3 + 1 = 4$ or $\frac{x}{2} = 3 + (-1) = 2$
   $x = 2 \cdot 4 = 8$ or $x = 2 \cdot (-2) = -4$

19. $|3 - 4x| = 5$
   $3 - 4x = 5$ or $3 - 4x = -5$
   $-4x = 5 - 3 = 2$ or $-4x = -5 - 3 = -8$
   $x = \frac{2}{-4} = \frac{-1}{2}$ or $x = \frac{-8}{-4} = 2$

21. $|3x + 24| = 0$
   $3x + 24 = 0$
   $3x = -24$
   $x = -8$

23. $\left|\frac{3x + 48}{3}\right| = 12$
   $\frac{3x + 48}{3} = 12$ or $\frac{3x + 48}{3} = -12$
   $3x + 48 = 36$ or $3x + 48 = -36$
   $3x = -12$ or $3x = -84$
   $x = -4$ or $x = -28$

25. $|x + 3| + 7 = 10$ Thus, $|x + 3| = 3$. Then
   $x + 3 = 3$ or $x + 3 = -3$
   $x = 0$ or $x = -6$

27. $\left|\frac{3}{5}x-4\right|-2=-2$

$\left|\frac{3}{5}x-4\right|=0$

$\frac{3}{5}x-4=0$

$\frac{3}{5}x=4$

$x=4\cdot\frac{5}{3}=\frac{20}{3}$

29. $|2x+1|=|3x+3|$

$2x+1=3x+3$ or $2x+1=-(3x+3)$

$-x=2$ or $2x+1=-3x-3$

$x=-2$ or $5x=-4$

$x=-2$ or $x=\frac{-4}{5}$

31. $|3x-1|=|x+5|$

$3x-1=x+5$ or $3x-1=-(x+5)$

$2x=6$ or $3x-1=-x-5$

$x=3$ or $4x=-4$

$x=3$ or $x=-1$

33. $|2-x|=|3x+2|$

$2-x=3x+2$ or $2-x=-(3x+2)$

$0=4x$ or $2-x=-3x-2$

$0=x$ or $2x=-4$

$x=0$ or $x=-2$

35. $\left|\frac{x}{2}+2\right|=\left|\frac{x}{2}-2\right|$

$\frac{x}{2}+2=\frac{x}{2}-2$ or $\frac{x}{2}+2=-\left(\frac{x}{2}-2\right)$

$2=-2$ or $\frac{x}{2}+2=2-\frac{x}{2}$

$\varnothing$ or $x+4=4-x$

$2x=0$ so that $x=0$

The only solution is $x = 0$.

37. $\left|x+\frac{1}{3}\right|=|x-3|$

$x+\frac{1}{3}=x-3$ or $x+\frac{1}{3}=-(x-3)$

$\frac{1}{3}=-3$ or $x+\frac{1}{3}=3-x$

$\varnothing$ or $2x=\frac{8}{3}$

$x=\frac{4}{3}$

The only solution is $x = 4/3$.

39. $|3x+7| = -|8x-2|$

There is no solution unless both sides are zero for the same value of x and this does not happen. Thus, no solution.

41. One example is a = 2, b = -3.

43. One example is a = 2, b = -3.

Review

45. a. $\frac{4}{5}+\frac{-8}{9}=\frac{4\cdot9}{5\cdot9}+\frac{-8\cdot5}{9\cdot5}=\frac{36}{45}+\frac{-40}{45}=\frac{-4}{45}$

b. $\left(-\frac{7}{8}\right)\left(-\frac{3}{5}\right)\div\left(\frac{-9}{20}\right)=\frac{21}{40}\div\frac{-9}{20}=\frac{21}{40}\cdot\frac{20}{-9}=$

$\frac{3\cdot7\cdot20}{2\cdot20\cdot3\cdot-3}=\frac{7}{-6}=-\frac{7}{6}$

47. $P(x) = -2x^2 - x + 2$. Thus, $P(-1) =$ $-2(-1)^2 - (-1) + 2 = -2\cdot1 + 1 + 2 = 1$

49.

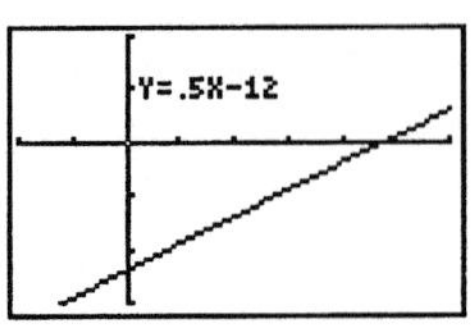

The window is [-10,30] by [-15,10].

Chapter 4 Summary

1. a. $\frac{-4}{5} = \frac{28}{-35}$ is a proportion since (-4)(-35) = 140 = 5(28)

b. $\frac{14}{19} = \frac{-42}{57}$ is not a proportion since 14(57) ≠ 19(-42)

c. $\frac{x+4}{x-2} = \frac{3x}{x+1}$ is not a proportion since $(x+1)(x+4) \neq 3x(x-2)$

d. $\frac{x+1}{x^2+x} = \frac{2}{2x}$ is a proportion since $2x(x+1) = 2x^2 + 2x = 2(x^2 + x)$

2. a. $\frac{x-4}{4} = \frac{x+4}{8}$

$8(x-4) = 4(x+4)$

$8x - 32 = 4x + 16$

$4x = 48$

$x = 12$

b. $\frac{x+3}{2} = \frac{x-3}{3}$

$3(x+3) = 2(x-3)$

$3x + 9 = 2x - 6$

$x = -15$

c. $\frac{x-5}{3} = \frac{2x+1}{2}$

$2(x-5) = 3(2x+1)$

$2x - 10 = 6x + 3$

$-13 = 4x$

$\frac{-13}{4} = x$

d. $\frac{3x-1}{2} = \frac{4x+5}{3}$

$3(3x-1) = 2(4x+5)$

$9x - 3 = 8x + 10$

$x = 13$

3. Let w = width of the painting. Then w satisfies $2w + 2(3+2w) = 84$

$2w + 6 + 4w = 84$

$6w + 6 = 84$

$6w = 78$

$w = 13$

The painting is 13 inches wide.

4. Let x = number of tons or iron ore needed.

Then $\frac{5}{3} = \frac{x}{18}$

$3x = 5 \cdot 18$

$3x = 90$

$x = 30$

Thus, 30 tons of iron ore is needed to make 18 tons of pig iron.

5. Let x = number of grams in a single dose.

Then $\frac{3}{300} = \frac{x}{5}$

$300x = 15$

$x = \frac{15}{300} = \frac{1}{20}$

There is 1/20 gram in a single 5 ml. dose.

6. Let h = height of the building. By similar triangles, $\frac{h}{5} = \frac{53}{2}$

$2h = 5 \cdot 53 = 265$

$h = 132.5$

The building is 132.5 feet tall.

7. Let x = break-even point. Then x satisfies

$3x = 1.25x + 668.50$

$1.75x = 668.50$

$x = 382$

Thus, 382 parts is the break-even point.

8. Let x = number of caps to be sold. Then

$3x + 85 = 2.5x + 105$

$.5x = 20$

$x = 40$

At 40 caps, the two machines give equal cost.

9. Let x = amount to be invested at 7%. Then 27000 – x = amount to be invested at 9% and x satisfies $.07x + .09(27000 - x) = 2110$

$$.07x + 2430 - .09x = 2110$$
$$-.02x = -320$$
$$x = 16,000$$

Thus, \$16,000 was invested at 7% and \$11,000 at 9%.

10. Let x = amount invested at 4 ½ %. Then 13,750 – x is invested at 5% and x satisfies

$$.045x + .05(13750 - x) = 647.50$$
$$.045 + 687.50 - .05x = 647.50$$
$$-.005x = -40$$
$$x = 8000$$

\$8,000 was invested at 4 ½ %.

11. Let x = number of liters of water to be added. Then x satisfies $0 \cdot x + .10(30) = .08(x + 30)$

$$3 = .08x + 2.4$$
$$.6 = .08x$$
$$7.5 = x$$

7.5 liters must be added to dilute it to an 8% solution.

12. Let x = number of pounds of \$.90/lb candy. Then 20 – x = number of pounds of \$1.65/lb candy and x satisfies the equation

$$.90x + 1.65(20 - x) = 20(1.20)$$
$$90x + 165(20 - x) = 20(120)$$
$$90x + 3300 - 165x = 2400$$
$$-75x = -900$$
$$x = 12$$

Thus, 12 pounds of the \$.90/lb candy must be mixed with 8 pounds of the \$1.65/lb. candy.

13. Let x = number of hours until they meet. Then x satisfies $3x + 12x = 5$

$$15x = 5$$
$$x = \frac{1}{3}$$

They will meet in 1/3 hour or 20 minutes.

14. a. $|3x + 1| = 10$

$$3x + 1 = 10 \text{ or } 3x + 1 = -10$$
$$3x = 9 \text{ or } 3x = -11$$
$$x = 3 \text{ or } x = \frac{-11}{3}$$

b. $\left|\frac{3}{2}x - 4\right| = 9$

$$\frac{3}{2}x - 4 = 9 \text{ or } \frac{3}{2}x - 4 = -9$$
$$\frac{3}{2}x = 13 \text{ or } \frac{3}{2}x = -5$$
$$x = \frac{26}{3} \text{ or } x = \frac{-10}{3}$$

c. $|x - 5| = -3$ No solution since absolute value is never negative.

d. $|2x - 3| + 2 = 5$

$$|2x - 3| = 3$$
$$2x - 3 = 3 \text{ or } 2x - 3 = -3$$
$$2x = 6 \text{ or } 2x = 0$$
$$x = 3 \text{ or } x = 0$$

e. $|4x - 1| - 7 = -3$

$$|4x - 1| = 4$$
$$4x - 1 = 4 \text{ or } 4x - 1 = -4$$
$$4x = 5 \text{ or } 4x = -3$$
$$x = \frac{5}{4} \text{ or } x = \frac{-3}{4}$$

15. a. $|3x + 2| = |2x - 3|$

$$3x + 2 = 2x - 3 \text{ or } 3x + 2 = -(2x - 3)$$
$$x + 2 = -3 \text{ or } 3x + 2 = -2x + 3$$
$$x = -5 \text{ or } 5x = 1$$
$$x = -5 \text{ or } x = \frac{1}{5}$$

b. $|5x - 4| = |4x - 5|$

$$5x - 4 = 4x - 5 \text{ or } 5x - 4 = -(4x - 5)$$
$$x - 4 = -5 \text{ or } 5x - 4 = -4x + 5$$
$$x = -1 \text{ or } 9x = 9$$
$$x = -1 \text{ or } x = 1$$

Chapter 4 Test

1. $\frac{-42}{22} = \frac{14}{-11}$ is not a proportion since $(-11)(-42) \neq 22(14)$

2. $\frac{2x-1}{3} = \frac{4x-1}{6}$ is not a proportion since $6(2x-1) \neq 3(4x-1)$.

3. $$\frac{4-x}{13} = \frac{11}{26}$$
$$26(4-x) = 13(11)$$
$$104 - 26x = 143$$
$$-26x = 39$$
$$x = \frac{-39}{26} = \frac{-3}{2}$$

4. $$\frac{2x+5}{4} = \frac{x-1}{6}$$
$$6(2x+5) = 4(x-1)$$
$$12x + 30 = 4x - 4$$
$$8x + 30 = -4$$
$$8x = -34$$
$$x = \frac{-34}{8} = \frac{-17}{4}$$

5. Let x = amount of altitude the plane will lose. Then $\frac{575}{\frac{1}{2}} = \frac{x}{7}$
$$7(575) = \frac{1}{2}x$$
$$14(575) = x$$
$$8050 = x$$
The plane will lose 8050 ft. in altitude.

6. Let x = number of hours for the truck to meet the car. Then the car travels 65x miles, the truck travels 55x miles and so x satisfies
$$65x + 55x = 72$$
$$120x = 72$$
$$x = \frac{72}{120} = \frac{6}{10}$$
They will meet after 6/10 hr. or 36 min.

7. Let x = amount invested at 6%. Then 50000 – x = amount invested at 5 ½ % and x satisfies $.06x + .055(50000 - x) = 2975$
$$.06x + 2750 - .055x = 2975$$
$$.005x = 225$$
$$x = 45000$$
$45,000 is in the certificate of deposit.

8. Let x = number of ounces of juice to be used. Then $\frac{x}{x+13} = \frac{35}{100}$
$$100x = 35(x+13)$$
$$100x = 35x + 455$$
$$65x = 455$$
$$x = 7$$
7 ounces of juice must be used.

9. Let x = length of the shortest side. The other sides are 2x + 4 and 2x. Then x satisfies
$$x + (2x+4) + 2x = 24$$
$$x + 2x + 4 + 2x = 24$$
$$5x + 4 = 24$$
$$5x = 20$$
$$x = 4$$
The short side is 4 units long.

10. Let x = number of envelopes she must stuff. Then $100 + .35x = .45x$
$$100 = .10x$$
$$1000 = x$$
She must stuff 1000 envelopes to break even.

11. $\left|\frac{2}{3}x-5\right|=6$

$\frac{2}{3}x-5=6$ or $\frac{2}{3}x-5=-6$

$\frac{2}{3}x=11$ or $\frac{2}{3}x=-1$

$x=\frac{33}{2}$ or $x=\frac{-3}{2}$

12. $\left|4x+\frac{2}{3}\right|=-1$ No solution since absolute value is never negative.

13. $|x+5|-3=2$

$|x+5|=5$

$x+5=5$ or $x+5=-5$

$x=0$ or $x=-10$

14. $|3x-2|+4=2$ so $|3x-2|=-2$. Thus, there is no solution since absolute value is never negative.

15. $|x-5|=-|x+3|$ No solution since both sides cannot be zero for the same value of x.

16. $|4x+1|=|2x-3|$

$4x+1=2x-3$ or $4x+1=-(2x-3)$

$2x+1=-3$ or $4x+1=-2x+3$

$2x=-4$ or $6x=2$

$x=-2$ or $x=\frac{1}{3}$

Cumulative Review: Chapters 3 - 4

Vocabulary/Concepts

1. for all; $\mathbb{R}$

2. no; { } or ϕ

3. ordered pairs

4. domain

5. range

6. a. quadrant II
   b. quadrant III
   c. quadrant IV
   d. quadrant IV

7. 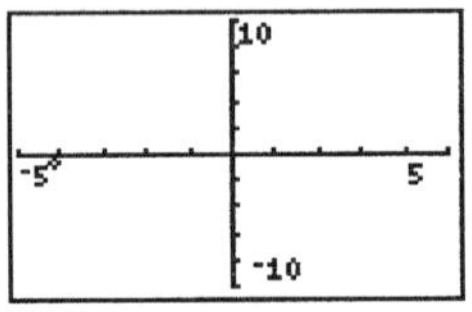

8. In ZInteger, all x-coordinates are integers.

9. A trip of 50 miles will cost $49.50 rental.

10. STOre ; TABLE

11. comparison

12. *18 to 24* means $\frac{18}{24}=\frac{3}{4}$

13. similar; proportion

14. Revenue = 16.99x

15. a. 6% = .06
    b. 7 ¾ % = 7.75% = .0775

16. There are x liters of alcohol.

17. a; -a

18. It is invalid because $|x| \geq 0$ always.

19. a. $-|-3|=-(3)=-3$
    b. $|2-\pi|=-(2-\pi)=\pi-2$

20. $i=prt$ so that $p=\frac{i}{rt}$

Practice

21. $3x + 4(2 - x) = 3(x + 4) - 5$

$$3x + 8 - 4x = 3x + 12 - 5$$
$$-x + 8 = 3x + 7$$
$$1 = 4x$$
$$\frac{1}{4} = x$$

22. $\frac{3(x-1)}{6} - 4 = \frac{x+3}{6}$

$$3(x-1) - 24 = x + 3$$
$$3x - 3 - 24 = x + 3$$
$$3x - 27 = x + 3$$
$$2x = 30$$
$$x = 15$$

23.

x	y
-3	12
-2	7
-1	4
0	3
1	4
2	7
3	12

$y = x^2 + 3$

$12 = (-3)^2 + 3$
$7 = (-2)^2 + 3$
$4 = (-1)^2 + 3$
$3 = 0^2 + 3$
$4 = 1^2 + 3$
$7 = 2^2 + 3$
$12 = 3^2 = 3$

24. b. (-2,-1) is not a solution of $2x + y = 5$ since $2(-2) + (-1) = -4 + -1 = -5 \neq 5$. All other pairs are solutions.

25. b. {(2,4), (2,-4),(3,1), (-3,1)} is not a function since two pairs have the same x coordinate

26. The x-intercepts are (-5,0) and (2,0) while the y-intercept is (0,-10).

27. $y = 4x - 1$

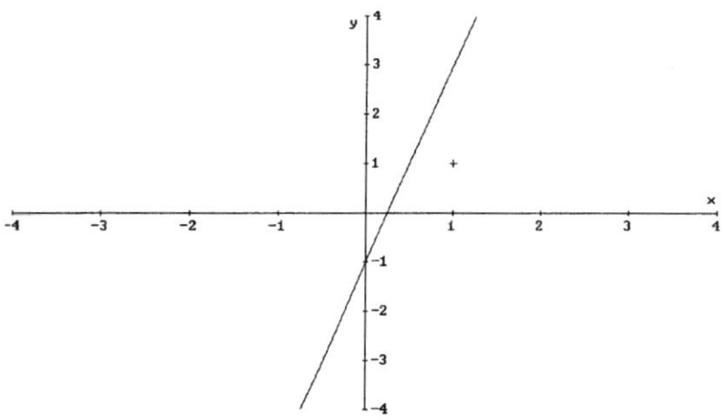

28. Given $y = x^2 - 8x + 7$. The x-intercepts are (1,0) and (7,0) while the y-intercept is (0,7).

29. ZInteger is a better window than ZStandard in exercise 28 above because ZStandard has no pixel with x-coordinates 1 or 7 or y-coordinate 7.

30. ZStandard is not a good window for solving the equation $3 + [x - (2 - x)] = -16$ since -16 is outside the y range in ZStandard. Ymin must be set to -17 or smaller to get a good window.

31. $\frac{4(x+1)}{3} = \frac{x-2}{4}$

$$4 \cdot 4 \cdot (x+1) = 3(x-2)$$
$$16(x+1) = 3(x-2)$$
$$16x + 16 = 3x - 6$$
$$13x = -22$$
$$x = \frac{-22}{13}$$

32. $|6x - 5| = 2$

$$6x - 5 = 2 \text{ or } 6x - 5 = -2$$
$$6x = 7 \text{ or } 6x = 3$$
$$x = \frac{7}{6} \text{ or } x = \frac{1}{2}$$

33. $\left|\frac{x}{2}+1\right|-5=3$

$\left|\frac{x}{2}+1\right|=8$

$\frac{x}{2}+1=8$ or $\frac{x}{2}+1=-8$

$x+2=16$ or $x+2=-16$

$x=14$ or $x=-18$

34. $|2x-1| = |3x+4|$

$2x-1=3x+4$ or $2x-1=-(3x+4)$

$-x-1=4$ or $2x-1=-3x-4$

$-x=5$ or $5x=-3$

$x=-5$ or $x=\frac{-3}{5}$

35. $\frac{1}{2}x-\frac{1}{3}y=6$

$-\frac{1}{3}y=6-\frac{1}{2}x$

$y=-3\cdot 6-(-3)\frac{1}{2}x$

$y=-18+\frac{3}{2}x$

36. $Ax+By=C$

$By=C-Ax$

$y=\frac{C}{B}-\frac{A}{B}x$

37. Solve graphically: $\left|\frac{1}{2}x-1\right|=6$

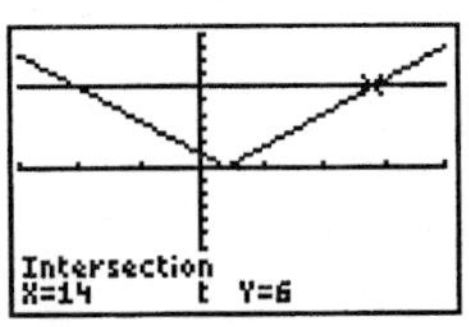

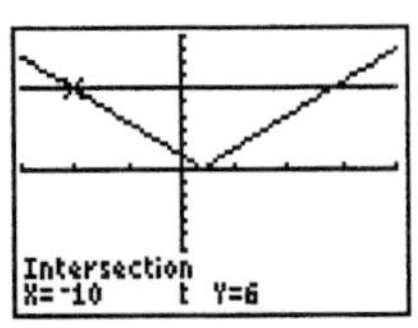

The solutions are x = 14 and x = -10.

38. Solve graphically: $-|x+2| = -3$

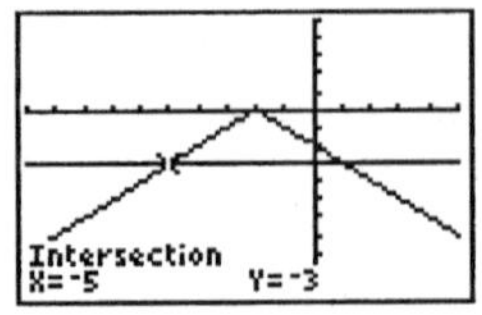

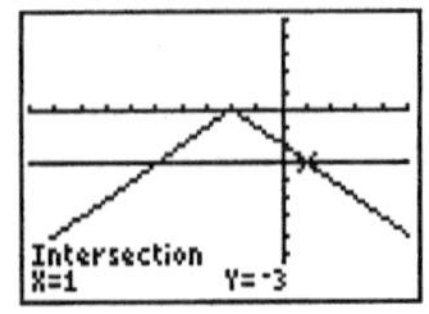

The solutions are x = -5 and x = 1.

39. Solve graphically: $|x+2| = -3$

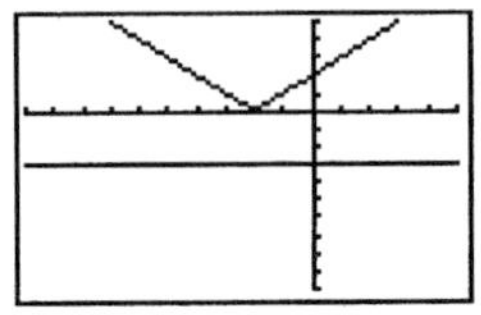

No solution since the graphs don't intersect.

40. Solve graphically: $|3x+2| = |x-4|$

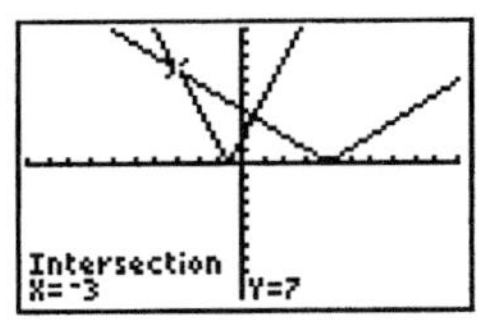

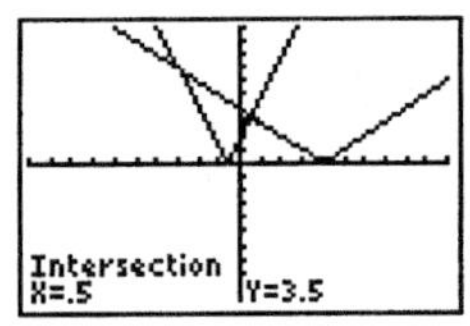

The solutions are x = -3 and x = .5.

Applications

41. The amounts of 3.75 oz. vinegar and 1.25 oz. alcohol are correct since 3.75 + 1.25 = 5 oz and $\frac{3.75}{1.25}=\frac{3}{1}$.

42. Let x and x + 1 represent the integers. Then

$x+3(x+1)=171$

$x+3x+3=171$

$4x=168$

$x=42$

The smaller integer is 42.

43. Let t = number of years for the accounts to have the same amount of money. Then

$$3000 + .07(3000)t = 4000 + .04(4000)t$$
$$3000 + 210t = 4000 + 160t$$
$$50t = 1000$$
$$t = 20$$

It will take 20 years for them to have the same amount of money.

44. Let x = number of 29¢ stamps and so 4x = number of 33¢ stamps. Then x satisfies

$$29x + 33(4x) = 2415$$
$$29x + 132x = 2415$$
$$161x = 2415$$
$$x = 15$$

She purchased 4(15) = 60 of the 33¢ stamps.

45. Let x = number of hours for the vehicle to overtake the transport. Then x + 1 = number of hours the transport will travel. Thus, when the vehicle overtakes the transport,

$$70x = 55(x+1)$$
$$70x = 55x + 55$$
$$15x = 55$$
$$x = \frac{55}{15} = \frac{11}{3}$$

The vehicle will overtake the transport in 11/3 hrs. or 3 hours and 40 minutes.

46. Let x = number of ounces of the 10% juice beverage to be used. The x will satisfy

$$.10x + .50(64) = .35(64 + x)$$
$$10x + 50(64) = 35(64 + x)$$
$$10x + 3200 = 2240 + 35x$$
$$960 = 25x$$
$$38.4 = x$$

38.4 ounces of the 10% juice beverage will be needed.

47. When r = 9 in., $V = \frac{4}{3}\pi r^3$

$$= \frac{4}{3}\pi(9^3)$$
$$\approx 3053.6 \text{ in.}^3 \text{ of air}$$

48. Let t = number of years the money was invested. Then t satisfies the equation

$$2937.50 = 2500(1 + .035t)$$
$$\frac{2937.50}{2500} = 1 + .035t$$
$$1.175 = 1 + .035t$$
$$.175 = .035t$$
$$5 = t$$

The money was invested for 5 years.

49. a. The revenue is \$20,700 when x = 230 CD players.

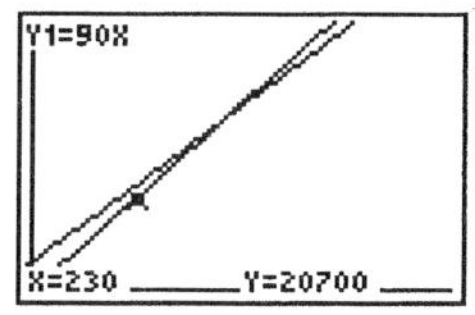

b. When revenue is \$15,300 the number of CD players is 170.

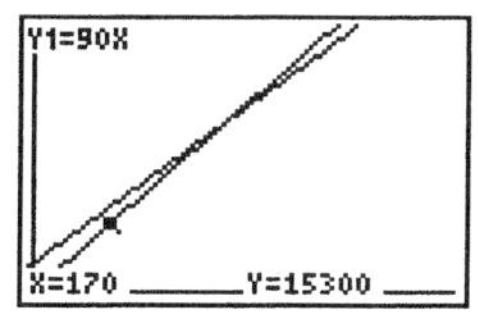

50. a. When 170 CD players are produced the cost is \$19,155.

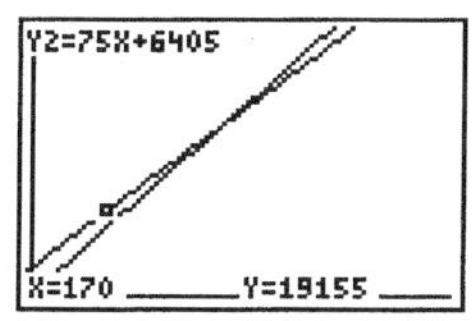

b. For 170 CD players, the cost is greater than the revenue.

c. If the cost is \$52,905 then 620 CD players are produced. The revenue is \$55,800 so there is a profit.

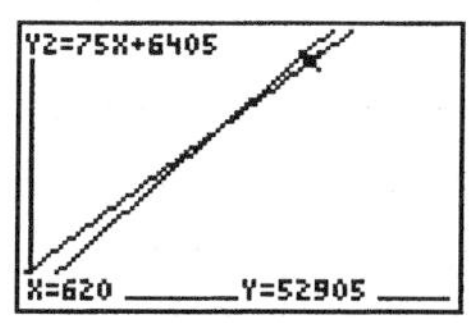

Chapter 5 Factoring

Section 5.1 Using the Distributive Property to Factor

Vocabulary

1. factors

3. prime

5. distributive

7. polynomial

9. grouping

Concepts

11. F 1 is not a prime

13. T $12 = 2^2 \cdot 3$ and $16 = 2^4$, so GCF $= 2^2 = 4$

15. T All factors are prime.

17. F $2xy - x$ can be factored further.

19. T

21. $36y = 2 \cdot 2 \cdot 3 \cdot 3 \cdot y$ and $6 = 2 \cdot 3$. The common factors are 2 and 3, so the GCF $= 2 \cdot 3 = 6$

23. $4x^3y^5 = 2 \cdot 2 \cdot x^3 \cdot y^5$
$2x^2y^4 = 2 \cdot x^2 \cdot y^4$
$6x^3y^3 = 2 \cdot 3 \cdot x^3 \cdot y^3$
The greatest common numerical factor is 2. The smallest powers on x and y are 2 and 3, respectively. Thus, the GCF $= 2x^2y^3$.

25. $20t^3v = 2^2 \cdot t^3 \cdot v$
$15t^2v^3 = 3 \cdot 5 \cdot t^2 \cdot v^3$
$3tv = 3 \cdot t \cdot v$
The greatest common numerical factor is 1. The smallest powers on t and v are both 1. Thus, the GCF $= 1 \cdot tv = tv$

27. $6x^3 + 3x = 3x(2x^2) + 3x(1) = 3x(\mathbf{2x^2 + 1})$

29. $5a^3b^2 + 7a^2b = a^2b(5ab) + a^2b(7) = a^2b(\mathbf{5ab + 7})$

31. $-4x^3y - 2xy = -2xy(2x^2) - 2xy(1) = -2xy(\mathbf{2x^2 + 1})$

Practice

33. The GCF is x. $xy - xz = x(y - z)$

35. The GCF is -1. $-2x + 5y = (-1)2x - (-1)5y = (-1)(2x - 5y) = -(2x - 5y)$

37. The GCF is r^2. $r^4 - r^2 = r^2 \cdot r^2 - r^2 \cdot 1 = r^2(r^2 - 1)$

39. The GCF is -1. $-3m - 4n + 1 = (-1)3m + (-1)4n - (-1)1 = (-1)(3m + 4n - 1) = -(3m + 4n - 1)$

41. The GCF is $8xy^2z^3$. $24x^2y^3z^4 + 8xy^2z^3 = 8xy^2z^3(3xyz) + 8xy^2z^3(1) = 8xy^2z^3(3xyz + 1)$

43. The GCF is 3. $3x + 3y - 6z = 3 \cdot x + 3 \cdot y - 3 \cdot 2z = 3(x + y - 2z)$

45. The GCF is -1. $-3ab - 5ac + 9bc = (-1)3ab + (-1)(5ac) - (-1)9bc = (-1)(3ab + 5ac - 9bc) = -(3ab + 5ac - 9bc)$

47. The GCF is 2y. $4y^2 + 8y - 2xy = 2y(2y) + 2y(4) - 2y \cdot x = 2y(2y + 4 - x)$

49. The GCF is $-4a^2b^2$. $-4a^2b^3 + 12a^3b^2 = (-4a^2b^2)(b) - (-4a^2b^2)(3a) = (-4a^2b^2)(b - 3a)$

51. The GCF is -7ab. $-14a^6b^6 + 49a^2b^3 - 21ab = -7ab(2a^5b^5) + (-7ab)(-7ab^2) - (-7ab)(-3) = -7ab(2a^5b^5 - 7ab^2 + 3)$

53. The GCF is $2xyz^2$. $4x^2y^2z^2 - 6xy^2z^2 + 12xyz^2 = 2xyz^2(2xy) - 2xyz^2(3y) + 2xyz^2(6) = 2xyz^2(2xy - 3y + 6)$

55. The GCF is a + 2b. $x(a + 2b) + y(a + 2b) = (a + 2b)(x + y)$.

57. The GCF is 3t + 5. $(3t + 5)^2 - (3t + 5) =$
$(3t + 5)(3t + 5) - (3t + 5)\cdot 1 =$
$(3t + 5)(3t + 5 - 1) = (3t + 5)(3t + 4)$

59. The GCF is 3ab(3x – 2y). $9a^2b^2(3x - 2y) -$
$6ab(3x - 2y) = 3ab(3x - 2y)(3ab) -$
$3ab(3x - 2y)\cdot 2 = 3ab(3x - 2y)(3ab - 2)$

61. The GCF is x – 5y. $(x - 5y)(a + 2) - (x - 5y)$
$= (x - 5y)(a + 2) - (x - 5y)\cdot 1 =$
$(x - 5y)(a + 2 - 1) = (x - 5y)(a + 1)$

63. $pm - pn + qm - qn = (pm - pn) + (qm - qn)$
$= p(m - n) + q(m - n) = (m - n)(p + q)$

65. $3ac + a + 3bc + b = (3ac + a) + (3bc + b) =$
$a(3c + 1) + b(3c + 1) = (3c + 1)(a + b)$

67. $3ab + 9a - 2b - 6 = (3ab + 9a) - (2b + 6) =$
$3a(b + 3) - 2(b + 3) = (b + 3)(3a - 2)$

69. $6x^2u - 3x^2v + 2yu - yv = (6x^2u - 3x^2v) +$
$(2yu - yv) = 3x^2(2u - v) + y(2u - v) =$
$(2u - v)(3x^2 + y)$

71. $p(m - n) - q(n - m) =$
$p(m - n) - q(-1)(m - n) =$
$p(m - n) + q(m - n) = (m - n)(p + q)$

73. $5x + ry + rx + 5y = (5x + 5y) + (rx + ry) =$
$5(x + y) + r(x + y) = (x + y)(5 + r)$

75. $mr + ns + ms + nr = (mr + ms) + (nr + ns) =$
$m(r + s) + n(r + s) = (r + s)(m + n)$

77. $sx - ry + rx - sy = sx + rx - sy - ry =$
$(sx + rx) - (sy + ry) = x(s + r) - y(s + r) =$
$(s + r)(x - y)$

79. The first student's factorization is not correct. Each side of the equation will evaluate to the same thing in a correct factorization.
The second student's factorization is probably correct, although it is possible that these expressions might <u>not</u> have the same value for other values of x and y.

Review

81. $3x + 4y - 2(x - 2y) = 3x + 4y - 2x + 4y =$
$(3x - 2x) + (4y + 4y) = x + 8y$

83. a. $3(x+2) - 4x = \frac{1}{2}(x-3) + 2$

$$3x + 6 - 4x = \frac{1}{2}(x-3) + 2$$
$$6 - x = \frac{1}{2}(x-3) + 2$$
$$2(6 - x) = 2\cdot\frac{1}{2}(x-3) + 2\cdot 2$$
$$12 - 2x = (x-3) + 4$$
$$12 - 2x = x - 3 + 4$$
$$12 - 2x = x + 1$$
$$11 = 3x$$
$$\frac{11}{3} = x$$

b. 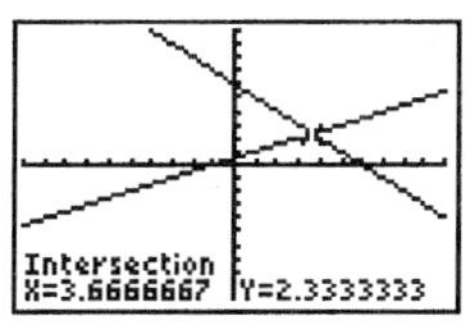

c. The solution is a rational number since it can be expressed as the quotient of two integers.

Section 5.2 Factoring Trinomials

Vocabulary

1. trinomials

3. multiplication

5. factors

Concepts

7. Given $x^2 + 5x + 6$
a. The leading coefficient is 1. $(1\cdot x^2)$

b. The coefficient of the second term is 5.
c. The coefficient of the first term is 1.
d. The constant term is 6.

9. $x^2 - 2x - 15$ a. $(x + 3)(x - 5)$

11. $6x^2 - x - 15$ c. $(2x + 3)(3x - 5)$

13. $2x^2 + 2x - 4$ b. $2(x + 2)(x - 1)$

15. Both factorizations are correct since real number multiplication is commutative.

Practice

17. $x^2 - 9x - 12$
$ac = 1(-12) = -12$ and no pair of factors of -12 add up to -9. Thus, it is prime.

19. $4 - 5x + x^2$ Rewrite as $x^2 - 5x + 4$.
$ac = 1(4) = 4 = (-1)(-4)$ and $-1 + (-4) = -5$.
Thus, $x^2 - 5x + 4 = x^2 - x - 4x + 4 = (x^2 - x) - (4x - 4) = x(x - 1) - 4(x - 1) = (x - 1)(x - 4)$

21. $x^2 + 3x + 2$
$ac = 1 \cdot 2 = 2$ and $1 + 2 = 3$. Thus, $x^2 + 3x + 2 = x^2 + x + 2x + 2 = (x^2 + x) + (2x + 2) = x(x + 1) + 2(x + 1) = (x + 1)(x + 2)$

23. $2x^2 + 10x + 12 = 2(x^2 + 5x + 6)$
$ac = 1 \cdot 6 = 6 = 3 \cdot 2$ and $3 + 2 = 5$. Thus,
$x^2 + 5x + 6 = x^2 + 3x + 2x + 6 = (x^2 + 3x) + (2x + 6) = x(x + 3) + 2(x + 3) = (x + 3)(x + 2)$
The complete factorization is $2(x + 3)(x + 2)$.

25. $-5t^2 + 25t - 30 = -5(t^2 - 5t + 6)$
$ac = 1 \cdot 6 = 6 = (-2)(-3)$ and $-2 + (-3) = -5$
Thus, $t^2 - 5t + 6 = t^2 - 2t - 3t + 6 = (t^2 - 2t) - (3t - 6) = t(t - 2) - 3(t - 2) = (t - 2)(t - 3)$.
The complete factorization is $-5(t - 2)(t - 3)$

27. $x^2 + 9x + 5$
$ac = 1 \cdot 5 = 5$ and no pair of factors of 5 will add up to 9. Thus, $x^2 + 9x + 5$ is prime.

29. $u^2 + 2uv - 15v^2$
$ac = 1(-15) = -15 = 5(-3)$ and $5 + (-3) = 2$
Thus, $u^2 + 2uv - 15v^2 = u^2 + 5uv - 3uv - 15v^2 = (u^2 + 5uv) - (3uv + 15v^2) = u(u + 5v) - 3v(u + 5v) = (u + 5v)(u - 3v)$

31. $-3rs + r^2 + 2s^2$ Rewrite as $r^2 - 3rs + 2s^2$.
$ac = 1 \cdot 2 = 2 = (-1)(-2)$ and $-1 + (-2) = -3$
Thus, $r^2 - 3rs + 2s^2 = r^2 - rs - 2rs + 2s^2 = (r^2 - rs) - (2rs - 2s^2) = r(r - s) - 2s(r - s) = (r - s)(r - 2s)$

33. $-4x^2y - 4x^3 + 24xy^2 = -4x(xy + x^2 - 6y^2) = -4x(x^2 + xy - 6y^2)$. In $x^2 + xy - 6y^2$, $ac = 1(-6) = -6 = 3(-2)$ and $3 + (-2) = 1$. Thus $x^2 + xy - 6y^2 = x^2 + 3xy - 2xy - 6y^2 = (x^2 + 3xy) - (2xy + 6y^2) = x(x + 3y) - 2y(x + 3y) = (x + 3y)(x - 2y)$. The complete factorization is $-4x(x + 3y)(x - 2y)$

35. $24y^2 + 14xy + 2x^2 = 2(12y^2 + 7xy + x^2)$ In $12y^2 + 7xy + x^2$, $ac = 12 \cdot 1 = 4 \cdot 3$ and $4 + 3 = 7$
Thus, $12y^2 + 7xy + x^2 = 12y^2 + 4xy + 3xy + x^2 = (12y^2 + 4xy) + (3xy + x^2) = 4y(3y + x) + x(3y + x) = (3y + x)(4y + x)$. The complete factorization is $2(3y + x)(4y + x)$.

37. $9y^3 + 3y^2 - 6y = 3y(3y^2 + y - 2)$ In $3y^2 + y - 2$, $ac = 3(-2) = -6$ and $3 + (-2) = 1$. Thus, $3y^2 + y - 2 = 3y^2 + 3y - 2y - 2 = (3y^2 + 3y) - (2y + 2) = 3y(y + 1) - 2(y + 1) = (y + 1)(3y - 2)$. The complete factorization is $3y(y + 1)(3y - 2)$.

39. $y^2 - 8y + 15$
$ac = 1 \cdot 15 = 15 = (-3)(-5)$ and $-3 + (-5) = -8$.
Thus, $y^2 - 8y + 15 = y^2 - 3y - 5y + 15 = (y^2 - 3y) - (5y - 15) = y(y - 3) - 5(y - 3) = (y - 3)(y - 5)$

41. $2y^2 + 5y - 2$
$ac = 2(-2) = -4$ and no pair of factors of -4 will add to 5. Thus, $2y^2 + 5y - 2$ is prime.

43. $6x^2 + 3xy - 18y^2 = 3(2x^2 + xy - 6y^2)$ In $2x^2 + xy - 6y^2$, $ac = 2(-6) = -12 = 4(-3)$ and $4 + (-3) = 1$. Thus, $2x^2 + xy - 6y^2 = 2x^2 + 4xy - 3xy - 6y^2 = (2x^2 + 4xy) - (3xy + 6y^2) =$

$2x(x + 2y) - 3y(x + 2y) = (x + 2y)(2x - 3y)$. The complete factorization is $3(x + 2y)\cdot(2x - 3y)$

45. a. $x^2 + 11x - 26$
$ac = 1(-26) = -26 = 13(-2)$ and $13 + (-2) = 11$
Thus, $x^2 + 11x - 26 = x^2 + 13x - 2x - 26 = (x^2 + 13x) - (2x + 26) = x(x + 13) - 2(x + 13) = (x + 13)(x - 2)$
b.

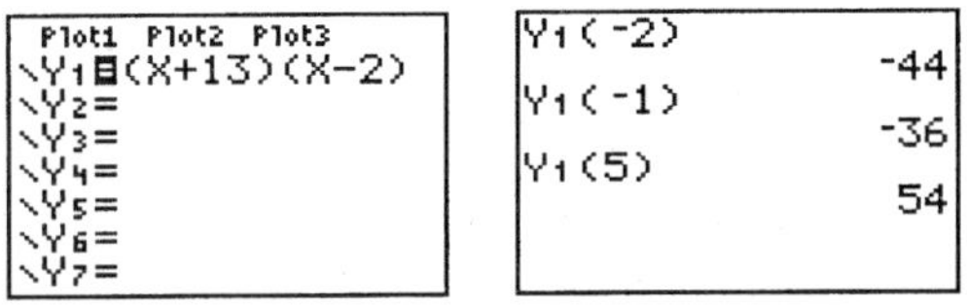

47. $6x^2 - 15x + 4x - 10 = (6x^2 - 15x) + (4x - 10) = 3x(2x - 5) + 2(2x - 5) = (2x - 5)(3x + 2)$

49. $2y^2 + 16y - 5y - 40 = (2y^2 + 16y) - (5y + 40) = 2y(y + 8) - 5(y + 8) = (y + 8)(2y - 5)$

51. $42x^2 + 48x - 7x - 8 = (42x^2 + 48x) - (7x + 8) = 6x(7x + 8) - 1\cdot(7x + 8) = (7x + 8)(6x - 1)$

53. $x^2 + 6x + 9$
$ac = 1\cdot 9 = 9 = 3\cdot 3$ and $3 + 3 = 6$. Thus, $x^2 + 6x + 9 = x^2 + 3x + 3x + 9 = (x^2 + 3x) + (3x + 9) = x(x + 3) + 3(x + 3) = (x + 3)(x + 3)$

55. $y^2 - 8y + 16$
$ac = 1\cdot 16 = 16 = (-4)(-4)$ and $-4 + (-4) = -8$.
Thus, $y^2 - 8y + 16 = y^2 - 4y - 4y + 16 = (y^2 - 4y) - (4y - 16) = y(y - 4) - 4(y - 4) = (y - 4)(y - 4) = (y - 4)^2$

57. $t^2 + 20t + 100$
$ac = 1\cdot 100 = 100 = 10\cdot 10$ and $10 + 10 = 20$.
Thus, $t^2 + 20t + 100 = t^2 + 10t + 10t + 100 = (t^2 + 10t) + (10t + 100) = t(t + 10) + 10(t + 10) = (t + 10)(t + 10) = (t + 10)^2$

59. $u^2 - 18u + 81$
$ac = 1\cdot 81 = 81 = (-9)(-9)$ and $-9 + (-9) = -18$.
Thus, $u^2 - 18u + 81 = u^2 - 9u - 9u + 81 = (u^2 - 9u) - (9u - 81) = u(u - 9) - 9(u - 9) = (u - 9)(u - 9) = (u - 9)^2$

61. $x^2 + 4xy + 4y^2$
$ac = 1\cdot 4 = 4 = 2\cdot 2$ and $2 + 2 = 4$. Thus, $x^2 + 4xy + 4y^2 = x^2 + 2xy + 2xy + 4y^2 = (x^2 + 2xy) + (2xy + 4y^2) = x(x + 2y) + 2y(x + 2y) = (x + 2y)(x + 2y) = (x + 2y)^2$

Applications

63. a. $V = h^3 + 7h^2 + 12h$

$$= h(h^2 + 7h + 12)$$
$$= h(h + 3)(h + 4)$$

The length is 4 greater than the height h and the width is 3 greater.
b. If h = 6", length = 6 + 4 = 10" and the width = 6 + 3 = 9"

65. Consider $6x^2 + bx + 6$.
$ac = 6\cdot 6 = 36 = (\pm 36)(\pm 1) = (\pm 18)(\pm 2) = (\pm 12)(\pm 3) = (\pm 9)(\pm 4) = (\pm 6)(\pm 6)$. Thus, $b = \pm(36 + 1) = \pm 37$ or $\pm(18 + 2) = \pm 20$ or $\pm(12 + 3) = \pm 15$ or $\pm(9 + 4) = \pm 13$ or $\pm(6 + 6) = \pm 12$.

Review

67. $(x + 3)^2 = x\cdot x + 2(3)x + 3\cdot 3 = x^2 + 6x + 9$

69. $\dfrac{a^4b^2}{b} = a^4b$

71. a. $x^2 + 5x - 3 = x(x - 4) + 3(2x - 6)$

$$x^2 + 5x - 3 = x^2 - 4x + 6x - 18$$
$$5x - 3 = 2x - 18$$
$$3x - 3 = -18$$
$$3x = -15$$
$$x = -5$$

b. Using a window of [-10,10] by [-20,10]:

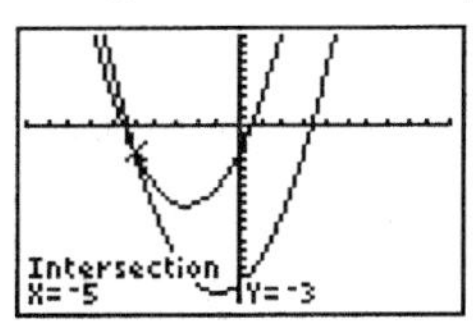

Section 5.3 Factoring the Difference of Two Squares and a Summary of Factoring Techniques

Vocabulary

1. difference

3. conjugate binomial

5. GCF(greatest common factor)

Concepts

7. F $2y^2$ is not a perfect square

9. T

11. F $x + y$ is prime, for example.

13. In $13x^2 + 220x - 17$, $ac = 13(-17) = -221$.
In $36x^2 - 117x + 80$, $ac = 36(80) = 2880$.
The number 2880 has many more pairs of factors to check than does -221.

Practice

15. $x^2 - 100 = x^2 - 10^2 = (x - 10)(x + 10)$

17. $4y^2 - 25 = (2y)^2 - 5^2 = (2y - 5)(2y + 5)$

19. $9x^2 - 81y^2 = 9(x^2 - 9y^2) = 9[x^2 - (3y)^2] =$
$9(x - 3y)(x + 3y)$

21. $x^4 - y^4 = (x^2)^2 - (y^2)^2 = (x^2 - y^2)(x^2 + y^2) =$
$(x - y)(x + y)(x^2 + y^2)$

23. $9x^4 - 16y^2 = (3x^2)^2 - (4y)^2 =$
$(3x^2 - 4y)(3x^2 + 4y)$

25. $4y^2 - 4 = 4(y^2 - 1) = 4(y - 1)(y + 1)$

27. $y^3 + xy^2 = y^2(y + x)$

29. $4x^2y - 16y^3 = 4y(x^2 - 4y^2) = 4y[x^2 - (2y)^2] =$
$4y(x - 2y)(x + 2y)$

31. $x^4 - 81 = (x^2)^2 - 9^2 = (x^2 - 9)(x^2 + 9) =$
$(x^2 - 3^2)(x^2 + 9) = (x - 3)(x + 3)(x^2 + 9)$

33. $x^3 - 8 = x^3 - 2^3 = (x - 2)(x^2 + 2x + 2^2) =$
$(x - 2)(x^2 + 2x + 4)$

35. $b^3 + 125 = b^3 + 5^3 = (b + 5)(b^2 - 5b + 5^2) =$
$(b + 5)(b^2 - 5b + 25)$

37. $27 - y^3 = 3^3 - y^3 = (3 - y)(3^2 + 3y + y^2) =$
$(3 - y)(9 + 3y + y^2)$

39. $8u^3 + w^3 = (2u)^3 + w^3 =$
$(2u + w)[(2u)^2 - 2uw + w^2] =$
$(2u + w)(4u^2 - 2uw + w^2)$

41. $x^3 - 27y^3 = x^3 - (3y)^3 =$
$(x - 3y)[x^2 + x \cdot 3y + (3y)^2] =$
$(x - 3y)(x^2 + 3xy + 9y^2)$

43. $27a^3 - b^3 = (3a)^3 - b^3 =$
$(3a - b)[(3a)^2 + (3a)b + b^2] =$
$(3a - b)(9a^2 + 3ab + b^2)$

45. $27x^3 + 125 = (3x)^3 + 5^3 =$
$(3x + 5)[(3x)^2 - (3x)5 + 5^2] =$
$(3x + 5)(9x^2 - 15x + 25)$

47. $64x^3 + 27y^3 = (4x)^3 + (3y)^3 =$
$(4x + 3y)[(4x)^2 - (4x)(3y) + (3y)^2] =$
$(4x + 3y)(16x^2 - 12xy + 9y^2)$

49.
```
11→A:7→B:A^3-B^3
                988
(A-B)(A²+AB+B²)
                988
```

51. Q(x) is not the correct factorization of P(x) since P and Q have different values for the same value of x.

53. The factorization is guaranteed to be correct if the polynomial is of degree 6 or less.

55. The factorization is not correct since $Y1 \neq Y2$ for at least one value of x.

57. $x^2 - 9 = x^2 - 3^2 = (x - 3)(x + 3)$

59. $x^3 + 8 = x^3 + 2^3 = (x + 2)(x^2 - 2x + 2^2) =$
$(x + 2)(x^2 - 2x + 4)$

61. $3rs^2 - 6r^2st = (3rs)s - (3rs)(2rt) = 3rs(s - 2rt)$

63. $ac + ad + bc + bd = (ac + ad) + (bc + bd) =$
$a(c + d) + b(c + d) = (c + d)(a + b)$

65. $6p^2 - 3p - 2$
$ac = 6(-2) = -12$ and no pair of factors of -12 will add to -3. Thus, $6p^2 - 3p - 2$ is prime.

67. $2x^2 - 32 = 2(x^2 - 16) = 2(x^2 - 4^2) =$
$2(x - 4)(x + 4$

69. $2xy^2 + 8xy - 24x = 2x(y^2 + 4y - 12)$. In the trinomial, $ac = 1(-12) = -12 = 6(-2)$ and $6 + (-2) = 4$. Thus, $y^2 + 4y - 12 =$
$y^2 + 6y - 2y - 12 = (y^2 + 6y) - (2y + 12) =$
$y(y + 6) - 2(y + 6) = (y + 6)(y - 2)$. The complete factorization is $2x(y + 6)(y - 2)$.

71. $-5x^2 - 10x + 40 = -5(x^2 + 2x - 8)$. In the trinomial, $ac = 1(-8) = -8 = 4(-2)$ and $4 + (-2) = 2$. Thus, $x^2 + 2x - 8 = x^2 + 4x - 2x - 8 =$
$(x^2 + 4x) - (2x + 8) = x(x + 4) - 2(x + 4) =$
$(x + 4)(x - 2)$. The complete factorization is $-5(x + 4)(x - 2)$.

73. $-8p^3q^7 - 4p^2q^3 = -4p^2q^3(2pq^4 + 1)$

75. $64y^3 - z^3 = (4y)^3 - z^3 =$
$(4y - z)[(4y)^2 + (4y)z + z^2] =$
$(4y - z)(16y^2 + 4yz + z^2)$

77. $x^2 + 7x + 1$
$ac = 1 \cdot 1 = 1$ and no pair of factors of 1 will add up to 7. Thus, $x^2 + 7x + 1$ is prime

79. $25x^2 + 16y^2$ is prime.

81. $14t^3 - 40t^2 + 6t^4 = 2t^2(7t - 20 + 3t^2) = 2t^2(3t^2 + 7t - 20)$. In the trinomial, $ac = 3(-20) = -60 = 12(-5)$ and $12 + (-5) = 7$. Thus, $3t^2 + 7t - 20 = 3t^2 + 12t - 5t - 20 = (3t^2 + 12t) - (5t + 20) = 3t(t + 4) - 5(t + 4) = (t + 4)(3t - 5)$. The complete factorization is $2t^2(t + 4)(3t - 5)$.

83. $-6x^2 - 7x + 20 = -1(6x^2 + 7x - 20)$
$ac = 6(-20) = -120 = 15(-8)$ and $15 + (-8) = 7$. Thus, $6x^2 + 7x - 20 = 6x^2 + 15x - 8x - 20 =$
$(6x^2 + 15x) - (8x + 20) =$
$3x(2x + 5) - 4(2x + 5) = (2x + 5)(3x - 4)$.
The complete factorization is $-(2x + 5)(3x - 4)$

85. $16x^3 - 54y^3 = 2(8x^3 - 27y^3) = 2[(2x)^3 - (3y)^3]$
$= 2(2x - 3y)[(2x)^2 + (2x)(3y) + (3y)^2] =$
$2(2x - 3y)(4x^2 + 6xy + 9y^2)$

87. $a^2(x - a) - b^2(x - a) = (x - a)(a^2 - b^2) = (x - a)$
$(a - b)(a + b)$

89. $8a^2x^3y - 2b^2xy = 2xy(4a^2x^2 - b^2) = 2xy$
$[(2ax)^2 - b^2] = 2xy(2ax - b)(2ax + b)$

91. $81p^4 - 16q^4 = (9p^2)^2 - (4q^2)^2 = (9p^2 - 4q^2)$
$(9p^2 + 4q^2) = [(3p)^2 - (2q)^2](9p^2 + 4q^2) =$
$(3p - 2q)(3p + 2q)(9p^2 + 4q^2)$

93. $4x^2 + 9y^2$ prime

95. $4x^2y^2 - y^4 = y^2(4x^2 - y^2) = y^2[(2x)^2 - y^2] =$
$y^2(2x - y)(2x + y)$

97. $10r^2 - 13r - 4$
$ac = 10(-4) = -40$ and no pair of factors of -40 will add up to -13 and so $10r^2 - 13r - 4$ is prime.

99. $21t^3 - 10t^2 + t = t(21t^2 - 10t + 1)$. In the trinomial, $ac = 21 \cdot 1 = 21 = (-3)(-7)$ and $-3 + (-7) = -10$. Thus, $21t^2 - 10t + 1 = 21t^2 - 3t - 7t + 1 = (21t^2 - 3t) - (7t - 1) =$
$3t(7t - 1) - (7t - 1) = (7t - 1)(3t - 1)$. The complete factorization is $t(7t - 1)(3t - 1)$.

Review

101. $l = f + (n-1)d = f + nd - d$

$l - f + d = nd$

$\dfrac{l - f + d}{d} = n$

Section 5.4 Using Factoring to Solve Equations

Vocabulary

1. transformation

3. root

5. at least one

7. root (or zero)

9. x; y

Concepts

11. $2x^2 - 10x + 3x - 15 = (2x^2 - 10x) + (3x - 15)$
$= 2x(x - 5) + 3(x - 5) = (x - 5)(2x + 3)$

13. $2x^2 - 7x - 15 = 0$

$(x-5)(2x+3) = 0$

$x - 5 = 0$ or $2x + 3 = 0$

$x = 5$ or $x = \dfrac{-3}{2}$

15. b. and c. are equations; a. and d. are expressions.

Practice

17. $(x+5)(x-2) = 0$

$x + 5 = 0$ or $x - 2 = 0$

$x = -5$ or $x = 2$

19. $15x^2 - 20x = 0$

$5x(3x - 4) = 0$

$5x = 0$ or $3x - 4 = 0$

$x = 0$ or $x = \dfrac{4}{3}$

21. $2x^2 - 5x + 2 = 0$

$(2x-1)(x-2) = 0$

$2x - 1 = 0$ or $x - 2 = 0$

$x = \dfrac{1}{2}$ or $x = 2$

23. $5x^2 - 6x + 1 = 0$

$(5x-1)(x-1) = 0$

$5x - 1 = 0$ or $x - 1 = 0$

$x = \dfrac{1}{5}$ or $x = 1$

25. $x(6x+5) = 6$

$6x^2 + 5x - 6 = 0$

$(3x-2)(2x+3) = 0$

$3x - 2 = 0$ or $2x + 3 = 0$

$x = \dfrac{2}{3}$ or $x = \dfrac{-3}{2}$

27. $(x+1)(8x+1) = 18x$

$8x^2 + 9x + 1 = 18x$

$8x^2 - 9x + 1 = 0$

$(8x-1)(x-1) = 0$

$8x - 1 = 0$ or $x - 1 = 0$

$x = \dfrac{1}{8}$ or $x = 1$

29. $6x^3 + x^2 = 12x$

$6x^3 + x^2 - 12x = 0$

$x(6x^2 + x - 12) = 0$

$x(3x-4)(2x+3) = 0$

$x = 0$ or $3x = 4$ or $2x + 3 = 0$

$x = 0$ or $x = \frac{4}{3}$ or $x = -\frac{3}{2}$

31. $9y^2 - 1 = 0$
$(3y-1)(3y+1) = 0$
$3y - 1 = 0$ or $3y + 1 = 0$
$y = \frac{1}{3}$ or $y = \frac{-1}{3}$

33. $2x^3 = 2x(x+2)$
$2x^3 = 2x^2 + 4x$
$2x^3 - 2x^2 - 4x = 0$
$2x(x^2 - x - 2) = 0$
$2x(x-2)(x+1) = 0$
$2x = 0$ or $x - 2 = 0$ or $x + 1 = 0$
$x = 0$ or $x = 2$ or $x = -1$

35. $x^2(x+10) = 2x(x-8)$
$x^3 + 10x^2 = 2x^2 - 16x$
$x^3 + 8x^2 + 16x = 0$
$x(x^2 + 8x + 16) = 0$
$x(x+4)(x+4) = 0$
$x = 0$ or $x + 4 = 0$
$x = 0$ or $x = -4$

37. $\frac{3}{2}x^2 + \frac{27}{2}x = -30$
$3x^2 + 27x = -60$
$3x^2 + 27x + 60 = 0$
$3(x^2 + 9x + 20) = 0$
$3(x+5)(x+4) = 0$
$x + 5 = 0$ or $x + 4 = 0$
$x = -5$ or $x = -4$

39. $4x^2 = 81$
$4x^2 - 81 = 0$
$(2x-9)(2x+9) = 0$
$2x - 9 = 0$ or $2x + 9 = 0$
$x = \frac{9}{2}$ or $x = \frac{-9}{2}$

41. $3x^2 - \frac{11}{2}x = 5$
$6x^2 - 11x = 10$
$6x^2 - 11x - 10 = 0$
$(3x+2)(2x-5) = 0$
$3x + 2 = 0$ or $2x - 5 = 0$
$x = \frac{-2}{3}$ or $x = \frac{5}{2}$

43. $(x-2)^2 - 3 = -3$
$(x-2)^2 = 0$
$x - 2 = 0$
$x = 2$

45. $\frac{1}{8}x^2 - \frac{1}{4}x - 6 = 0$
$x^2 - 2x - 48 = 0$
$(x-8)(x+6) = 0$
$x - 8 = 0$ or $x + 6 = 0$
$x = 8$ or $x = -6$

47. $(x-2)(x+3) = 0$
a. C
b. The solutions are 2 and -3.
c. $x^2 + x - 6 = 0$

49. $x(x-3)(x+3) = 0$
a. B
b. The solutions are 0, 3 and -3.
c. $x^3 - 9x = 0$

51. $x(x-3)(x+2) = 0$
a. H
b. The solutions are 0, 3 and -2.

c. $x^3 - x^2 - 6x = 0$

53. $(x + 4)(x - 2) = 0$
 a. D
 b. The solutions are x = 2 and -4.
 c. $x^2 + 2x - 8 = 0$

55. $x^2 + 3.44x - .21 = 0$

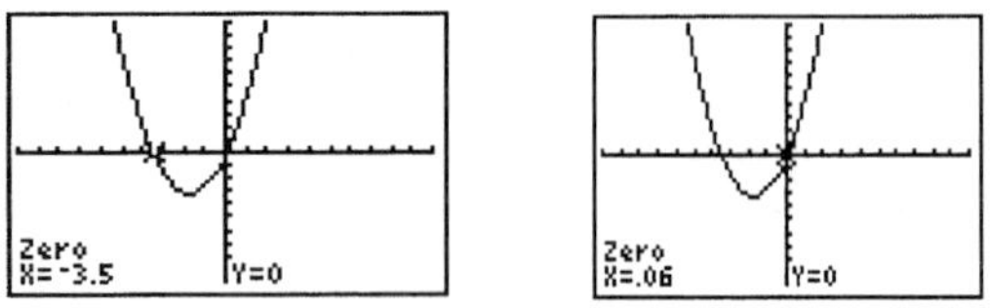

The solutions are -3.5 and 0.06

57. $x^2 + 4.9x - 1.56 = 0$

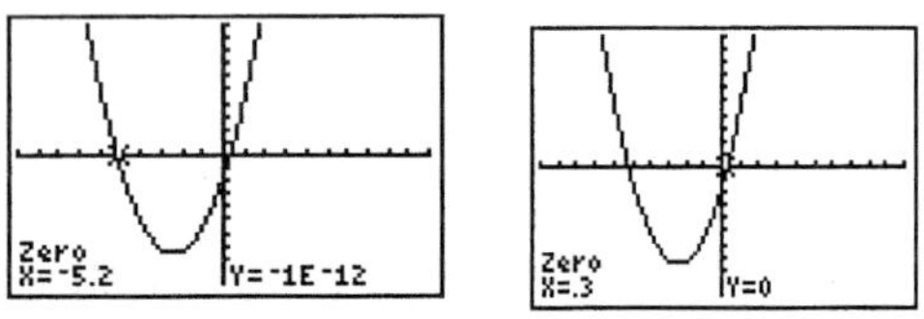

The solutions are -5.2 and 0.3.

59. $x^3 + 1.2x^2 - 15x + 2.944 = 0$

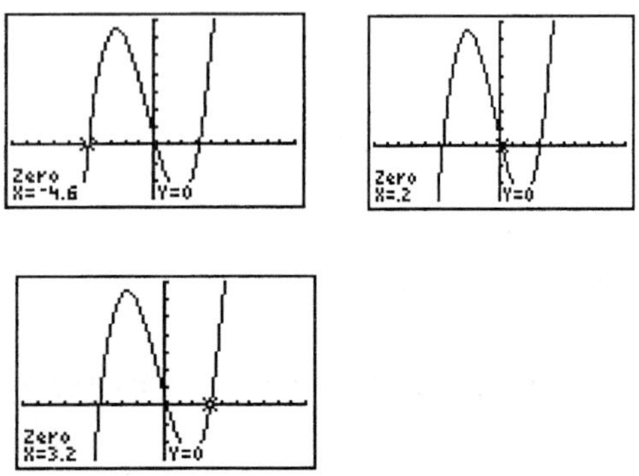

The solutions are - 4.6, 0.2 and 3.2.

Applications

61. Here v = 144, so $h = 144t - 16t^2$. On the ground means h = 0, so $144t - 16t^2 = 0$.
 Thus, $16t(9 - t) = 0$
 $16t = 0$ or $9 - t = 0$
 $t = 0$ or $t = 9$
 It will be on the ground after 9 seconds.

63. Here v = 220 and h = 600. Thus,
 $600 = 220t - 16t^2$
 $16t^2 - 220t + 600 = 0$
 $4(4t^2 - 55t + 150) = 0$
 $4(4t - 15)(t - 10) = 0$
 $4t - 15 = 0$ or $t - 10 = 0$
 $t = \frac{15}{4}$ or $t = 10$
 At 15/4 sec. and 10 sec. it will be at a height of 600 ft.

65. $h = -16t^2 + 64$. The dive ends when h = 0.
 Thus, $-16t^2 + 64 = 0$
 $-16(t^2 - 4) = 0$
 $-16(t - 2)(t + 2) = 0$
 $t = 2$ or $t = -2$
 The dive ends after t = 2 seconds.

67. Let x and x + 2 represent the integers. Then
 $x(x + 2) = 35$
 $x^2 + 2x = 35$
 $x^2 + 2x - 35 = 0$
 $(x + 7)(x - 5) = 0$
 $x + 7 = 0$ or $x - 5 = 0$
 $x = -7$ or $x = 5$
 The positive integers are 5 and 5 + 2 = 7.

69. Let x = the composite integer ($\neq 1$ or a prime)
 Then, $4 + x^2 = 10x - 5$
 $x^2 - 10x + 9 = 0$
 $(x - 9)(x - 1) = 0$
 $x - 9 = 0$ or $x - 1 = 0$
 $x = 9$ or $x = 1$
 The composite integer is 9.

71. Let w and 2w + 1 represent the dimensions.

Then, $w(2w+1)=36$

$$2w^2+w=36$$
$$2w^2+w-36=0$$
$$(2w+9)(w-4)=0$$
$$2w+9=0 \text{ or } w-4=0$$
$$w=\frac{-9}{2} \text{ or } w=4$$

The dimensions are 4 m. by $2\cdot 4+1=9$ m.

73. Let x = width. Then x + 2 = length and

$$x(x+2)=143$$
$$x^2+2x=143$$
$$x^2+2x-143=0$$
$$(x+13)(x-11)=0$$
$$x=-13 \text{ or } x=11$$

The perimeter P = 2(11) + 2(11 + 2) = 22 + 26 = 48. It will take 48' of crown molding.

75. Let h = height of the tent. Then 2 + 2h is the base and for a triangle, area = $\frac{1}{2}\cdot$ base $\cdot$ height. Thus, $\frac{1}{2}h(2+2h)=30$

$$h+h^2-30=0$$
$$(h+6)(h-5)=0$$
$$h=-6 \text{ or } h=5$$

The height is 5', the base 2 + 2·5 = 12'.

77. Let x = area. Then x – 6 = height and x – 3 = base and so $\frac{1}{2}(x-3)(x-6)=x$

$$(x-3)(x-6)=2x$$
$$x^2-9x+18=2x$$
$$x^2-11x+18=0$$
$$(x-9)(x-2)=0$$
$$x=9 \text{ or } x=2$$

The area is 9 square units.

79. Let h = height and so 2h = base. Then

$$h(2h)=200$$
$$2h^2-200=0$$
$$2(h^2-100)=0$$
$$2(h-10)(h+10)=0$$
$$h=10 \text{ or } h=-10$$

The base is 2(10) = 20 cm.

81. Let h = height of the trapezoid. Thus, one base = h and the other 8 m. Then the area is

$$\frac{h(8+h)}{2}=24$$
$$h(8+h)=48$$
$$8h+h^2=48$$
$$h^2+8h-48=0$$
$$(h+12)(h-4)=0$$
$$h=4 \text{ or } h=-12$$

The height is 4 m.

83. Let x and x + 2 represent the length and width of the base. Since the height is 12 cm. and the volume is 192 cm.^{3}, x satisfies

$$192=\frac{x(x+2)\cdot 12}{3}$$
$$192=4x(x+2)$$
$$192=4x^2+8x$$
$$0=4x^2+8x-192$$
$$0=4(x^2+2x-48)$$
$$0=4(x+8)(x-6)$$
$$x=6 \text{ or } x=-8$$

The rectangular base is 6 cm. by 8 cm.

85. Let x represent the length so that x – 3 = width. The height is 4 cm. and the volume is 72 cm.^{3}, so x satisfies the equation

$$72=x(x-3)\cdot 4$$
$$72=4x^2-12x$$
$$0=4x^2-12x-72$$
$$0=4(x^2-3x-18)$$
$$0=4(x-6)(x+3)$$
$$x=6 \text{ or } x=-3$$

The length is 6 cm., the width 3 cm. and so the sum is 9 cm.

87. Solutions: 3, - 4
$(x - 3)(x + 4) = 0$ or $x^2 + x - 12 = 0$

89. Solutions: -3, 0, $\frac{7}{3}$.

$(x + 3)(x - 0)(x - \frac{7}{3}) = 0$

$(x^2 + 3x)(3x - 7) = 0$ or $3x^3 + 2x^2 - 21x = 0$

91. Solutions: $\frac{3}{4}, \frac{2}{3}$

$(x - \frac{3}{4})(x - \frac{2}{3}) = 0$

$(4x - 3)(3x - 2) = 0$

or $12x^2 - 17x + 6 = 0$

Review

93. $2q^2 - 9 = q(q + 3) + q^2$

$2q^2 - 9 = q^2 + 3q + q^2$

$2q^2 - 9 = 2q^2 + 3q$

$-9 = 3q$

$-3 = q$

95.
$$1 + \frac{2}{x - 3} = \frac{1}{3}$$
$$3(x - 3) \cdot 1 + 3(x - 3)\frac{2}{x - 3} = 3(x - 3)\frac{1}{3}$$
$$3x - 9 + 6 = x - 3$$
$$3x - 3 = x - 3$$
$$2x = 0$$
$$x = 0$$

97. The pair (6,-2) is a solution of $2x + 3y = 6$ since $2 \cdot 6 + 3(-2) = 12 - 6 = 6$

Chapter 5 Summary

1. GCF = 3
$3x + 9y = 3(x + 3y)$

2. GCF = 5a
$5ax^2 + 15a = 5a(x^2 + 3)$

3. GCF = 7x
$7x^2 + 14x = 7x(x + 2)$

4. GCF = 3x
$3x^2 - 3x = 3x(x - 3)$

5. GCF = 2x
$2x^3 + 4x^2 - 8x = 2x(x^2 + 2x - 4)$

6. GCF = a
$ax + ay - az = a(x + y - z)$

7. GCF = a
$ax + ay - a = a(x + y - 1)$

8. GCF = xyz
$x^2yz + xy^2z = xyx(x + y)$

9. GCF = a
$5a^2 + 6ab^2 + 10acd - 15a =$
$a(5a + 6b^2 + 10cd - 15)$

10. GCF = 7xy
$7axy + 21x^2y - 35x^3y + 7xy^2 =$
$7xy(a + 3x - 5x^2 + y)$

11. GCF = x + y
$(x + y)a + (x + y)b = (x + y)(a + b)$

12. GCF = x + y
$(x + y)^2 + (x + y) = (x + y)(x + y + 1)$

13. GCF = 2x(x + 2)
$2x^2(x + 2) + 6x(x + 2) = 2x(x + 2) \cdot x +$
$2x(x + 2) \cdot 3 = 2x(x + 2)(x + 3)$

14. GCF = 3x(y + z)
$3x(y + z) - 9x(y + z)^2 = 3x(y + z) \cdot 1 +$
$3x(y + z) \cdot 3(y + z)$
$= 3x(y + z)[1 - 3(y + z)] =$
$3x(y + z)(1 - 3y - 3z)$

15. $3p + 9q + ap + 3aq = (3p + 9q) + (ap + 3aq)$
$= 3(p + 3q) + a(p + 3q) = (p + 3q)(3 + a)$

16. $ar - 2as + 7r - 14s = (ar - 2as) + (7r - 14s) =$
$a(r - 2s) + 7(r - 2s) = (r - 2s)(a + 7)$

17. $x^2 + ax + bx + ab = (x^2 + ax) + (bx + ab) =$
$x(x + a) + b(x + a) = (x + a)(x + b)$

18. $xy + 2x - 2y - 4 = (xy + 2x) - (2y + 4) =$
$x(y + 2) - 2(y + 2) = (y + 2)(x - 2)$

19. $3x^2y - xy^2 - 6xy + 2y^2 = y(3x^2 - xy - 6x + 2y)$
$= y[(3x^2 - xy) - (6xy - 2y^2)] =$
$y[x(3x - y) - 2(3x - y)] = y(3x - y)(x - 2)$

20. $5x^2 + 10x - 15xy - 30y = 5(x^2 + 2x - 3xy -$
$6y) = 5[(x^2 + 2x) - (3xy + 6y)] =$
$5[x(x + 2) - 3y(x + 2)] = 5(x + 2)(x - 3y)$

21. $x^2 + 10x + 21$
$ac = 1(21) = 21 = 3 \cdot 7$ and $3 + 7 = 10$. Thus,
$x^2 + 10x + 21 = x^2 + 3x + 7x + 21 =$
$(x^2 + 3x) + (7x + 21) = x(x + 3) + 7(x + 3) =$
$(x + 3)(x + 7)$

22. $x^2 + 4x - 21$
$ac = 1(-21) = -21 = 7(-3)$ and $7 + (-3) = 4$.
Thus, $x^2 + 4x - 21 = x^2 + 7x - 3x - 21 =$
$(x^2 + 7x) - (3x + 21) = x(x + 7) - 3(x + 7) =$
$(x + 7)(x - 3)$

23. $x^2 + 2x - 24$
$ac = 1(-24) = -24 = 6(-4)$ and $6 + (-4) = 2$.
Thus, $x^2 + 2x - 24 = x^2 + 6x - 4x - 24 =$
$(x^2 + 6x) - (4x + 24) = x(x + 6) - 4(x + 6) =$
$(x + 6)(x - 4)$

24. $x^2 - 4x - 12$
$ac = 1(-12) = -12 = 2(-6)$ and $2 + (-6) = -4$.
Thus, $x^2 - 4x - 12 = x^2 - 6x + 2x - 12 =$
$(x^2 - 6x) + (2x - 12) = x(x - 6) + 2(x - 6) =$
$(x - 6)(x + 2)$

25. $2x^2 - 5x - 3$
$ac = 2(-3) = -6 = 1(-6)$ and $1 + (-6) = -5$.
Thus, $2x^2 - 5x - 3 = 2x^2 + x - 6x - 3 =$
$(2x^2 + x) - (6x + 3) = x(2x + 1) - 3(2x + 1) =$
$(2x + 1)(x - 3)$

26. $3x^2 - 14x - 5$
$ac = 3(-5) = -15 = 1(-15)$ and $1 + (-15) = -14$.
Thus, $3x^2 - 14x - 5 = 3x^2 + x - 15x - 5 =$
$(3x^2 + x) - (15x + 5) = x(3x + 1) - 5(3x + 1)$
$= (3x + 1)(x - 5)$

27. $6x^2 + 7x - 3$
$ac = 6(-3) = -18 = 9(-2)$ and $9 + (-2) = 7$.
Thus, $6x^2 + 7x - 3 = 6x^2 + 9x - 2x - 3 =$
$(6x^2 + 9x) - (2x + 3) = 3x(2x + 3) - (2x + 3)$
$= (2x + 3)(3x - 1)$

28. $6x^2 + 3x - 3 = 3(2x^2 + x - 1)$. In the trinomial, $ac = 2(-1) = -2$ and $2 + (-1) = 1$. Thus, $2x^2 + x - 1 = 2x^2 + 2x - x - 1 =$ $(2x^2 + 2x) - (x + 1) = 2x(x + 1) - (x + 1) =$ $(x + 1)(2x - 1)$. The complete factorization is $3(x + 1)(2x - 1)$.

29. $6x^3 + 17x^2 - 3x = x(6x^2 + 17x - 3)$. In the trinomial, $ac = 6(-3) = -18 = 18(-1)$ and $18 + (-1) = 17$. Thus, $6x^2 + 17x - 3 = 6x^2 + 18x - x - 3 = (6x^2 + 18x) - (x + 3) = 6x(x + 3) - (x + 3) = (x + 3)(6x - 1)$. The complete factorization is $x(x + 3)(6x - 1)$.

30. $4x^3 - 5x^2 - 6x = x(4x^2 - 5x - 6)$. In the trinomial, $ac = 4(-6) = -24 = 3(-8)$ and $3 + (-8) = -5$. Thus, $4x^2 - 5x - 6 = 4x^2 + 3x - 8x - 6 = (4x^2 + 3x) - (8x + 6) = x(4x + 3) - 2(4x + 3) = (4x + 3)(x - 2)$. The complete factorization is $x(4x + 3)(x - 2)$.

31. $x^2 - 9 = x^2 - 3^2 = (x - 3)(x + 3)$

32. $x^2y^2 - 16 = (xy)^2 - 4^2 = (xy - 4)(xy + 4)$

33. $(x + 2)^2 - y^2 = (x + 2 - y)(x + 2 + y)$

34. $z^2 - (x + y)^2 = [z - (x + y)][z + (x + y)] =$
$(z - x - y)(z + x + y)$

35. $6x^2y - 24y^3 = 6y(x^2 - 4y^2) = 6y[x^2 - (2y)^2] =$
$6y(x - 2y)(x + 2y)$

36. $(x + y)^2 - z^2 = (x + y - z)(x + y + z)$

37. $64a^3 - 1 = (4a)^3 - 1^3 =$
$(4a - 1)[(4a)^2 + 4a \cdot 1 + 1^2] =$
$(4a - 1)(16a^2 + 4a + 1)$

38. $5 - 40y^3 = 5(1 - 8y^3) = 5[1^3 - (2y)^3] =$
$5(1 - 2y)[1^2 + 1 \cdot 2y + (2y)^2] =$
$5(1 - 2y)(1 + 2y + 4y^2)$

39. $x^3 + 125 = x^3 + 5^3 = (x + 5)(x^2 - 5 \cdot x + 5^2) =$
$(x + 5)(x^2 - 5x + 25)$

40. $3x + 81x^4 = 3x(1 + 27x^3) = 3x[1^3 + (3x)^3] =$
$3x(1 + 3x)[1 - 1 \cdot 3x + (3x)^2] =$
$3x(1 + 3x)(1 - 3x + 9x^2)$

41. $24xy^2 - 6xy = 6xy(4y - 1)$

42. $x^3 + 27 = x^3 + 3^3 = (x + 3)(x^2 - 3x + 9)$

43. $6x^2 + x - 2$
$ac = 6(-2) = -12 = 4(-3)$ and $4 + (-3) = 1$.
Thus, $6x^2 + x - 2 = 6x^2 + 4x - 3x - 2 = (6x^2 + 4x) - (3x + 2) = 2x(3x + 2) - (3x + 2) = (3x + 2)(2x - 1)$.

44. $9x^2 - 49 = (3x)^2 - 7^2 = (3x - 7)(3x + 7)$

45. $16x^2 - 25 = (4x)^2 - 5^2 = (4x - 5)(4x + 5)$

46. $2x^2 - 10x + 3x - 15 = (2x^2 - 10x) + (3x - 15)$
$= 2x(x - 5) + 3(x - 5) = (x - 5)(2x + 3)$

47. $4x^2 + 8x - 12 = 4(x^2 + 2x - 3)$. In the trinomial, $ac = 1(-3) = -3 = 3(-1)$ and $3 + (-1) = 2$. Thus, $x^2 + 2x - 3 = x^2 + 3x - x - 3 = (x^2 + 3x) - (x + 3) = x(x + 3) - (x + 3) = (x + 3)(x - 1)$. The complete factorization is $4(x + 3)(x - 1)$.

48. $4x^2 + 12x + 9 = (2x)^2 + 2(2x) \cdot 3 + 3^2 =$
$(2x + 3)^2$

49. $x^3 - 1 = x^3 - 1^3 = (x - 1)(x^2 + 1 \cdot x + 1^2) =$
$(x - 1)(x^2 + x + 1)$

50. $2x^3 - 16 = 2(x^3 - 8) = 2(x^3 - 2^3) =$ $2(x - 2)(x^2 + 2 \cdot x + 2^2) = 2(x - 2)(x^2 + 2x + 4)$

51. $x^2 + 2x = 0$
$x(x + 2) = 0$
$x = 0$ or $x + 2 = 0$
$x = 0$ or $x = -2$

52. $2x^2 - 6x = 0$
$2x(x - 3) = 0$
$2x = 0$ or $x - 3 = 0$
$x = 0$ or $x = 3$

53. $x^2 - 9 = 0$
$(x - 3)(x + 3) = 0$
$x - 3 = 0$ or $x + 3 = 0$
$x = 3$ or $x = -3$

54. $x^2 - 25 = 0$
$(x - 5)(x + 5) = 0$
$x - 5 = 0$ or $x + 5 = 0$
$x = 5$ or $x = -5$

55. $a^2 - 7a + 12 = 0$
$(a - 4)(a - 3) = 0$
$a - 4 = 0$ or $a - 3 = 0$
$a = 4$ or $a = 3$

56. $x^2 - 2x - 15 = 0$
$(x - 5)(x + 3) = 0$
$x - 5 = 0$ or $x + 3 = 0$
$x = 5$ or $x = -3$

57. $2x - x^2 + 24 = 0$
$x^2 - 2x - 24 = 0$
$(x - 6)(x + 4) = 0$
$x - 6 = 0$ or $x + 4 = 0$
$x = 6$ or $x = -4$

58. $16 + x^2 - 10x = 0$

$x^2 - 10x + 16 = 0$

$(x-8)(x-2) = 0$

$x - 8 = 0$ or $x - 2 = 0$

$x = 8$ or $x = 2$

59. $2x^2 - 5x - 3 = 0$

$(2x+1)(x-3) = 0$

$2x + 1 = 0$ or $x - 3 = 0$

$x = \frac{-1}{2}$ or $x = 3$

60. $2x^2 + x - 3 = 0$

$(2x+3)(x-1) = 0$

$2x + 3 = 0$ or $x - 1 = 0$

$x = \frac{-3}{2}$ or $x = 1$

61. $4x^2 = 1$

$4x^2 - 1 = 0$

$(2x-1)(2x+1) = 0$

$2x - 1 = 0$ or $2x + 1 = 0$

$x = \frac{1}{2}$ or $x = \frac{-1}{2}$

62. $9x^2 = 4$

$9x^2 - 4 = 0$

$(3x-2)(3x+2) = 0$

$3x - 2 = 0$ or $3x + 2 = 0$

$x = \frac{2}{3}$ or $x = \frac{-2}{3}$

63. $x^3 - 7x^2 + 12x = 0$

$x(x^2 - 7x + 12) = 0$

$x(x-4)(x-3) = 0$

$x = 0$ or $x - 4 = 0$ or $x - 3 = 0$

$x = 0$ or $x = 4$ or $x = 3$

64. $x^3 + 5x^2 + 6x = 0$

$x(x^2 + 5x + 6) = 0$

$x(x+2)(x+3) = 0$

$x = 0$ or $x + 2 = 0$ or $x + 3 = 0$

$x = 0$ or $x = -2$ or $x = -3$

65. $2x^3 + 5x^2 = 3x$

$2x^3 + 5x^2 - 3x = 0$

$x(2x^2 + 5x - 3) = 0$

$x(2x-1)(x+3) = 0$

$x = 0$ or $2x - 1 = 0$ or $x + 3 = 0$

$x = 0$ or $x = \frac{1}{2}$ or $x = -3$

66. $3x^3 - 2x = x^2$

$3x^3 - x^2 - 2x = 0$

$x(3x^2 - x - 2) = 0$

$x(3x+2)(x-1) = 0$

$x = 0$ or $3x + 2 = 0$ or $x - 1 = 0$

$x = 0$ or $x = \frac{-2}{3}$ or $x = 1$

67. Let x = one integer. Then 12 – x is the other integer and $x(12 - x) = 35$

$21x - x^2 = 35$

$0 = x^2 - 21x + 35$

$0 = (x-7)(x-5)$

$x = 7$ or $x = 5$

The integers are 5 and 7.

68. Let x = the positive number. Then x satisfies

$3x^2 + 5x = 2$

$3x^2 + 5x - 2 = 0$

$(3x-1)(x+2) = 0$

$x = \frac{1}{3}$ or $x = -2$

The number is $\frac{1}{3}$.

69. Let x = height, so 2x + 3 = base. Area = ½ · base · height so $45 = \frac{1}{2}x(2x+3)$

$$90 = x(2x+3)$$
$$90 = 2x^2 + 3x$$
$$0 = 2x^2 + 3x - 90$$
$$0 = (2x+15)(x-6)$$
$$x = \frac{-15}{2} \text{ or } x = 6$$

The base is 2·6 + 3 = 15 ft.

70. The height h must be 0 at impact. Thus,

$$0 = 3000 + 40t - 16t^2$$
$$16t^2 - 40t - 3000 = 0$$
$$8(2t^2 - 5t - 375) = 0$$
$$8(2t+25)(t-15) = 0$$
$$t = \frac{-25}{2} \text{ or } t = 15$$

The bomb will hit its target in 15 sec.

71. Let x = width of the bed so 2x + 3 = length. The area is 27 ft.^{2}, so x satisfies

$$x(2x+3) = 27$$
$$2x^2 + 3x = 27$$
$$2x^2 + 3x - 27 = 0$$
$$(2x+9)(x-3) = 0$$
$$x = \frac{-9}{2} \text{ or } x = 3$$

The rectangular bed is 3' by 2·3 + 3 = 9'

72. Let x = width, so x + 3 = length. Then its area is x(x + 3) and its perimeter is 2x + 2(x + 3) = 4x + 6. Since area = perimeter, x satisfies $x(x+3) = 4x+6$

$$x^2 + 3x = 4x + 6$$
$$x^2 - x - 6 = 0$$
$$(x-3)(x+2) = 0$$
$$x = 3 \text{ or } x = -2$$

The rectangle is 3' by (3 + 3) = 6'.

Chapter 5 Test

1. GCF = 2x
$4xy^2 + 12x^2y^3 - 6xz = 2x(2y^2) + 2x(6xy^3) - 2x(3z) = 2x(2y^2 + 6xy^3 - 3z)$

2. GCF = 3x(a + b)
$3x^2(a + b) - 6xy(a + b) = 3x(a + b)\cdot x - 3x(a + b)\cdot 2y = 3x(a + b)(x - 2y)$

3. $ax + ay + bx + by = (ax + ay) + (bx + by) = a(x + y) + b(x + y) = (x + y)(a + b)$

4. $x^2 - 25 = x^2 - 5^2 = (x - 5)(x + 5)$

5. $3a^2 - 27b^2 = 3(a^2 - 9b^2) = 3[a^2 - (3b)^2] = 3(a - 3b)(a + 3b)$

6. $16x^4 - 81y^4 = (4x^2)^2 - (9y^2)^2 = (4x^2 - 9y^2)(4x^2 + 9y^2) = [(2x)^2 - (3y)^2](4x^2 + 9y^2) = (2x - 3y)(2x + 3y)(4x^2 + 9y^2)$

7. $x^2 + 4x + 3$
ac = 1·3 = 3 and 1 + 3 = 4. Thus, $x^2 + 4x + 3 = x^2 + x + 3x + 3 = (x^2 + x) + (3x + 3) = x(x + 1) + 3(x + 1) = (x + 1)(x + 3)$

8. $x^2 - 9x - 22$
ac = 1(-22) = -22 = 2(-11) and 2 + (-11) = -9. Thus, $x^2 - 9x - 22 = x^2 + 2x - 11x - 22 = (x^2 + 2x) - (11x + 22) = x(x + 2) - 11(x + 2) = (x + 2)(x - 11)$

9. $27 - x^3 = 3^3 - x^3 = (3 - x)(3^2 + 3 \cdot x + x^2) = (3 - x)(9 + 3x + x^2)$

10. $3x^2 + 13x + 4$
ac = 3(4) = 12 = 1(12) and 1 + 12 = 13. Thus, $3x^2 + 13x + 4 = 3x^2 + x + 12x + 4 = (3x^2 + x) + (12x + 4) = x(3x + 1) + 4(3x + 1) = (3x + 1)(x + 4)$

11. $2y^2 + 5y - 12$
ac = 2(-12) = -24 = 8(-3) and 8 + (-3) = 5. Thus, $2y^2 + 5y - 12 = 2y^2 + 8y - 3y - 12 = (2y^2 + 8y) - (3y + 12) = 2y(y + 4) - 3(y + 4) = (y + 4)(2y - 3)$

12. $2x^2 + 3xy - 2y^2$
ac = 2(-2) = - 4 = 4(-1) and 4 + (-1) = 3.
Thus, $2x^2 + 3xy - 2y^2 = 2x^2 + 4xy - xy - 2y^2$
$= (2x^2 + 4xy) - (xy + 2y^2) =$
$2x(x + 2y) - y(x + 2y) = (x + 2y)(2x - y)$

13. $12 - 25x + 12x^2$ Rewrite as $12x^2 - 25x + 12$.
ac = 12(12) = 144 = (-9)(-16) and -9 + (-16)
= -25. Thus, $12x^2 - 25x + 12 = 12x^2 - 9x -$
$16x + 12 = (12x^2 - 9x) - (16x - 12) =$
$3x(4x - 3) - 4(4x - 3) = (4x - 3)(3x - 4)$

14. $2y^4 - 16y = 2y(y^3 - 8) = 2y(y^3 - 2^3) = 2y\cdot$
$(y - 2)(y^2 + 2y + 2^2) = 2y(y - 2)(y^2 + 2y + 4)$

15. $(x-5)(x+5)=0$

$x-5=0$ or $x+5=0$

$x=5$ or $x=-5$

16. $2x^2+5x+3=0$

$(2x+3)(x+1)=0$

$2x+3=0$ or $x+1=0$

$x=\frac{-3}{2}$ or $x=-1$

17. $10x^2-13x=9$

$10x^2-13x-9=0$

$(5x-9)(2x+1)=0$

$5x-9=0$ or $2x+1=0$

$x=\frac{9}{5}$ or $x=\frac{-1}{2}$

18. $-3(y-6)+2=y^2+2$

$-3y+18+2=y^2+2$

$0=y^2+3y-18$

$0=(y+6)(y-3)$

$y+6=0$ or $y-3=0$

$y=-6$ or $y=3$

19. The cannonball is on the ground when h = 0.

Since v = 192 ft./sec., $0=192t-16t^2$

$16t^2-192=0$

$16t(t-12)=0$

$t=0$ or $t=12$

It will be on the ground after 12 sec.

20. Let x = height of the triangle, so that x + 2 = base. The area is 40 m.^{2} so x satisfies

$40=\frac{1}{2}x(x+2)$

$80=x(x+2)$

$80=x^2+2x$

$0=x^2+2x-80$

$0=(x+10)(x-8)$

$x=-10$ or $x=8$

The base is 8 + 2 = 10 m.

Chapter 6 Rational Expressions and Equations

Section 6.1 Simplifying Rational Expressions

Vocabulary

1. rational

Concepts

3. To find the excluded value(s) set the denominator equal to 0 and solve for x.

   a. $\frac{3+x}{x}$, x = 0. The excluded value is 0.

   b. $\frac{5+x}{2x}$, 2x = 0. The excluded value is 0.

   c. $\frac{x+2}{x-5}$, x − 5 = 0. The excluded value is 5.

   d. $\frac{2(x+4)}{x+6}$, x + 6 = 0.
   The excluded value is -6.

5. a. The excluded value is 5.
   b. The excluded values are -2 and 3.
   c. The excluded values are -2.5 and -0.5.

7. a. $\frac{8}{52}=\frac{4\cdot 2}{4\cdot 13}=\frac{2}{13}$

   b. $\frac{15}{21}=\frac{3\cdot 5}{3\cdot 7}=\frac{5}{7}$

   c. $\frac{-6x}{18}=\frac{6(-x)}{6\cdot 3}=\frac{-x}{3}$

   d. $\frac{-25y}{5}=\frac{5(-5y)}{5}=-5y$

   e. $\frac{x+3}{3(x+3)}=\frac{1}{3}$

   f. $\frac{2(x+7)}{x+7}=\frac{2}{1}=2$

9. $-\frac{x+y}{x-y}=\frac{x+y}{-(x-y)}=\frac{x+y}{y-x}=\frac{-(x+y)}{x-y}=$ $\frac{-x-y}{x-y}$ Only the expression in b. is equivalent.

Practice

11. Set the denominator x^2+5 equal to 0. Since there are no real solutions, there are no excluded values.

13. Set the denominator $x^2-2x-24=0$ and solve for x. $x^2-2x-24=0$
    $(x-6)(x+4)=0$
    $x=6$ or $x=-4$
    The excluded values are 6 and - 4.

15. $\frac{15x^2y^2}{5xy^2}=\frac{5xy^2(3x)}{5xy^2}=3x$

17. $\frac{5x+35}{x+7}=\frac{5(x+7)}{x+7}=\frac{5}{1}=5$

19. $\frac{x^2+3x}{2x+6}=\frac{x(x+3)}{2(x+3)}=\frac{x}{2}$

21. $\frac{15x-3x^2}{25y-5xy}=\frac{3x(5-x)}{5y(5-x)}=\frac{3x}{5y}$

23. $\frac{x-7}{7-x}=\frac{-(x-7)}{x-7}=\frac{-1}{1}=-1$

25. $\frac{6x-3y}{3y-6x}=\frac{-(3y-6x)}{3y-6x}=\frac{-1}{1}=-1$

27. $\frac{a+b-c}{c-a-b}=\frac{(a+b)-c}{c-(a+b)}=\frac{-[c-(a+b)]}{c-(a+b)}=$
    $\frac{-1}{1}=-1$

29. $\frac{x^2+3x+2}{x^2+x-2}=\frac{(x+2)(x+1)}{(x+2)(x-1)}=\frac{x+1}{x-1}$

31. $\frac{x^2-8x+15}{x^2-x-6}=\frac{(x-5)(x-3)}{(x+2)(x-3)}=\frac{x-5}{x+2}$

33. $\frac{2x^2-8x}{x^2-6x+8}=\frac{2x(x-4)}{(x-2)(x-4)}=\frac{2x}{x-2}$

35. $\frac{xy+2x^2}{2xy+y^2}=\frac{x(y+2x)}{y(2x+y)}=\frac{x}{y}$

37. $\frac{x^2+3x+2}{x^3+x^2}=\frac{(x+2)(x+1)}{x^2(x+1)}=\frac{x+2}{x^2}$

39. $\frac{x^2-8x+16}{x^2-16}=\frac{(x-4)(x-4)}{(x-4)(x+4)}=\frac{x-4}{x+4}$

41. $\frac{x^2-2x-15}{x^2+2x-15}=\frac{(x-5)(x+3)}{(x+5)(x-3)}$, simplified.

43. $\frac{x^2-3(2x-3)}{9-x^2}=\frac{x^2-6x+9}{9-x^2}=\frac{(x-3)(x-3)}{(3-x)(3+x)}$

$=\frac{(x-3)(x-3)}{-(x-3)(3+x)}=-\frac{x-3}{3+x}=\frac{3-x}{3+x}$

45. $\frac{4(x+3)+4}{3(x+2)+6}=\frac{4x+12+4}{3x+6+6}=\frac{4x+16}{3x+12}=$

$\frac{4(x+4)}{3(x+4)}=\frac{4}{3}$

47. $\frac{x^2-9}{(2x+3)-(x+6)}=\frac{(x+3)(x-3)}{2x+3-x-6}=$

$\frac{(x+3)(x-3)}{x-3}=x+3$

49. a. 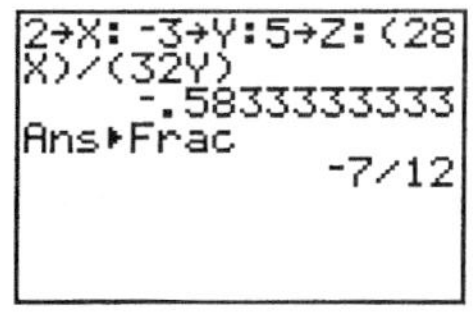

b. $\frac{28x}{32y}=\frac{4(7x)}{4(8y)}=\frac{7x}{8y}$

c. For x = 2, y = -3,

$\frac{7x}{8y}=\frac{7\cdot 2}{8(-3)}=-\frac{14}{24}=-\frac{7}{12}$

51. a. 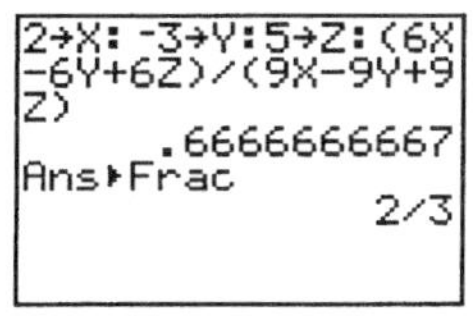

b. $\frac{6x-6y+6z}{9x-9y+9z}=\frac{6(x-y+z)}{9(x-y+z)}=\frac{6}{9}=\frac{2}{3}$

c. The expression is always 2/3.

53. a. 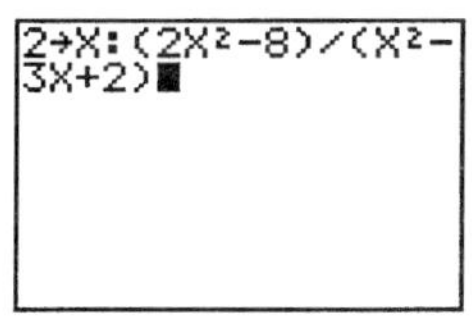

Pressing ENTER gives a 'divide by 0' error, since the denominator is 0 when x = 2.

b. $\frac{2x^2-8}{x^2-3x+2}=\frac{2(x-2)(x+2)}{(x-2)(x-1)}=\frac{2(x+2)}{x-1}$

c. 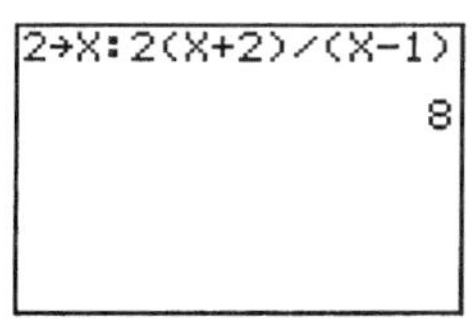

The original and simplified forms of an expression will have the same value only where they both are defined.

55. $\dfrac{3x}{x+1} = \dfrac{3x(x+1)}{(x+1)^2}$

57. $\dfrac{2y}{x} = \dfrac{2y(x+1)}{x(x+1)} = \dfrac{2y(x+1)}{x^2+x}$

59. $\dfrac{z}{z-1} = \dfrac{z(z+1)}{(z-1)(z+1)} = \dfrac{z(z+1)}{z^2-1}$

61. $\dfrac{x+2}{x-2} = \dfrac{(x+2)(x+2)}{(x-2)(x+2)} = \dfrac{(x+2)^2}{x^2-4}$

63. $\dfrac{2}{x+1} = \dfrac{2(x+2)}{(x+1)(x+2)} = \dfrac{2(x+2)}{x^2+3x+2}$

65. a.

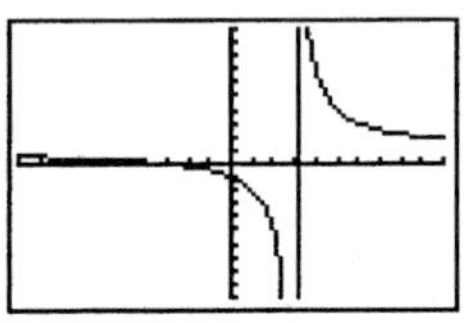

b.

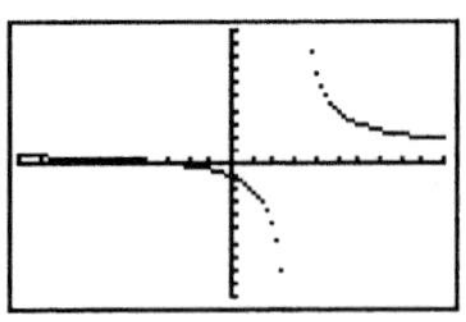

c. There should be a break in the graph at the excluded value of 3, since y has no value when x = 3.

67. a. $\dfrac{x(x+3)-3(x-1)}{x^2+3} = \dfrac{x^2+3x-3x+3}{x^2+3} =$

$\dfrac{x^2+3}{x^2+3} = 1$

b. You get a horizontal line at height 1, since y will equal 1 for every value of x.

c. y will be 1 for every value of x.

Review

69. a. $4x^3 - 10x^2 + 6x = 2x(2x^2 - 5x + 3) =$
$2x(2x-3)(x-1)$

b. $y^2 + 2xy - 63x^2 = (y+9x)(y-7x)$

c. $15 + 3b - 5a^2 - a^2b = (15 + 3b) - (5a^2 + a^2b)$
$= 3(5+b) - a^2(5+b) = (5+b)(3-a^2)$

71. $y = \dfrac{2}{3}x - 5$

y-intercept: (0,-5) x-intercept: (7.5,0)

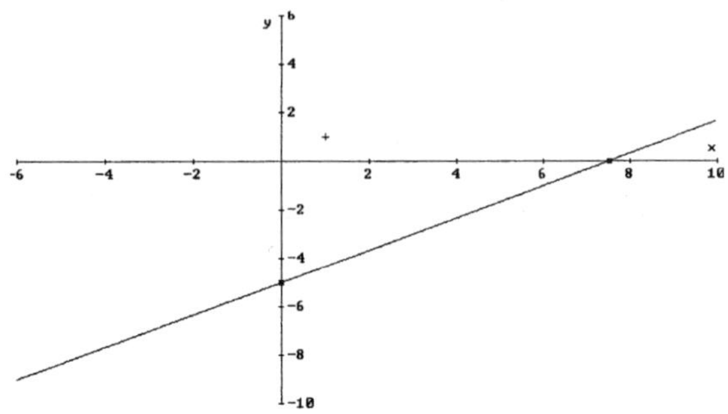

73. Solve ½ (x – 1) + 2 = 3 – 4x

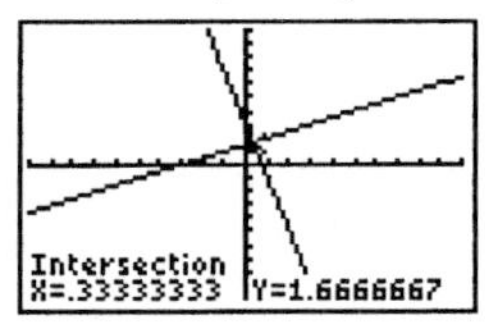

The solution is x = ⅓.

Section 6.2 Adding and Subtracting Rational Expressions

Concepts

1. $\dfrac{8}{7} + \dfrac{6}{7} = \dfrac{8+6}{7} = \dfrac{14}{7} = 2$

3. $\dfrac{9y}{3x} - \dfrac{6y}{3x} = \dfrac{9y-6y}{3x} = \dfrac{3y}{3x} = \dfrac{y}{x}$

5. $\dfrac{15z}{22} - \dfrac{-15z}{22} = \dfrac{15z-(-15z)}{22} = \dfrac{30z}{22} = \dfrac{15z}{11}$

7. Factor each denominator.
$21 = 3\cdot 7$, $42 = 2\cdot 3\cdot 7$
The LCD $= 2\cdot 3\cdot 7 = 42$

9. Factor each denominator.
$3x = 3\cdot x$, $9xy = 3^2\cdot x\cdot y$

The LCD = $3^2 \cdot x \cdot y$

11. LCD = 2·3 = 6, $\frac{1}{2}+\frac{2}{3}=\frac{3}{6}+\frac{4}{6}=\frac{7}{6}$

13. LCD = 3·3 = 9,

$$\frac{2y}{9}+\frac{y}{3}=\frac{2y}{9}+\frac{3y}{9}=\frac{2y+3y}{9}=\frac{5y}{9}$$

15. LCD = 2·7·3 = 42,

$$\frac{21x}{14}-\frac{5x}{21}=\frac{3(21x)}{42}-\frac{2(5x)}{42}=\frac{63x-10x}{42}=\frac{53x}{42}$$

Practice

17. $\frac{y+2}{2z}-\frac{y-4}{2z}=\frac{y+2-(y-4)}{2z}=\frac{y+2-y+4}{2z}$

$$=\frac{6}{2z}=\frac{3}{z}$$

19. $\frac{y+2}{5z}+\frac{y+4}{5z}=\frac{y+2+y+4}{5z}=\frac{2y+6}{5z}$

21. $\frac{3x-5}{x-2}+\frac{6x-13}{x-2}=\frac{3x-5+6x-13}{x-2}=\frac{9x-18}{x-2}$

$$\frac{9(x-2)}{x-2}=9$$

23. $\frac{3y-2}{y+3}-\frac{2y-5}{y+3}=\frac{3y-2-(2y-5)}{y+3}=$

$$\frac{3y-2-2y+5}{y+3}=\frac{y+3}{y+3}=1$$

25. $\frac{3x}{y+2}-\frac{3y}{y+2}+\frac{x+y}{y+2}=\frac{3x-3y+(x+y)}{y+2}=$

$$\frac{3x-3y+x+y}{y+2}=\frac{4x-2y}{y+2}$$

27. $\frac{x+1}{x-2}-\frac{2(x-3)}{x-2}+\frac{3(x+1)}{x-2}=$

$$\frac{x+1-2(x-3)+3(x+1)}{x-2}=$$

$$\frac{x+1-2x+6+3x+3}{x-2}=\frac{2x+10}{x-2}$$

29. Factor each denominator.
$x^2-1=(x-1)(x+1)$, $\quad x+1=x+1$
LCD = $(x-1)(x+1)$

31. Factor each denominator.
$x^2+6x=x(x+6)$, $\quad x+6=x+6$
LCD = $x(x+6)$

33. Factor each denominator.
$x^2-x-2=(x-2)(x+1)$,
$(x-2)^2=(x-2)(x-2)$
LCD = $(x-2)^2(x+1)$

35. Factor each denominator.
$x^2-4x-5=(x-5)(x+1)$,
$x^2-25=(x-5)(x+5)$
LCD = $(x-5)(x+1)(x+5)$

37. LCD = 15y

$$\frac{y+2}{5y}+\frac{y+4}{15y}=\frac{3(y+2)}{15y}+\frac{y+4}{15y}=$$

$$\frac{3(y+2)+(y+4)}{15y}=\frac{3y+6+y+4}{15y}=\frac{4y+10}{15y}$$

39. LCD = x^2y

$$\frac{x+5}{xy}-\frac{x-1}{x^2y}=\frac{x(x+5)}{x^2y}-\frac{x-1}{x^2y}=$$

$$\frac{x(x+5)-(x-1)}{x^2y}=\frac{x^2+5x-x+1}{x^2y}=$$

$$\frac{x^2+4x+1}{x^2y}$$

41. LCD = x − 2

$$\frac{3}{x-2}-(x-1)=\frac{3}{x-2}-\frac{(x-2)(x-1)}{x-2}=$$

$$\frac{3-(x-1)(x-2)}{x-2}=\frac{3-(x^2-3x+2)}{x-2}=$$

$$\frac{3-x^2+3x-2}{x-2}=\frac{-x^2+3x+1}{x-2}$$

43. LCD = (x – 2)(x + 2)

$$\frac{2x+2}{x-2}-\frac{2x}{x+2}=\frac{(2x+2)(x+2)}{(x-2)(x+2)}-\frac{2x(x-2)}{x^2-4}$$

$$=\frac{(2x+2)(x+2)-2x(x-2)}{x^2-4}=$$

$$\frac{2x^2+6x+4-2x^2+4x}{x^2-4}=\frac{10x+4}{x^2-4}$$

45. LCD = $(x-2)^2(x+2)$

$$\frac{x}{(x-2)^2}+\frac{x-4}{(x+2)(x-2)}=\frac{x(x+2)}{(x-2)^2(x+2)}+$$

$$\frac{(x-4)(x-2)}{(x-2)^2(x+2)}=\frac{x(x+2)+(x-4)(x-2)}{(x-2)^2(x+2)}=$$

$$\frac{x^2+2x+x^2-6x+8}{(x-2)^2(x+2)}=\frac{2x^2-4x+8}{(x-2)^2(x+2)}$$

47. LCD = (x – 1)(x – 2)

$$\frac{2x}{x^2-3x+2}+\frac{2x}{x-1}-\frac{x}{x-2}=\frac{2x}{(x-1)(x-2)}+$$

$$\frac{2x(x-2)}{(x-1)(x-2)}-\frac{x(x-1)}{(x-1)(x-2)}=$$

$$\frac{2x+2x(x-2)-x(x-1)}{(x-1)(x-2)}=\frac{2x+2x^2-4x-x^2+x}{(x-1)(x-2)}=$$

$$\frac{x^2-x}{(x-1)(x-2)}=\frac{x(x-1)}{(x-1)(x-2)}=\frac{x}{x-2}$$

49. LCD = 2(x + 2)(x – 2)

$$\frac{x+1}{2x+4}-\frac{x^2}{2x^2-8}=\frac{(x+1)(x-2)}{2(x+2)(x-2)}-\frac{x^2}{2(x+2)(x-2)}$$

$$=\frac{(x+1)(x-2)-x^2}{2(x+2)(x-2)}=\frac{x^2-x-2-x^2}{2(x+2)(x-2)}=$$

$$\frac{-(x+2)}{2(x+2)(x-2)}=\frac{-1}{2(x-2)}$$

51. LCD = (x + 2)(x – 3)

$$\frac{x-1}{x+2}+\frac{x}{3-x}+\frac{9x+3}{x^2-x-6}=\frac{x-1}{x+2}-\frac{x}{x-3}+\frac{9x+3}{x^2-x-6}=$$

$$\frac{(x-1)(x-3)}{(x+2)(x-3)}-\frac{x(x+2)}{(x+2)(x-3)}+\frac{9x+3}{(x+2)(x-3)}=$$

$$\frac{(x-1)(x-3)-x(x+2)+9x+3}{(x+2)(x-3)}=$$

$$\frac{x^2-4x+3-x^2-2x+9x+3}{(x+2)(x-3)}=\frac{3x+6}{(x+2)(x-3)}=$$

$$\frac{3(x+2)}{(x+2)(x-3)}=\frac{3}{x-3}$$

53. a. Perimeter = $4\left(\frac{x}{x-6}\right)=\frac{4x}{x-6}$

b. If x = 8, perimeter = $\frac{4\cdot 8}{8-6}=\frac{32}{2}=16$ ft.

c. If x = 6, the length of a side is undefined.

Review

55. $2x^3-7x^2-15x=x(2x^2-7x-15)=$
$x(2x+3)(x-5)$

57. The solutions of $2x^3-7x^2-15x=0$, using a viewing window of [-5,8] by [-55,10] are 0, 5 and -1.5.

Section 6.3 Multiplying and Dividing Rational Expressions

Concepts

1. a. $35=5\cdot 7$ b. $24=2^3\cdot 3$
   c. $56=2^3\cdot 7$ d. $27=3^3$

3. $\frac{25}{35}\cdot\frac{21}{55}=\frac{5^2\cdot 3\cdot 7}{5\cdot 7\cdot 5\cdot 11}=\frac{(5^2\cdot 7)\cdot 3}{(5^2\cdot 7)\cdot 11}=\frac{3}{11}$

5. $\frac{2}{3}\cdot\frac{15}{2}\cdot\frac{1}{7}=\frac{2\cdot 3\cdot 5}{3\cdot 2\cdot 7}=\frac{5}{7}$

7. $\frac{5y}{7}\cdot\frac{7x}{5z}=\frac{(5\cdot 7)xy}{(5\cdot 7)z}=\frac{xy}{z}$

9. $\frac{35}{2}\div\frac{15}{2}=\frac{35}{2}\cdot\frac{2}{15}=\frac{5\cdot 7\cdot 2}{2\cdot 3\cdot 5}=\frac{7}{3}$

11. $\frac{x^2}{3} \div \frac{2x}{4} = \frac{x^2}{3} \cdot \frac{4}{2x} = \frac{x \cdot x \cdot 2 \cdot 2}{3 \cdot 2 \cdot x} = \frac{2x}{3}$

Practice

13. $\frac{8x^2y^2}{4x^2} \cdot \frac{2xy}{2y} = \frac{2^4x^3y^3}{2^3x^2y} = 2xy^2$

15. $\frac{-2xy}{x^2} \cdot \frac{3xy}{2} = \frac{-2 \cdot 3 \cdot x^2y^2}{2x^2} = -3y^2$

17. $\frac{z+7}{7} \cdot \frac{z+2}{z} = \frac{(z+7)(z+2)}{7z}$

19. $\frac{x-2}{2} \cdot \frac{2x}{x-2} = \frac{2x(x-2)}{2(x-2)} = x$

21. $\frac{(x+1)^2}{x+1} \cdot \frac{x+2}{x+1} = \frac{(x+1)^2(x+2)}{(x+1)^2} = x+2$

23. $\frac{2x+6}{x+3} \cdot \frac{3}{4x} = \frac{2(x+3) \cdot 3}{4x(x+3)} = \frac{3}{2x}$

25. $\frac{x^2+x-6}{5x} \cdot \frac{5x-10}{x+3} = \frac{(x+3)(x-2) \cdot 5(x-2)}{5x(x+3)}$

$= \frac{(x-2)^2}{x}$

27. $\frac{x^2+7xy+12y^2}{x^2+2xy-8y^2} \cdot \frac{x^2-xy-2y^2}{x^2+4xy+3y^2} =$

$\frac{(x+4y)(x+3y)}{(x+4y)(x-2y)} \cdot \frac{(x-2y)(x+y)}{(x+3y)(x+y)} =$

$\frac{(x+4y)(x+3y)(x-2y)(x+y)}{(x+4y)(x-2y)(x+3y)(x+y)} = 1$

29. $\frac{3r^2+15rs+18s^2}{6r^2-24s^2} \cdot \frac{2r-4s}{3r+9s} =$

$\frac{3(r^2+5rs+6s^2)}{6(r^2-4s^2)} \cdot \frac{2(r-2s)}{3(r+3s)} =$

$\frac{3(r+3s)(r+2s)}{2 \cdot 3(r-2s)(r+2s)} \cdot \frac{2(r-2s)}{3(r+3s)} =$

$\frac{2 \cdot 3(r+3s)(r+2s)(r-2s)}{2 \cdot 3^2(r+3s)(r+2s)(r-2s)} = \frac{1}{3}$

31. $\frac{x^2+5x+6}{x^2} \cdot \frac{x^2-2x}{x^2-9} \cdot \frac{x^2-3x}{x^2-4} =$

$\frac{(x+2)(x+3)}{x^2} \cdot \frac{x(x-2)}{(x+3)(x-3} \cdot \frac{x(x-3)}{(x+2)(x-2} =$

$\frac{x^2(x+3)(x+2)(x-3)(x-2)}{x^2(x+3)(x+2)(x-3)(x-2)} = 1$

33. $\frac{x+2}{3x} \div \frac{x+2}{2} = \frac{x+2}{3x} \cdot \frac{2}{x+2} = \frac{2}{3x}$

35. $\frac{(z-2)^2}{3z^2} \div \frac{z-2}{6z} = \frac{(z-2)^2}{3z^2} \cdot \frac{6z}{z-2} = \frac{2(z-2)}{z}$

37. $\frac{5x^2+13x-6}{x+3} \div \frac{5x^2-17x+6}{x-2} =$

$\frac{(5x-2)(x+3)}{x+3} \cdot \frac{x-2}{(5x-2)(x-3)} = \frac{x-2}{x-3}$

39. $\frac{2}{3} \cdot \frac{15}{5} \div \frac{10}{5} = \frac{2}{3} \cdot \frac{15}{5} \cdot \frac{5}{10} = \frac{2 \cdot 3 \cdot 5^2}{3 \cdot 2 \cdot 5^2} = 1$

41. $\frac{x^2}{18} \div \frac{x^3}{6} \div \frac{12}{x^2} = \frac{x^2}{18} \cdot \frac{6}{x^3} \cdot \frac{x^2}{12} = \frac{2 \cdot 3x^4}{2^3 3^3 x^3} = \frac{x}{36}$

43. $\frac{x^2-1}{x^2-9} \cdot \frac{x+3}{x+2} \div \frac{5}{x+2} =$

$\frac{(x-1)(x+1)(x+3)}{(x-3)(x+3)(x+2)} \cdot \frac{x+2}{5} = \frac{(x-1)(x+1)}{5(x-3)}$

45. $\dfrac{x-x^2}{x^2-4}\left(\dfrac{2x+4}{x+2}\div\dfrac{5}{x+2}\right)=$

$\dfrac{x(1-x)}{(x-2)(x+2)}\left(\dfrac{2(x+2)}{x+2}\cdot\dfrac{x+2}{5}\right)=$

$\dfrac{2x(1-x)(x+2)^2}{5(x-2)(x+2)^2}=\dfrac{2x(1-x)}{5(x-2)}$

47. $\dfrac{y^2}{x+1}\cdot\dfrac{x^2+2x+1}{x^2-1}\div\dfrac{3y}{xy-y}=$

$\dfrac{y^2}{x+1}\cdot\dfrac{(x+1)^2}{(x-1)(x+1)}\cdot\dfrac{y(x-1)}{3y}=$

$\dfrac{y^3(x+1)^2(x-1)}{3y(x+1)^2(x-1)}=\dfrac{y^2}{3}$

Review

49. $\dfrac{x^3+2x^2-25x-50}{x^2-10x+25}=\dfrac{(x^3+2x^2)-(25x+50)}{(x-5)^2}=$

$\dfrac{x^2(x+2)-25(x+2)}{(x-5)^2}=\dfrac{(x+2)(x^2-25)}{(x-5)^2}=$

$\dfrac{(x+2)(x-5)(x+5)}{(x-5)^2}=\dfrac{(x+2)(x+5)}{x-5}$

51. Solve $4x-3y=7$ for y.

$4x-3y=7$

$-3y=-4x+7$

$y=\dfrac{4}{3}x-\dfrac{7}{3}$

Section 6.4 Mixed Operations and Complex Fractions

Vocabulary

1. complex

Concepts

3. $\dfrac{\frac{3}{2}}{5}=3\cdot\dfrac{5}{2}=\dfrac{15}{2}$ while $\dfrac{\frac{3}{2}}{5}=\dfrac{3}{2}\cdot\dfrac{1}{5}=\dfrac{3}{10}$ They are not equal since division is not associative in general.

5. a. $\dfrac{x+2}{3x}\div\dfrac{x+2}{2}=\dfrac{x+2}{3x}\cdot\dfrac{2}{x+2}=\dfrac{2}{3x}$

b. $\dfrac{\frac{x+2}{3x}}{\frac{x+2}{2}}$

c. They would be the same since the main fraction bar is the same as ÷.

7. a. 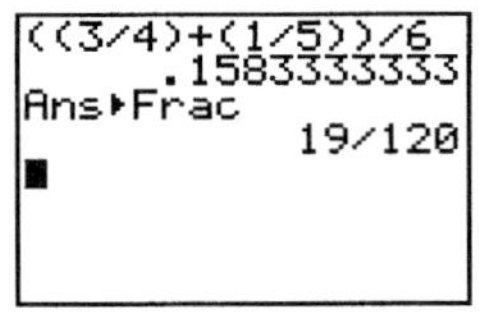

b. $\left(\dfrac{3}{4}+\dfrac{1}{5}\right)\div 6$

c. $\left(\dfrac{15}{20}+\dfrac{4}{20}\right)\div 6=\dfrac{19}{20}\cdot\dfrac{1}{6}=\dfrac{19}{120}$ This is equal to the result in part a. since the two expressions are equivalent.

Practice

9. $\dfrac{\frac{3}{5}}{\frac{2}{7}}=\dfrac{3}{5}\cdot\dfrac{7}{2}=\dfrac{21}{10}$

11. $\dfrac{\frac{7}{8}}{\frac{49}{4}}=\dfrac{7}{8}\cdot\dfrac{4}{49}=\dfrac{1}{14}$

13. $\dfrac{\frac{3}{5}-2}{\frac{2}{5}-2}=\dfrac{5\left(\frac{3}{5}-2\right)}{5\left(\frac{2}{5}-2\right)}=\dfrac{3-10}{2-10}=\dfrac{-7}{-8}=\dfrac{7}{8}$

15. LCD = 6

$\dfrac{\frac{2}{3}-\frac{5}{2}}{\frac{2}{3}-\frac{3}{2}}=\dfrac{6\left(\frac{2}{3}-\frac{5}{2}\right)}{6\left(\frac{2}{3}-\frac{3}{2}\right)}=\dfrac{4-15}{4-9}=\dfrac{-11}{-5}=\dfrac{11}{5}$

17. $\dfrac{\frac{y}{x}}{\frac{x}{xy}}=\dfrac{y}{x}\cdot\dfrac{xy}{x}=\dfrac{y^2}{x}$

19. $\dfrac{\frac{5w^2}{4tz}}{\frac{15wt}{z^2}}=\dfrac{5w^2}{4tz}\cdot\dfrac{z^2}{15wt}=\dfrac{5w^2z^2}{4\cdot 3\cdot 5wzt^2}=\dfrac{wz}{12t^2}$

21. $\dfrac{\frac{1}{y}+3}{\frac{3}{y}-2}=\dfrac{y\left(\frac{1}{y}+3\right)}{y\left(\frac{3}{y}-2\right)}=\dfrac{1+3y}{3-2y}$

23. $\dfrac{\frac{3}{x}-3}{\frac{9}{x}-3}=\dfrac{x\left(\frac{3}{x}-3\right)}{x\left(\frac{9}{x}-3\right)}=\dfrac{3-3x}{9-3x}=\dfrac{3(1-x)}{3(3-x)}=\dfrac{1-x}{3-x}$

25. $\dfrac{\frac{y}{x}+3y}{y+\frac{2y}{x}}=\dfrac{x\left(\frac{y}{x+3y}\right)}{x\left(y+\frac{2y}{x}\right)}=\dfrac{y+3xy}{xy+2y}=\dfrac{y(1+3x)}{y(x+2)}=\dfrac{1+3x}{x+2}$

27. $\dfrac{\frac{1}{x-1}}{1-\frac{1}{x-1}}=\dfrac{(x-1)\left(\frac{1}{x-1}\right)}{(x-1)\left(1-\frac{1}{x-1}\right)}=\dfrac{1}{(x-1)-1}=\dfrac{1}{x-2}$

29. $\dfrac{\frac{2}{x-2}}{\frac{2}{x-2}-1}=\dfrac{(x-2)\left(\frac{2}{x-2}\right)}{(x-2)\left(\frac{2}{x-2}-1\right)}=\dfrac{2}{2-(x-2)}=$

$\dfrac{2}{2-x+2}=\dfrac{2}{4-x}$

31. LCD = ab $\dfrac{1}{\frac{b}{a}-\frac{a}{b}}=\dfrac{ab(1)}{ab\left(\frac{b}{a}-\frac{a}{b}\right)}=\dfrac{ab}{b^2-a^2}$

33. LCD = 3y

$\dfrac{\frac{2y}{3}}{\frac{2y}{3}-\frac{8}{y}}=\dfrac{3y\left(\frac{2y}{3}\right)}{3y\left(\frac{2y}{3}-\frac{8}{y}\right)}=\dfrac{2y^2}{2y^2-24}=\dfrac{2y^2}{2(y^2-12)}=$

$\dfrac{y^2}{y^2-12}$

35. LCD = x(x + 1)

$\dfrac{2-\frac{2}{x+1}}{2+\frac{2}{x}}=\dfrac{x(x+1)\left(2-\frac{2}{x+1}\right)}{x(x+1)\left(2+\frac{2}{x}\right)}=\dfrac{2x(x+1)-2x}{2x(x+1)+2(x+1)}$

$=\dfrac{2x^2+2x-2x}{2(x+1)(x+1)}=\dfrac{2x^2}{2(x+1)^2}=\dfrac{x^2}{(x+1)^2}$

37. LCD = y(y – 3)

$\dfrac{\frac{5}{y-3}-\frac{2}{y}}{\frac{1}{y}+\frac{2}{y-3}}=\dfrac{y(y-3)\left(\frac{5}{y-3}-\frac{2}{y}\right)}{y(y-3)\left(\frac{1}{y}+\frac{2}{y-3}\right)}=\dfrac{5y-2(y-3)}{(y-3)+2y}=$

$\dfrac{5y-2y+6}{y-3+2y}=\dfrac{3y+6}{3y-3}=\dfrac{3(y+2)}{3(y-1)}=\dfrac{y+2}{y-1}$

39. LCD = y(y + 1)

$\dfrac{\frac{5}{y}+\frac{4}{y+1}}{\frac{4}{y}-\frac{5}{y+1}}=\dfrac{y(y+1)\left(\frac{5}{y}+\frac{4}{y+1}\right)}{y(y+1)\left(\frac{4}{y}-\frac{5}{y+1}\right)}=\dfrac{5(y+1)+4y}{4(y+1)-5y}=$

$\dfrac{5y+5+4y}{4y+4-5y}=\dfrac{9y+5}{4-y}$

41. $\dfrac{\dfrac{2}{b^2-1}-\dfrac{3}{ab-a}}{\dfrac{3}{ab-a}-\dfrac{2}{b^2-1}}=-1$ This is the pattern

$\dfrac{a-c}{c-a}$ where $a=\dfrac{2}{b^2-1}$ and $c=\dfrac{3}{ab-a}$.

Review

43. $\dfrac{x+5}{x-3}=\dfrac{4}{5}$

$$5(x+5)=4(x-3)$$
$$5x+25=4x-12$$
$$x=-37$$

45. Let x = height of the bell tower. Then

$$\frac{3\text{ft.}}{x\text{ft.}}=\frac{4\text{ft.}}{50\text{ft.}}$$
$$4x=150$$
$$x=37.5$$

He finds the tower to be 37.5 ft. tall.

Section 6.5 Algebraic and Graphical Solutions of Rational Equations

Vocabulary

1. extraneous

Concepts

3. a. Expression $\dfrac{x}{3}+\dfrac{x}{2}=\dfrac{2x}{6}+\dfrac{3x}{6}=\dfrac{5x}{6}$

b. Equation $\dfrac{x}{3}+\dfrac{x}{2}=6$

$$6\left(\frac{x}{3}+\frac{x}{2}\right)=6\cdot 6$$
$$2x+3x=36$$
$$5x=36$$
$$x=\frac{36}{5}$$

c. Expression

$$\frac{3x}{x-1}-\frac{x}{x^2-1}=$$
$$\frac{3x(x+1)}{(x-1)(x+1)}-\frac{x}{(x-1)(x+1)}=$$
$$\frac{3x(x+1)-x}{(x-1)(x+1)}=\frac{3x^2+3x-x}{(x-1)(x+1)}=$$
$$\frac{3x^2+2x}{(x-1)(x+1)}$$

d. Expression

$$\frac{3}{x}+\frac{2}{4x}=\frac{3\cdot 4}{4x}+\frac{2}{4x}=\frac{14}{4x}=\frac{7}{2x}$$

e. Equation

$$\frac{3x}{x-1}-\frac{x}{x^2-1}=3$$
$$(x+1)(x-1)\left(\frac{3x}{x-1}-\frac{x}{x^2-1}\right)=3(x+1)(x-1)$$
$$3x(x+1)-x=3(x^2-1)$$
$$3x^2+3x-x=3x^2-3$$
$$2x=-3$$
$$x=\frac{-3}{2}$$

f. Equation

$$\frac{3}{x}+\frac{2}{4x}=7$$
$$4x\left(\frac{3}{x}+\frac{2}{4x}\right)=7(4x)$$
$$12+2=28x$$
$$14=28x$$
$$x=\frac{14}{28}=\frac{1}{2}$$

Practice

5. $\frac{y}{3}+6=\frac{4y}{3}$

$3\left(\frac{y}{3}+6\right)=3\left(\frac{4y}{3}\right)$

$y+18=4y$

$-3y=-18$

$y=6$

7. $\frac{x}{2}-3=\frac{x}{5}$ (LCD = 10)

$10\left(\frac{x}{2}-3\right)=10\left(\frac{x}{5}\right)$

$5x-30=2x$

$3x=30$

$x=10$

9. $\frac{x+2}{2}-3x=x+8$

$2\left(\frac{x+2}{2}-3x\right)=2(x+8)$

$(x+2)-6x=2x+16$

$2-5x=2x+16$

$-7x=14$

$x=-2$

11. $\frac{b+2}{3}=b-2$

$3\left(\frac{b+2}{3}\right)=3(b-2)$

$b+2=3b-6$

$-2b=-8$

$b=4$

13. $\frac{3(x-1)}{2}+2=x$

$2\left(\frac{3(x-1)}{2}+2\right)=x$

$3(x-1)+4=2x$

$3x-3+4=2x$

$3x+1=2x$

$x=-1$

15. $\frac{11}{x}+\frac{3}{x}=12$

$\frac{24}{x}=12$

$24=12x$

$2=x$

17. $\frac{3}{p+6}-2=\frac{7}{p+6}$

$(p+6)\left(\frac{3}{p+6}-2\right)=(p+6)\left(\frac{7}{p+6}\right)$

$3-2(p+6)=7$

$3-2p-12=7$

$-2p=16$

$p=-8$

19. $\frac{z^2}{z+1}+2=\frac{1}{z+1}$

$(z+1)\left(\frac{z^2}{z+1}+2\right)=(z+1)\left(\frac{1}{z+1}\right)$

$z^2+2(z+1)=1$

$z^2+2z+2-1=0$

$z^2+2z+1=0$

$(z+1)^2=0$

$z+1=0$

$z=-1$

No solution since -1 is a restricted value.

21. $$\frac{3}{y-2}+1=\frac{3}{y-2}$$
$$(y-2)\left(\frac{3}{y-2}+1\right)=(y-2)\left(\frac{3}{y-2}\right)$$
$$3+(y-2)=3$$
$$y-2=0$$
$y=2$, no solution

23. $$\frac{2p}{3}-\frac{1}{p}=\frac{2p-1}{3} \quad (\text{LCD}=3p)$$
$$3p\left(\frac{2p}{3}-\frac{1}{p}\right)=3p\left(\frac{2p-1}{3}\right)$$
$$p(2p)-3=p(2p-1)$$
$$2p^2-3=2p^2-p$$
$$-3=-p$$
$$3=p$$

25. $$\frac{3}{5}+\frac{7}{x+2}=2 \quad (\text{LCD}=5(x+2))$$
$$5(x+2)\left(\frac{3}{5}+\frac{7}{x+2}\right)=5(x+2)\cdot 2$$
$$3(x+2)+7\cdot 5=2\cdot 5(x+2)$$
$$3x+6+35=10(x+2)$$
$$3x+41=10x+20$$
$$-7x=-21$$
$$x=3$$

27. $\frac{-5}{x^2+x-2}+\frac{3}{x+2}=\frac{1}{x-1}$ The LCD is $(x+2)(x-1)=x^2+x-2$. Multiplying both sides by the LCD gives
$$-5+3(x-1)=x+2$$
$$-5+3x-3=x+2$$
$$2x=10$$
$$x=5$$

29. $\frac{x-3}{4x-4}+\frac{1}{9}=\frac{x-5}{6x-6}$ The LCD is $36(x-1)$.
Multiplying both sides by the LCD gives
$$9(x-3)+4(x-1)=6(x-5)$$
$$9x-27+4x-4=6x-30$$
$$13x-31=6x-30$$
$$7x=1$$
$$x=\frac{1}{7}$$

31. $$\frac{3}{5x-20}-\frac{4}{5}=\frac{3}{5x-20}-\frac{x}{5}$$
Subtracting $\frac{3}{5x-20}$ from both sides:
$$-\frac{4}{5}=-\frac{x}{5}$$
$$-5x=-20$$
$$x=4$$
No solution since 4 is a restricted value.

33. $1-\frac{3}{x}=\frac{-8x}{x^2+3x}$ The LCD is $x(x+3)=x^2+3x$. Multiplying both sides by the LCD gives $(x^2+3x)-3(x+3)=-8x$
$$x^2+3x-3x-9=-8x$$
$$x^2+8x-9=0$$
$$(x+9)(x-1)=0$$
$$x=-9 \text{ or } x=1$$

35. $\frac{x-3}{x-2}=\frac{1}{x}+\frac{x-3}{x}$ The LCD is $x(x-2)$.
Multiplying both sides by the LCD gives
$$x(x-3)=(x-2)+(x-2)(x-3)$$
$$x^2-3x=x-2+x^2-5x+6$$
$$-3x=-4x+4$$
$$x=4$$

37. $\frac{3x}{4}-6=\frac{x}{4}$ Use [-5,15] by [-10,10]

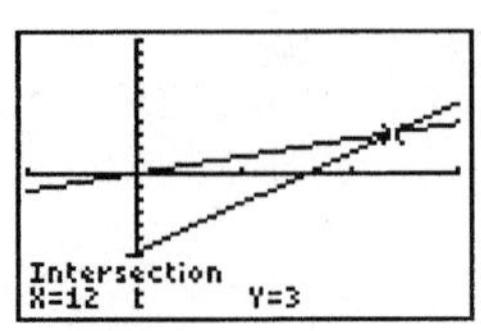

The solution is 12.

39. $\frac{1}{x-3}+\frac{2}{x-3}=1$ Use ZDecimal x 2.

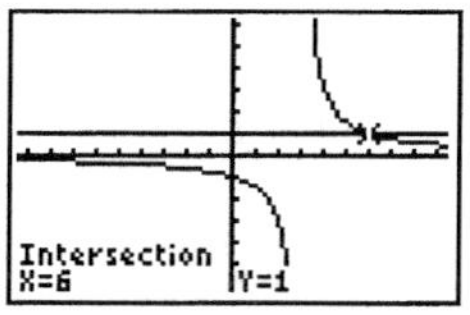

The solution is 6.

41. $x+\frac{3}{4}=\frac{3x-50}{4x-24}$ Use [1,5] by [2,5].

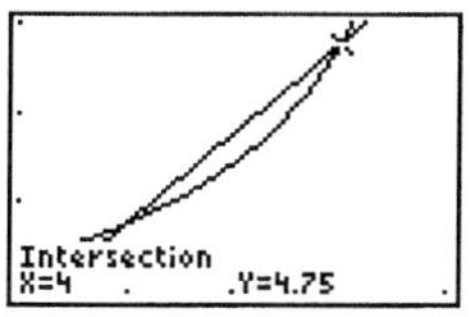

The solutions are 4 (shown above) and 2.

43. $\frac{x-5}{x-4}+\frac{x-3}{x-4}=x-4$

a. The LCD is x – 4. Multiplying both sides by the LCD gives

$(x-5)+(x-3)=(x-4)^2$

$2x-8=x^2-8x+16$

$0=x^2-10x+24$

$0=(x-6)(x-4)$

$x=4$ or $x=6$, but 4 is extraneous.

b. $\frac{(x-5)+(x-3)}{x-4}=x-4$

$\frac{2x-8}{x-4}=x-4$

$\frac{2(x-4)}{(x-4)}=x-4$

$2=x-4$

$6=x$

c.

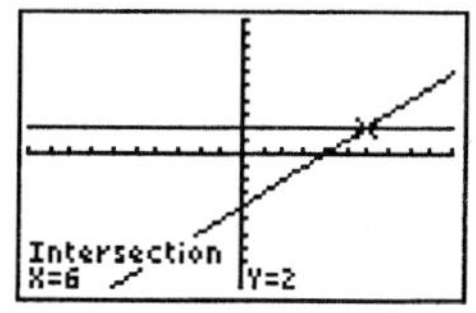

e. The graph would help as it clearly indicates that there is only <u>one</u> solution.

Review

45. $\frac{4x^2+6xy-12x^3}{4x}=\frac{4x^2}{4x}+\frac{6xy}{4x}-\frac{12x^3}{4x}=$

$x+\frac{3y}{2}-3x^2$

47. a. y = 75 + 15x, y in dollars, x in hours.

b. Using the window [0,10] by [0,300] gives

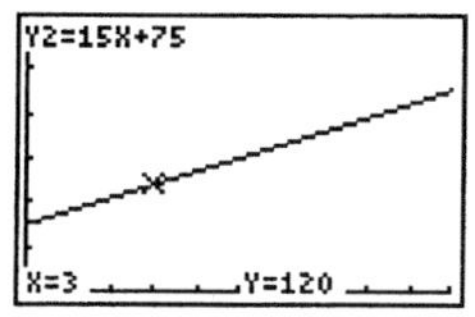

c. You might use TABLE, STOre or, as above, value.

Section 6.6 Applications of Rational Equations

Concepts

1. The reciprocals are $\frac{1}{x}$ and $\frac{1}{x+1}$.

3. The new rational expression is

$\frac{(x+2)-3}{x+3}=\frac{x-1}{x+3}$

Applications

5. Let x = the desired number. Then x satisfies

$$\frac{7+x}{8-x}=2$$
$$7+x=2(8-x)$$
$$7+x=16-2x$$
$$3x=9$$
$$x=3$$

The desired number is 3.

7. Let x = the desired number. Then x satisfies

$$\frac{5+x}{7-2x}=8$$
$$5+x=8(7-2x)$$
$$5+x=56-16x$$
$$17x=51$$
$$x=3$$

The desired number is 3.

9. Let x and x + 2 be the consecutive even integers. Then $\frac{1}{x}+\frac{1}{x+2}=\frac{7}{24}$.

Multiplying both sides by the LCD = 24x(x + 2) gives

$$24(x+2)+24x=7x(x+2)$$
$$24x+48+24x=7x^2+14x$$
$$0=7x^2-34x-48$$
$$0=(7x+8)(x-6)$$
$$x=\frac{-8}{7} \text{ or } x=6$$

Thus x = 6 and x + 2 = 8 are the integers.

11. Let x = rate of the plane.

	r	d
plane	x	300
car	x − 90	120

Time for the plane is $\frac{d}{r}=\frac{300}{x}$.

Time for the car is $\frac{d}{r}=\frac{120}{x-90}$.

Since these two times are the same, x satisfies $\frac{300}{x}=\frac{120}{x-90}$

$$300(x-90)=120x$$
$$300x-27{,}000=120x$$
$$180x=27{,}000$$
$$x=150$$

The plane flies at 150 m.p.h.

13. Let x = wind speed.

	r	d
upwind	255 − x	210
downwind	255 + x	300

The time upwind is $\frac{d}{r}=\frac{210}{255-x}$.

The time downwind is $\frac{d}{r}=\frac{300}{255+x}$. Since these times are the same, x satisfies

$$\frac{210}{255-x}=\frac{300}{255+x}$$
$$210(255+x)=300(255-x)$$
$$53{,}550+210x=76{,}500-300x$$
$$510x=22{,}950$$
$$x=45$$

The wind speed is 45 m.p.h.

15. Let x = speed of the smaller plane.

	r	d
jet	2x	680
smaller plane	x	680

The time going (jet) is $\frac{d}{r}=\frac{680}{2x}=\frac{340}{x}$.

The time returning (smaller plane) is $\frac{d}{r}=\frac{680}{x}$

The sum of the times is 6 hours so x satisfies

$$\frac{340}{x}+\frac{680}{x}=6$$
$$\frac{1020}{x}=6$$
$$1020=6x$$
$$170=x$$
The planes fly at 170 and 340 m.p.h.

17. Let x = time to fill the pool.

	hours	rate
inlet 1	5	1/5
inlet 2	4	1/4
together	x	1/x

The rate together is $\frac{1}{5}+\frac{1}{4}=\frac{4}{20}+\frac{5}{20}=\frac{9}{20}$.

Thus, $\frac{1}{x}=\frac{9}{20}$ and so $x=\frac{20}{9}=2\frac{2}{9}$. The time for both pipes together is $2\frac{2}{9}$ hours.

19. Let x = mowing time for the son alone.

	hours	rate
Blaine	3	1/3
son	x	1/x
together	2	1/3 + 1/x

The rate together is $\frac{1}{3}+\frac{1}{x}=\frac{x+3}{3x}$. Thus, the time together is $\frac{3x}{x+3}$. Then x satisfies

$$\frac{3x}{x+3}=2$$
$$3x=2(x+3)$$
$$3x=2x+6$$
$$x=6$$
It will take his son 6 hours alone.

21. Let x = time to process the forms together.

	days	rate
Becky	5/2	2/5
Sherry	2	½
together	x	1/x

Their rate together is $\frac{2}{5}+\frac{1}{2}=\frac{4}{10}+\frac{5}{10}=\frac{9}{10}$.

Thus, $\frac{1}{x}=\frac{9}{10}$ and so $x=\frac{10}{9}=1\frac{1}{9}$ days. It will take them $1\frac{1}{9}$ days together.

23. Let x = lower interest rate so that x + .02 = higher rate.

	I	r	t
1st fund	315	x	1
2nd fund	385	x +.02	1

For one year, I = Pr(1) = Pr so that

Principal in 1st fund = $\frac{I}{r}=\frac{315}{x}$.

Principal in 2nd fund = $\frac{I}{r}=\frac{385}{x+.02}$.

The two principals are the same so x satisfies

$$\frac{315}{x}=\frac{385}{x+.02}$$
$$315(x+.02)=385x$$
$$315x+6.3=385x$$
$$6.3=70x$$
$$.09=x$$
The rates are 9% and (9 + 2) = 11%

25. a. Let x = number of texts. The total cost of producing x texts is 55x + 50,000. Since the average cost per text is to be $75, x satisfies

$$\frac{55x+50000}{x}=75$$
$$55x+50000=75x$$
$$50000=20x$$
$$2500=x$$

The average production cost will be $75 per text if 2500 texts are produced.

b. When 1600 texts per year are produced the average cost of production per text is $86.25.

c. When 4000 texts per year are produced the average cost is $67.50 per text.

d. The average cost per text is -$25 if -625 texts are produced. It is not applicable since a negative number of texts cannot be produced.

Chapter 6 Summary

1. a. $-\frac{51}{153}=-\frac{3\cdot17}{3\cdot3\cdot17}=-\frac{1}{3}$

   b. $-\frac{105}{45}=-\frac{5\cdot3\cdot7}{5\cdot3\cdot3}=-\frac{7}{3}$

   c. $\frac{3x^2}{6x^3}=\frac{3\cdot x\cdot x}{2\cdot3\cdot x\cdot x\cdot x}=\frac{1}{2x}$

   Excluded value: $x=0$

   d. $\frac{5xy^2}{2x^2y^2}=\frac{5\cdot x\cdot y^2}{2\cdot x\cdot x\cdot y^2}=\frac{5}{2x}$

   Excluded values: $x=0$, $y=0$

2. a. $\frac{x^2}{x^2+x}=\frac{x\cdot x}{x(x+1)}=\frac{x}{x+1}$

   Excluded values: $x=0$, -1

   b. $\frac{x+2}{x^2+2x}=\frac{x+2}{x(x+2)}=\frac{1}{x}$

   Excluded values: $x=0$, -2

   c. $\frac{x^2+4x+3}{x^2-4x-5}=\frac{(x+1)(x+3)}{(x+1)(x-5)}=\frac{x+3}{x-5}$

   Excluded values: $x=-1$, 5

   d. $\frac{x^2-x-56}{x^2-5x-24}=\frac{(x-8)(x+7)}{(x-8)(x+3)}=\frac{x+7}{x+3}$

   Excluded values: $x=8$, -3

3. a. $\frac{2x^2-16x}{2x^2-18x+16}=\frac{2x(x-8)}{2(x^2-9x+8)}=$

   $\frac{2x(x-8)}{2(x-8)(x-1)}=\frac{x}{x-1}$

   Excluded values: $x=1$, 8

   b. $\frac{x^2+x-2}{x^2-x-2}=\frac{(x+2)(x-1)}{(x-2)(x+1)}$,in lowest terms

   Excluded values: $x=-1$, 2

4. a. $\frac{x}{x+y}+\frac{y}{x+y}=\frac{x+y}{x+y}=1$

   b. $\frac{3x}{x-7}-\frac{x-2}{x-7}=\frac{3x-(x-2)}{x-7}=\frac{3x-x+2}{x-7}=$

   $\frac{2x+2}{x-7}$

   c. LCD $=x(x-1)$

   $\frac{x}{x-1}+\frac{1}{x}=\frac{x\cdot x}{x(x-1)}+\frac{1(x-1)}{x(x-1)}=\frac{x^2+x-1}{x(x-1)}$

   d. LCD $=7x$

   $\frac{1}{7}-\frac{1}{x}=\frac{x}{7x}-\frac{7}{7x}=\frac{x-7}{7x}$

5. a. LCD $=x(x+1)$

   $\frac{3}{x+1}-\frac{2}{x}=\frac{3x}{x(x+1)}-\frac{2(x+1)}{x(x+1)}=$

   $\frac{3x-2(x+1)}{x(x+1)}=\frac{3x-2x-2}{x(x+1)}=\frac{x-2}{x(x+1)}$

   b. LCD $=2x^2$

   $\frac{x+2}{2x}-\frac{2-x}{x^2}=\frac{x(x+2)}{2x^2}-\frac{2(2-x)}{2x^2}=$

   $\frac{x(x+2)-2(2-x)}{2x^2}=\frac{x^2+2x-4+2x}{2x^2}=$

   $\frac{x^2+4x-4}{2x^2}$

   c. LCD $=x(x+2)=x^2+2x$

$$\frac{x}{x+2}+\frac{3}{x}-\frac{4}{x^2+2x}=\frac{x\cdot x}{x(x+2)}+\frac{3(x+2)}{x(x+2)}-\frac{4}{x^2+2x}$$

$$=\frac{x^2+3(x+2)-4}{x(x+2)}=\frac{x^2+3x+6-4}{x(x+2)}=\frac{x^2+3x+2}{x(x+2)}=$$

$$\frac{(x+1)(x+2)}{x(x+2)}=\frac{x+1}{x}$$

d. LCD = $(x-1)(x+1)=x^2-1$

$$\frac{2}{x-1}-\frac{3}{x+1}+\frac{x-5}{x^2-1}=\frac{2(x+1)}{x^2-1}-\frac{3(x-1)}{x^2-1}+\frac{x-5}{x^2-1}=$$

$$\frac{2(x+1)-3(x-1)+(x-5)}{x^2-1}=\frac{2x+2-3x+3+x-5}{x^2-1}$$

$$=\frac{0}{x^2-1}=0$$

6. a. $\frac{3xy}{2x}\cdot\frac{4x}{2y^2}=\frac{3\cdot 2^2\cdot x^2y}{2^2\cdot xy^2}=\frac{3x}{y}$

b. $\frac{3x}{x^2-x}\cdot\frac{2x-2}{x^2}=\frac{3x\cdot 2(x-1)}{x(x-1)x^2}=\frac{6}{x^2}$

c. $\frac{3x^2}{5x^2y}\div\frac{6x}{15xy}=\frac{3x^2}{5x^2y}\cdot\frac{15xy}{6x}=\frac{3^2 5x^3y}{5\cdot 3\cdot 2x^3y}=\frac{3y}{2}$

d. $\frac{x^2+5x}{x^2+4x-5}\div\frac{x^2}{x-1}=\frac{x(x+5)}{(x+5)(x-1)}\cdot\frac{x-1}{x^2}=\frac{1}{x}$

7. a. $\frac{x^2+3x+2}{x^2+2x}\cdot\frac{x}{x+1}=\frac{(x+2)(x+1)x}{x(x+2)(x+1)}=1$

b. $\frac{x^2+4x+4}{x^2-x-6}\cdot\frac{x^2-9}{x^2+5x+6}=$

$$\frac{(x+2)^2(x-3)(x+3)}{(x-3)(x+2)(x+3)(x+2)}=1$$

c. $\frac{x^2-x-6}{2x-1}\div\frac{x^2-2x-3}{2x^2+x-1}=$

$$\frac{(x-3)(x+2)}{2x-1}\cdot\frac{(2x-1)(x+1)}{(x-3)(x+1)}=x+2$$

d. $\frac{3n^2+5n-2}{12n^2-13n+3}\div\frac{n^2+3n+2}{4n^2+5n-6}=$

$$\frac{(3n-1)(n+2)}{(4n-3)(3n-1)}\cdot\frac{(4n-3)(n+2)}{(n+2)(n+1)}=\frac{n+2}{n+1}$$

8. a. $\frac{\frac{1}{x}+1}{\frac{1}{x}-1}=\frac{x\left(\frac{1}{x}+1\right)}{x\left(\frac{1}{x}-1\right)}=\frac{1+x}{1-x}$

b. The LCD = x^2.

$$\frac{1+\frac{3}{x}}{2-\frac{1}{x^2}}=\frac{x^2\left(1+\frac{3}{x}\right)}{x^2\left(2-\frac{1}{x^2}\right)}=\frac{x^2+3x}{2x^2-1}$$

c. The LCD = $(x-1)(x+1)=x^2-1$

$$\frac{\frac{2}{x-1}+\frac{x-1}{x+1}}{\frac{1}{x^2-1}}=\frac{(x-1)(x+1)\left(\frac{2}{x-1}+\frac{x-1}{x+1}\right)}{(x-1)(x+1)\left(\frac{1}{x^2-1}\right)}$$

$$=\frac{2(x+1)+(x-1)^2}{1}=2x+2+x^2-2x+1=$$

$$x^2+3$$

d. The LCD = ab.

$$\frac{\frac{a}{b}+c}{\frac{b}{a}+c}=\frac{ab\left(\frac{a}{b}+c\right)}{ab\left(\frac{b}{a}+c\right)}=\frac{a^2+abc}{b^2+abc}$$

9. a. $\frac{3}{x}=\frac{2}{x-1}$

$$3(x-1)=2x$$
$$3x-3=2x$$
$$x=3$$

b. $\frac{5}{x+4}=\frac{3}{x+2}$

$$5(x+2)=3(x+4)$$
$$5x+10=3x+12$$
$$2x=2$$
$$x=1$$

c. $\frac{2}{3x}+\frac{1}{x}=\frac{5}{9}$ The LCD = 9x. Multiplying both sides by the LCD gives

$3 \cdot 2 + 9 = 5x$

$15 = 5x$

$3 = x$

d. $\frac{2x}{x+4} = \frac{3}{x-1}$

$2x(x-1) = 3(x+4)$

$2x^2 - 2x = 3x + 12$

$2x^2 - 5x - 12 = 0$

$(2x+3)(x-4) = 0$

$x = \frac{-3}{2}$ or $x = 4$

e. $\frac{2}{x-1} + \frac{3}{x+4} = \frac{-5}{x^2+3x-4}$ The LCD is $(x-1)(x+4) = x^2 + 3x - 4$. Multiplying both sides by the LCD gives

$2(x+4) + 3(x-1) = -5$

$2x + 8 + 3x - 3 = -5$

$5x + 5 = -5$

$5x = -10$

$x = -2$

f. $\frac{4}{x+2} - \frac{3}{x+3} = \frac{6}{x^2+5x+6}$ The LCD is $(x+2)(x+3) = x^2 + 5x + 6$. Multiplying both sides by the LCD gives

$4(x+3) - 3(x+2) = 6$

$4x + 12 - 3x - 6 = 6$

$x + 6 = 6$

$x = 0$

10. $\frac{1}{x-1} + \frac{3}{x-1} = 1$ Use [-5,8] by [-5,5].

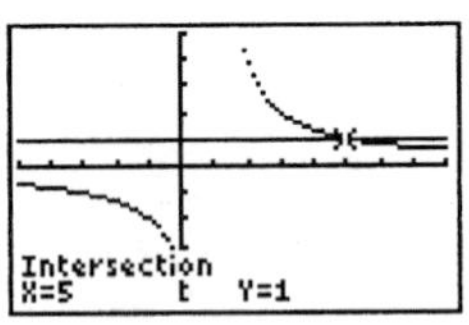

The solution is 5.

11. Let x = combined time to empty the basement.

	hours	rate
pump 1	18	1/18
pump 2	20	1/20
together	x	1/x

The combined rate is $\frac{1}{18} + \frac{1}{20} = \frac{10+9}{180} = \frac{19}{180} = \frac{1}{x}$. Thus, the combined time is $x = \frac{180}{19} = 9\frac{9}{19}$ hours.

12. Let x = combined time to paint the house.

	days	rate
painter	10	1/10
owner	14	1/14
together	x	1/x

The combined rate is $\frac{1}{10} + \frac{1}{14} = \frac{7+5}{70} = \frac{12}{70} = \frac{6}{35} = \frac{1}{x}$. Thus, the combined time $x = \frac{35}{6} = 5\frac{5}{6}$ days.

13. Let x = jogging rate, x + 10 = cycling rate

	r	d
jogging	x	10
cycling	x +10	30

The jogging time is $\frac{d}{r} = \frac{10}{x}$.

The cycling time is $\frac{d}{r} = \frac{30}{x+10}$. These times are equal so $\frac{10}{x} = \frac{30}{x+10}$

$$10(x+10) = 30x$$
$$10x + 100 = 30x$$
$$100 = 20x$$
$$5 = x$$

He jogs at 5 m.p.h.

14. Let x = wind velocity. The plane's velocity in still air is 360 m.p.h. Thus,

	r	d
upwind	360 – x	320
downwind	360 + x	400

The time upwind is $\frac{d}{r} = \frac{320}{360 - x}$.

The time downwind is $\frac{d}{r} = \frac{400}{360 + x}$.

The times are equal so that x satisfies

$$\frac{320}{360 - x} = \frac{400}{360 + x}$$
$$320(360 + x) = 400(360 - x)$$
$$115{,}200 + 320x = 144{,}000 - 400x$$
$$720x = 28{,}800$$
$$x = 40$$

The wind speed is 40 m.p.h.

15. Let x = lower interest rate, so that x + .005 = higher rate.

	I	r	t
Acc't 1	108	x	1
Acc't 2	121.50	x + .005	1

Since I = Pr(1) = Pr for one year,

Principal in account 1 = $\frac{I}{r} = \frac{108}{x}$.

Principal in account 2 = $\frac{I}{r} = \frac{121.50}{x + .005}$.

Since the two principals are the same

$$\frac{108}{x} = \frac{121.50}{x + .005}$$
$$108(x + .005) = 121.50x$$
$$108x + .54 = 121.50x$$
$$.54 = 13.50x$$
$$.04 = x$$

The lower rate is 4%.

16. Let x = desired number of children. The total monthly expenditure for x children is 150x + 6000. The average monthly cost per child is to be \$400 so x satisfies

$$\frac{150x + 6000}{x} = 400$$
$$150x + 6000 = 400x$$
$$6000 = 250x$$
$$24 = x$$

An enrollment of 24 children will give an average cost per child or \$400 per month.

Chapter 6 Test

1. $\frac{48x^2y}{54xy^2} = \frac{2^4 \cdot 3 \cdot x^2y}{2 \cdot 3^3 \cdot xy^2} = \frac{2^3x}{3^2y} = \frac{8x}{9y}$

2. $\frac{2x^2 - x - 3}{4x^2 - 9} = \frac{(2x-3)(x+1)}{(2x-3)(2x+3)} = \frac{x+1}{2x+3}$

   The excluded values are $\pm\frac{3}{2}$.

3. $\frac{3(x+2) - 3}{2x - 4 - (x - 5)} = \frac{3x + 6 - 3}{2x - 4 - x + 5} = \frac{3x + 3}{x + 1} =$

   $\frac{3(x+1)}{x+1} = 3$ The excluded value is -1.

4. $\frac{12x^2y}{15xyz} \cdot \frac{25y^2z}{16xt} = \frac{2^2 \cdot 3 \cdot 5^2 x^2 y^3 z}{2^4 \cdot 3 \cdot 5 x^2 yzt} = \frac{5y^2}{4t}$

5. $\dfrac{x^2+3x+2}{3x+9}\cdot\dfrac{x+3}{x^2-4}=\dfrac{(x+2)(x+1)(x+3)}{3(x+3)(x+2)(x-2)}=$
$\dfrac{x+1}{3(x-2)}$

6. $\dfrac{8x^2y}{25xt}\div\dfrac{16x^2y^3}{30xyt^3}=\dfrac{2^3x^2y}{5^2xt}\cdot\dfrac{2\cdot3\cdot5xyt^3}{2^4x^2y^3}=$
$\dfrac{2^4\cdot3\cdot5x^3y^2t^3}{2^45^2x^3y^3t}=\dfrac{3t^2}{5y}$

7. $\dfrac{x^2-x}{3x^2+6x}\div\dfrac{3x-3}{3x^3+6x^2}=\dfrac{x(x-1)}{3x(x+2)}\cdot\dfrac{3x^2(x+2)}{3(x-1)}=\dfrac{x^2}{3}$

8. $\dfrac{5x-4}{x-1}+\dfrac{5x+3}{x-1}=\dfrac{5x-4+5x+3}{x-1}=\dfrac{10x-1}{x-1}$

9. $\dfrac{3y+7}{2y+3}-\dfrac{3(y-2)}{2y+3}=\dfrac{3y+7-3(y-2)}{2y+3}=$
$\dfrac{3y+7-3y+6}{2y+3}=\dfrac{13}{2y+3}$

10. LCD = x(x + 1)
$\dfrac{x+1}{x}+\dfrac{x-1}{x+1}=\dfrac{(x+1)(x+1)}{x(x+1)}+\dfrac{x(x-1)}{x(x+1)}=$
$\dfrac{(x^2+2x+1)+(x^2-x)}{x(x+1)}=\dfrac{2x^2+x+1}{x(x+1)}$

11. $\dfrac{\dfrac{8x^2}{xy^3}}{\dfrac{4y^3}{x^2y^3}}=\dfrac{8x^2}{xy^3}\cdot\dfrac{x^2y^3}{4y^3}=\dfrac{8x^4y^3}{4xy^6}=\dfrac{2x^3}{y^3}$

12. $\dfrac{1+\dfrac{y}{x}}{\dfrac{y}{x}-1}=\dfrac{x\left(1+\dfrac{y}{x}\right)}{x\left(\dfrac{y}{x}-1\right)}=\dfrac{x+y}{y-x}$

13. $\dfrac{x}{10}-\dfrac{1}{2}=\dfrac{x}{5}$ The LCD is 10. Multiplying both sides of the equation by 10 gives
$x-5=2x$
$-5=x$

14. $3x-\dfrac{2(x+3)}{3}=16-\dfrac{x+2}{2}$ The LCD is 6.
Multiplying both sides by 6 gives
$6(3x)-2\cdot2(x+3)=6(16)-3(x+2)$
$18x-4x-12=96-3x-6$
$14x-12=90-3x$
$17x=102$
$x=6$

15. $\dfrac{7}{x+4}-\dfrac{1}{2}=\dfrac{3}{x+4}$ The LCD is 2(x + 4).
Multiplying both sides by 2(x + 4) gives
$2\cdot7-(x+4)=2\cdot3$
$14-x-4=6$
$10-x=6$
$4=x$

16. $\dfrac{y}{y-1}=\dfrac{y-2}{y}$
$y^2=(y-1)(y-2)$
$y^2=y^2-3y+2$
$3y=2$
$y=\dfrac{2}{3}$

17. Let x = combined time to pick up the trash.

	hours	rate
worker 1	7	1/7
worker 2	9	1/9
together	x	1/x

The combined rate is $\frac{1}{7}+\frac{1}{9}=\frac{9+7}{63}=\frac{16}{63}=\frac{1}{x}$.

The time together is $x=\frac{63}{16}=3\frac{15}{16}$ hours.

18. Let x = speed of the current. The speed of the boat in still water is 23 m.p.h. so we have

	r	d
upstream	23 – x	28
downstream	23 + x	28

The time upstream is $\frac{d}{r}=\frac{28}{x+23}$.

The time downstream is $\frac{d}{r}=\frac{18}{23-x}$.

The two times are equal, so x satisfies

$$\frac{28}{x+23}=\frac{18}{23-x}$$
$$28(23-x)=18(x+23)$$
$$644-28x=18x+414$$
$$230=46x$$
$$5=x$$

The speed of the current is 5 m.p.h.

19. Let x = rate of the novice cyclist. Then x + 10 = rate of the veteran cyclist.

	r	d
novice	x	20
veteran	x + 10	35

The novice's time is $\frac{d}{r}=\frac{20}{x}$.

The veteran's time is $\frac{d}{r}=\frac{35}{x+10}$.

The times are equal so that x satisfies

$$\frac{20}{x}=\frac{35}{x+10}$$
$$20(x+10)=35x$$
$$20x+200=35x$$
$$200=15x$$
$$\frac{40}{3}=x$$

The novice's rate is $\frac{40}{3}$ m.p.h. and so his

time is $\frac{20}{\frac{40}{3}}=\frac{60}{40}=1.5$ hours.

20. Let x = Betsy's jogging rate. Then x + 10 is Betsy's cycling rate.

	r	d
jogging	x	5
cycling	x + 10	20

Betsy's jogging time is $\frac{d}{r}=\frac{5}{x}$.

Betsy's cycling time is $\frac{d}{r}=\frac{20}{x+10}$.

The times are the same so x satisfies

$$\frac{5}{x}=\frac{20}{x+10}$$
$$5(x+10)=20x$$
$$5x+50=20x$$
$$50=15x$$
$$\frac{10}{3}=x$$

Her jogging rate is $\frac{10}{3}$ m.p.h. and her time is

$$\frac{d}{r}=\frac{5}{\frac{10}{3}}=\frac{15}{10}=1.5 \text{ hours.}$$

Cumulative Review: Chapters 5 and 6

Vocabulary/Concepts

1. F The completely factored form is $(x-2)(4+x)$.
2. T
3. trinomial
4. The leading coefficient is -1.

5. The trinomial $x^3 + x^2 + 1$ is not factorable since there is no common factor and the higher power of x is not double the smaller power.

6. $x^2 - y^2$
7. prime
8. two

9. Factoring $x^2 + 5x + 6$ means to rewrite it as a product.
   Solving $x^2 + 5x + 6 = 0$ means to find all values of x which make the left side evaluate to 0.

10. a. quadratic
    $4x(x-5) = 2x(x+1) - 3$
    $4x^2 - 20x = 2x^2 + 2x - 3$
    $2x^2 - 22x + 3 = 0$
    b. linear
    $x^2 - 5x = 2x(x+1) - x^2$
    $x^2 - 5x = 2x^2 + 2x - x^2$
    $x^2 - 5x = x^2 + 2x$
    $-7x = 0$
    c. quadratic
    $(x+2)(x-5) = (2x+1)(x-4)$
    $x^2 - 3x - 10 = 2x^2 - 7x - 4$
    $0 = x^2 - 4x + 6$

11. rational

12. To find the excluded values, set each denominator equal to 0 and solve for x.
    a. $\frac{x+5}{x-2}$, $x - 2 = 0$ gives $x = 2$ as the excluded value.
    b. $\frac{3x}{4x+1}$, $4x + 1 = 0$ gives $x = -\frac{1}{4}$ as the excluded value.
    c. $\frac{5}{6x}$, $6x = 0$ gives $x = 0$ as the excluded value.

13. $\frac{14xy}{15} - \frac{-6xy}{15} = \frac{14xy - (-6xy)}{15} = \frac{20xy}{15} = \frac{4xy}{3}$

14. Factor each denominator.
    $6x^2y = 2 \cdot 3 \cdot x^2 \cdot y$, $9xy^2 = 3^2 \cdot x \cdot y^2$
    The LCD is $2 \cdot 3^2 \cdot x^2 \cdot y^2 = 18x^2y^2$

15. a. $\frac{3xy}{5x^2} \cdot \frac{10}{6xy^2} = \frac{2 \cdot 3 \cdot 5 \cdot x \cdot y}{5 \cdot 2 \cdot 3 \cdot x^3 y^2} = \frac{1}{x^2 y}$
    b. $\frac{-8y^2}{6x} \div \frac{2}{5y^2x} = \frac{-2^3 y^2}{2 \cdot 3 \cdot x} \cdot \frac{5y^2x}{2} =$
    $\frac{-2 \cdot 5y^4}{3} = \frac{-10y^4}{3}$

16. complex

17. No, it is equivalent to $\left(\frac{1}{2} + \frac{2}{3}\right) \div \frac{3}{4}$.

18. extraneous solutions

19. a. $\frac{1}{x-3} + \frac{2}{x-3} = 1$
    $\frac{3}{x-3} = 1$
    $x - 3 = 3$
    $x = 6$
    The excluded value is 3.
    b. $\frac{1}{x-3} + \frac{2}{x-2} = \frac{1+2}{x-3} = \frac{3}{x-3}$
    The excluded value is 3.

20. If the consecutive odd integers are x and x + 2, then their reciprocals are $\frac{1}{x}$ and $\frac{1}{x+2}$

Practice

21. $x^2 - 49 = x^2 - 7^2 = (x - 7)(x + 7)$

22. $x^2 - 8x - 9$
$ac = 1(-9) = -9$ and $1 + (-9) = -8$. Thus, $x^2 - 8x - 9 = x^2 + x - 9x - 9 = (x^2 + x) - 9(x + 1) = x(x + 1) - 9(x + 1) = (x + 1)(x - 9)$

23. $18x^2 - 27x - 5$
$ac = 18(-5) = -90 = 3(-30)$ and $3 + (-30) = -27$ Thus, $18x^2 - 27x - 5 = 18x^2 + 3x - 30x - 5 = 18x^2 + 3x - (30x + 5) = 3x(6x + 1) - 5(6x + 1) = (6x + 1)(3x - 5)$

24. $x^3 + 2x^2 = x^2(x + 2)$

25. $x^2 - 18x + 81 = x^2 - 2 \cdot 9 \cdot x + 9^2 = (x - 9)^2$

26. $5x^3 - 125x = 5x(x^2 - 25) = 5x(x - 5)(x + 5)$

27. $ac - ad + bc - bd = (ac - ad) + (bc - bd) = a(c - d) + b(c - d) = (c - d)(a + b)$

28. $a^3 - 8y^3 = a^3 - (2y)^3 = (a - 2y)[a^2 + a(2y) + (2y)^2] = (a - 2y)(a^2 + 2ay + 4y^2)$

29. $x(x+8) - 5x = 5(1-x) - 12$
$x^2 + 8x - 5x = 5 - 5x - 12$
$x^2 + 3x = -5x - 7$
$x^2 + 8x + 7 = 0$
$(x+1)(x+7) = 0$
$x + 1 = 0$ or $x + 7 = 0$
$x = -1$ or $x = -7$

30. $x^3 + x^2 - 12x = 0$
$x(x^2 + x - 12) = 0$
$x(x+4)(x-3) = 0$
$x = 0$ or $x + 4 = 0$ or $x - 3 = 0$
$x = 0$ or $x = -4$ or $x = 3$

31. 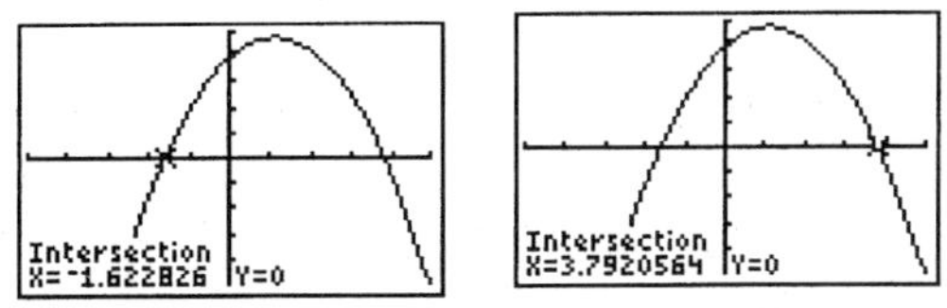

The solutions are -1.62 and 3.79.

32. The solutions are x = -2 and x = 5 and thus the factors would be (x + 2) and (x – 5). The quadratic equation is $(x + 2)(x - 5) = 0$ or $x^2 - 3x - 10 = 0$.

33. The factorization is incorrect. A polynomial and its factored form should evaluate to the same number for every value of x.

34. $\frac{4a + 4b}{a^2 + ab} = \frac{4(a+b)}{a(a+b)} = \frac{4}{a}$

35. $\frac{x}{x+5} = \frac{x(x-5)}{(x+5)(x-5)} = \frac{x^2 - 5x}{x^2 - 25}$

36. The LCD is (y + 2)(y – 2).
$$\frac{3}{y+2} - \frac{2}{y-2} = \frac{3(y-2)}{(y+2)(y-2)} - \frac{2(y+2)}{(y+2)(y-2)}$$
$$= \frac{3(y-2) - 2(y+2)}{(y+2)(y-2)} = \frac{3y - 6 - 2y - 4}{(y+2)(y-2)} =$$
$$\frac{y - 10}{(y+2)(y-2)}$$

37. $\frac{a-b}{b-a} = \frac{-(b-a)}{b-a} = -1$

38. $\frac{3-a}{b^2-b}\cdot\frac{b^2-2b+1}{a^2-9}=\frac{-(a-3)}{b(b-1)}\cdot\frac{(b-1)^2}{(a+3)(a-3)}$

$=\frac{-(b-1)}{b(a+3)}$

39. $\frac{(x+3)^5}{x^2+4x+3}\div\frac{x^2+6x+9}{x^2-3x}\cdot\frac{2x^2+8x+6}{x^2-9}=$

$\frac{(x+3)^5}{(x+3)(x+1)}\cdot\frac{x(x-3)}{(x+3)^2}\cdot\frac{2(x^2+4x+3)}{(x-3)(x+3)}=$

$\frac{(x+3)^5x(x-3)(x+1)(x+3)}{(x+3)^4(x+1)(x-3)}=2x(x+3)^2$

40. The LCD is 15.

$\frac{\frac{2}{3}+\frac{4}{5}}{\frac{2}{3}-\frac{4}{5}}=\frac{15\left(\frac{2}{3}+\frac{4}{5}\right)}{15\left(\frac{2}{3}-\frac{4}{5}\right)}=\frac{10+12}{10-12}=\frac{22}{-2}=-11$

41. LCD = a^2b^2

$\frac{\frac{a}{b}-\frac{b}{a}}{\frac{a}{b^2}-\frac{b}{a^2}}=\frac{a^2b^2\left(\frac{a}{b}-\frac{b}{a}\right)}{a^2b^2\left(\frac{a}{b^2}-\frac{b}{a^2}\right)}=\frac{a^3b-ab^3}{a^3-b^3}=$

$\frac{ab(a-b)(a+b)}{(a-b)(a^2+ab+b^2)}=\frac{ab(a+b)}{a^2+ab+b^2}$

42. $\frac{5}{a-3}+\frac{30}{9-a^2}=1$ or $\frac{5}{a-3}-\frac{30}{a^2-9}=1$ The LCD is a^2-9. Multiplying both sides by a^2-9 gives

$5(a+3)-30=a^2-9$

$5a+15-30=a^2-9$

$0=a^2-5a+6$

$0=(a-2)(a-3)$

$a=2$ or $a=3$, but 3 is extraneous.

The only solution is 2.

43. $\frac{6}{5x}+\frac{4}{5}-2x=0$ The LCD is 5x.

Multiplying both sides by 5x gives

$6+4x-2x(5x)=0$

$6+4x-10x^2=0$

$10x^2-4x-6=0$

$2(5x^2-2x-3)=0$

$2(5x+3)(x-1)=0$

$x=\frac{-3}{5}$ or $x=1$

44. The line is the result of graphing $y=\frac{x-1}{x+1}$ in connected mode. It should not be there.

45. From the table the excluded value is 3 and the solution is 5.

Applications

46. Since the initial velocity is 80 ft./sec. the equation is $h = 80t - 16t^2$. The object will be on the ground when $h = 0$, so t satisfies

$0=80t-16t^2$

$16t^2-80t=0$

$16t(t-5)=0$

$t=0$ or $t=5$

The object is on the ground after 5 seconds.

47. Let x = width. Then 2x + 8 = length and x satisfies $x(2x+8)=10$

$2x^2+8x-10=0$

$2(x^2+4x-5)=0$

$2(x+5)(x-1)=0$

$x=-5$ or $x=1$

The width is 1 yard.

48. Let x = combined time to enter the data.

	hours	rate
Becky	4	1/4
Sherry	6	1/6
together	x	1/x

Their rate together is $\frac{1}{4}+\frac{1}{6}=\frac{6+4}{24}=\frac{10}{24}=\frac{1}{x}$

Thus, time together is $x=\frac{24}{10}=2.4$ hours.

49. Let x = rate of the boat in still water, with the speed of the current equal to 1.5 m.p.h.

	r	d
upstream	x − 1.5	8
downstream	x + 1.5	12

The time upstream is $\frac{d}{r}=\frac{8}{x-1.5}$.

The time downstream is $\frac{d}{r}=\frac{12}{x+1.5}$.

Since the times are the same, x satisfies

$$\frac{8}{x-1.5}=\frac{12}{x+1.5}$$
$$8(x+1.5)=12(x-1.5)$$
$$8x+12=12x+18$$
$$-4x=-30$$
$$x=7.5$$

The rate of the current is 7.5 m.p.h.

Chapter 7 Making the Transition from Elementary to Intermediate Algebra

Section 7.1 Negative Integral Exponents

Vocabulary

1. scientific notation

Concepts

3. 3, -3 · 3 (= -9)

5. 1

7. $\frac{1}{x^n}$

9. a. $x^4 \cdot x^7 = x^{4+7} = x^{11}$
   b. $(x^4)^7 = x^{4\cdot 7} = x^{28}$
   c. $\frac{x^7}{x^4} = x^{7-4} = x^3$
   d. $\left(\frac{x}{y}\right)^7 = \frac{x^7}{y^7}$
   e. $(xy)^3 = x^3y^3$
   f. $(x^2y^3)^4 = (x^2)^4(y^3)^4 = x^{2\cdot 4}y^{3\cdot 4} = x^8y^{12}$

Practice

11. $2^5 \cdot 2^{-2} = 2^{5+(-2)} = 2^3 = 8$

13. $\frac{4^{-5}}{4^{-3}} = 4^{-5-(-3)} = 4^{-2} = \frac{1}{4^2} = \frac{1}{16}$

15. $\frac{2^3 \cdot 2^{-1}}{2^2} = 2^{3+(-1)-2} = 2^0 = 1$

17. $x^{-2} = \frac{1}{x^2}$

19. $(2y)^{-4} = \frac{1}{(2y)^4} = \frac{1}{2^4y^4} = \frac{1}{16y^4}$

21. $(ab^2)^{-3} = \frac{1}{(ab^2)^3} = \frac{1}{a^3(b^2)^3} = \frac{1}{a^3b^6}$

23. $\frac{y^4y^3}{y^{-4}y^{-2}} = \frac{y^{4+3}}{y^{-4+(-2)}} = \frac{y^7}{y^{-6}} = y^7y^6 = y^{7+6} = y^{13}$

25. $(x^{-4}x^3)^3 = (x^{-4+3})^3 = (x^{-1})^3 = x^{-3} = \frac{1}{x^3}$

27. $(-2x^3y^{-2})^{-5} = \frac{1}{(-2x^3y^{-2})^5} = \frac{1}{(-2)^5(x^3)^5(y^{-2})^5}$
$= \frac{1}{-32x^{15}y^{-10}} = \frac{y^{10}}{-32x^{15}}$

29. $\left(\frac{a^3}{a^{-4}}\right)^2 = (a^3a^4)^2 = (a^7)^2 = a^{7\cdot 2} = a^{14}$

31. $\left(\frac{x^5}{y^{-2}}\right)^{-2} = (x^5y^2)^{-2} = \frac{1}{(x^5y^2)^2} = \frac{1}{(x^5)^2(y^2)^2} =$
$\frac{1}{x^{10}y^4}$

33. $\left(\frac{4x^2}{3x^{-5}}\right)^4 = \left(\frac{4x^2x^5}{3}\right)^4 = \left(\frac{4x^7}{3}\right)^4 = \frac{4^4x^{28}}{3^4} =$
$\frac{256x^{28}}{81}$

35. $\left(\frac{12y^3z^{-2}}{3y^{-4}z^3}\right)^2 = \left(\frac{4y^3y^4}{z^3z^2}\right)^2 = \left(\frac{4y^7}{z^5}\right)^2 = \frac{4^2(y^7)^2}{(z^5)^2}$
$= \frac{16y^{14}}{z^{10}}$

37. $\left(\frac{9u^2v^3}{18u^{-3}v}\right)^4 = \left(\frac{u^{2-(-3)}v^{3-1}}{2}\right)^4 = \left(\frac{u^5v^2}{2}\right)^4 =$
$\frac{(u^5)^4(v^2)^4}{2^4} = \frac{u^{20}v^8}{16}$

39. $\dfrac{16(x^{-2}yz)^{-2}}{(2x^{-3}z^0)^4} = \dfrac{16x^{(-2)(-2)}y^{-2}z^{-2}}{2^4x^{(-3)4}\cdot 1} = \dfrac{x^{4-(-12)}}{y^2z^2} =$
$\dfrac{x^{16}}{y^2z^2}$

41. $\dfrac{1+x^{-1}}{x^{-1}-1} = \dfrac{x(1+x^{-1})}{x(x^{-1}-1)} = \dfrac{x+1}{1-x}$

43. $\dfrac{a^{-2}+a}{a+1} = \dfrac{a^2(a^{-2}+a)}{a^2(a+1)} = \dfrac{1+a^3}{a^2(a+1)} =$
$\dfrac{(1+a)(1-a+a^2)}{a^2(a+1)} = \dfrac{1-a+a^2}{a^2}$

45. $\dfrac{2x^{-1}+4x^{-2}}{2x^{-2}+x^{-1}} = \dfrac{x^2(2x^{-1}+4x^{-2})}{x^2(2x^{-2}+x^{-1})} = \dfrac{2x+4}{2+x} =$
$\dfrac{2(x+2)}{x+2} = 2$

47. 23,000 = 2.3 x 10^4

49. 0.062 = 6.2 x 10^{-2}

51. 42.5 x 10^2 = 4.25 x 10^3

53. 0.25 x 10^{-2} = 2.5 x 10^{-3}

55. 2.3 x 10^2 = 230

57. 1.14 x 10^{-3} = 0.00114

59. 25 x 10^6 = 25,000,000

61. 0.51 x 10^{-3} = 0.00051

63. a. 4.2 x 10^{-5} b. 0.000042

65. a. -1.0000001 x 10^{-11}
b. -0.000000000010000001

67. (3.4 x 10^2)(2.1 x 10^3) = (3.4)(2.1) x 10^{2+3} =
7.14 x 10^5 = 714,000

69. $\dfrac{9.3\times10^2}{3.1\times10^{-2}} = \dfrac{9.3}{3.1}\times10^{2-(-2)} = 3.0\times10^4 = 30{,}000$

71. $\dfrac{(12{,}000)(3600)}{0.0003} = \dfrac{(1.2\times10^4)(3.6\times10^3)}{3.0\times10^{-4}} =$
$\dfrac{(1.2)(3.6)}{3}\times10^{4+3-(-4)} = 1.44\times10^{11} =$
144,000,000,000

Applications

73. 25,700,000,000,000 = 2.57 x 10^{13} miles

75. 33,100 = 3.31 x 10^4 cm./sec.

77. 1.14 x 10^8 = 114,000,000 miles

79. (3.6 x 10^7) mi. $\dfrac{5.28\times10^3\text{ft.}}{1\text{ mi.}}$ = (3.6)(5.28) x
10^{10} ft. = 19.008 x 10^{10} ft. = 1.9008 x 10^{11} ft.

81. The mass is 10^6(1.7 x 10^{-24}) g. = 1.7 x 10^{-18} g.

Section 7.2 Review of Factoring

Vocabulary

1. factor

3. irreducible (or prime)

Concepts

5. a. $(3y^2)y$, $(3y^2)\cdot1$ GCF = $3y^2$
b. $(9x^2y^2)(7x)$, $(9x^2y^2)(9y^2)$ GCF = $9x^2y^2$

Practice

7. $2x + 8 = 2x + 2\cdot4 = 2(x + 4)$

9. $2x^2 - 6x = 2x(x) - 2x(3) = 2x(x - 3)$

11. $15x^2y - 10x^2y^2 = (5x^2y)(3) - (5x^2y)(2y) =$
$5x^2y(3 - 2y)$

13. $27z^3 + 12z^2 + 3z = (3z)(9z^2) + (3z)(4z) + (3z)1 = (3z)(9z^2 + 4z + 1)$

15. $-3a - 6 = (-3)a + (-3)\cdot 2 = -3(a + 2)$

17. $-6x^2 - 3xy = (-3x)(2x) + (-3x)y = -3x(2x + y)$

19. $ax + bx + ay + by = (ax + bx) + (ay + by) = x(a + b) + y(a + b) = (a + b)(x + y)$

21. $x^2 + xy + 2x + 2y = (x^2 + xy) + (2x + 2y) = x(x + y) + 2(x + y) = (x + y)(x + 2)$

23. $3c - cd + 3d - c^2 = (3c + 3d) - (cd + c^2) = 3(c + d) - c(d + c) = (c + d)(3 - c)$

25. $x^2 + 5x + 6$
$ac = 1(6) = 6 = 2\cdot 3$ and $3 + 2 = 5$. Thus, $x^2 + 5x + 6 = x^2 + 2x + 3x + 6 = (x^2 + 2x) + (3x + 6) = x(x + 2) + 3(x + 2) = (x + 2)(x + 3)$

27. $3x^2 + 12x - 63 = 3(x^2 + 4x - 21)$ In the trinomial, $ac = 1(-21) = -21 = 7(-3)$ and $7 + (-3) = 4$. Thus, $x^2 + 4x - 21 = x^2 + 7x - 3x - 21 = (x^2 + 7x) - (3x + 21) = x(x + 7) - 3(x + 7) = (x + 7)(x - 3)$. The complete factorization is $3(x + 7)(x - 3)$.

29. $-a^2 + 4a + 32 = (-1)(a^2 - 4a - 32)$. In the trinomial, $ac = 1(-32) = -32 = 4(-8)$ and $4 + (-8) = -4$. Thus, $a^2 - 4a - 32 = a^2 + 4a - 8a - 32 = (a^2 + 4a) - (8a + 32) = a(a + 4) - 8(a + 4) = (a + 4)(a - 8)$. The complete factorization is $(-1)(a + 4)(a - 8) = (a + 4)(8 - a)$.

31. $6y^2 + 7y + 2$
$ac = 6\cdot 2 = 12 = 4\cdot 3$ and $4 + 3 = 7$. Thus, $6y^2 + 7y + 2 = 6y^2 + 4y + 3y + 2 = (6y^2 + 4y) + (3y + 2) = 2y(3y + 2) + (3y + 2) = (3y + 2)(2y + 1)$.

33. $8a^2 + 6a - 9$
$ac = 8(-9) = -72 = 12(-6)$ and $12 + (-6) = 6$. Thus, $8a^2 + 6a - 9 = 8a^2 + 12a - 6a - 9 = (8a^2 + 12a) - (6a + 9) = 4a(2a + 3) - 3(2a + 3) = (2a + 3)(4a - 3)$

35. $a^2 - 3ab - 4b^2$
$ac = 1(-4) = -4$ and $1 + (-4) = -3$. Thus, $a^2 - 3ab - 4b^2 = a^2 + ab - 4ab - 4b^2 = (a^2 + ab) - (4ab + 4b^2) = a(a + b) - 4b(a + b) = (a + b)(a - 4b)$

37. $-3a^2 + ab + 2b^2 = (-1)(3a^2 - ab - 2b^2)$. In the trinomial, $ac = 3(-2) = -6 = 2(-3)$ and $2 + (-3) = -1$. Thus, $3a^2 - ab - 2b^2 = 3a^2 + 2ab - 3ab - 2b^2 = (3a^2 + 2ab) - (3ab + 2b^2) = a(3a + 2b) - b(3a + 2b) = (3a + 2b)(a - b)$. The complete factorization is $(-1)(3a + 2b)(a - b)$.

39. $3x^3 - 10x^2 + 3x = x(3x^2 - 10x + 3)$. In the trinomial, $ac = 3\cdot 3 = 9 = (-1)(-9)$ and $(-1) + (-9) = -10$. Thus, $3x^2 - 10x + 3 = 3x^2 - x - 9x + 3 = (3x^2 - x) - (9x - 3) = x(3x - 1) - 3(3x - 1) = (3x - 1)(x - 3)$. The complete factorization is $x(3x - 1)(x - 3)$.

41. $x^4 + 8x^2 + 15 = (x^2)^2 + 8x^2 + 15$
$ac = 1(15) = 15 = 3\cdot 5$ and $3 + 5 = 8$. Thus, $x^4 + 8x^2 + 15 = x^4 + 3x^2 + 5x^2 + 15 = (x^4 + 3x^2) + (5x^2 + 15) = x^2(x^2 + 3) + 5(x^2 + 3) = (x^2 + 3)(x^2 + 5)$

43. $y^4 - 13y^2 + 30 = (y^2)^2 - 13y^2 + 30$
$ac = 1(30) = 30 = (-10)(-3)$ and $(-10) + (-3) = -13$. Thus, $y^4 - 13y^2 + 30 = y^4 - 10y^2 - 3y^2 + 30 = (y^4 - 10y^2) - (3y^2 - 30) = y^2(y^2 - 10) - 3(y^2 - 10) = (y^2 - 10)(y^2 - 3)$.

45. $x^2 - 4 = x^2 - 2^2 = (x - 2)(x + 2)$

47. $9y^2 - 64 = (3y)^2 - 8^2 = (3y - 8)(3y + 8)$

49. $81a^4 - 49b^2 = (9a^2)^2 - (7b)^2 = (9a^2 - 7b)(9a^2 + 7b)$

51. $(x + y)^2 - z^2 = [(x + y) - z][(x + y) + z] = (x + y - z)(x + y + z)$

53. $x^4 - y^4 = (x^2)^2 - (y^2)^2 = (x^2 - y^2)(x^2 + y^2) = (x - y)(x + y)(x^2 + y^2)$

55. $2x^2 - 288 = 2(x^2 - 144) = 2(x^2 - 12^2) = 2(x - 12)(x + 12)$

57. $2x^3 - 32x = 2x(x^2 - 16) = 2x(x^2 - 4^2) =$
$2x(x - 4)(x + 4)$

59. $x^3 - 8 = x^3 - 2^3 = (x - 2)(x^2 + 2x + 4)$

61. $y^3 - 64 = y^3 - 4^3 = (y - 4)(y^2 + 4y + 16)$

63. $125 + a^3 = 5^3 + a^3 = (5 + a)(5^2 - 5a + a^2) =$
$(5 + a)(25 - 5a + a^2)$

65. $x^3 + 8y^3 = x^3 + (2y)^3 =$
$(x + 2y)[x^2 - x(2y) + (2y)^2] =$
$(x + 2y)(x^2 - 2xy + 4y^2)$

67. $2x^3y - 16y = 2y(x^3 - 8) =$
$2y(x - 2)(x^2 + 2x + 4)$

69. $(a - b)^3 - 8 = (a - b)^3 - 2^3 =$
$[(a - b) - 2][(a - b)^2 + 2(a - b) + 2^2] =$
$(a - b - 2)(a^2 - 2ab + b^2 + 2a - 2b + 4)$

71. $-3a^2 - 6ab - 3b^2 = -3(a^2 + 2ab + b^2) =$
$-3(a + b)^2$

73. $-x^3 - 2x^2 - x = -x(x^2 + 2x + 1) = -x(x + 1)^2$

75. $x^2 - 4x + 3$
$ac = 1 \cdot 3 = 3 = (-1)(-3)$ and $(-1) + (-3) = -4$.
Thus, $x^2 - 4x + 3 = x^2 - x - 3x + 3 =$
$(x^2 - x) - (3x - 3) = x(x - 1) - 3(x - 1) =$
$(x - 1)(x - 3)$

77. $y^3 - 216 = y^3 - 6^3 = (y - 6)(y^2 + 6y + 36)$

79. $-21x^7 + 14x^5 - 7x^4 = -7x^4(3x^3 - 2x + 1)$

Review

81. $4^{-5} \cdot 4^{-3} = 4^{-5+(-3)} = 4^{-8} = \dfrac{1}{4^8} = \dfrac{1}{65{,}536}$

83. a. $9.76 \times 10^{-5} = 0.0000976$
b. $8.12 \times 10^5 = 812{,}000$

Section 7.3 Equation Review

Vocabulary

1. solve

3. identity

5. literal equation

7. $\mathbb{R}$

Concepts

9. Both are solutions.

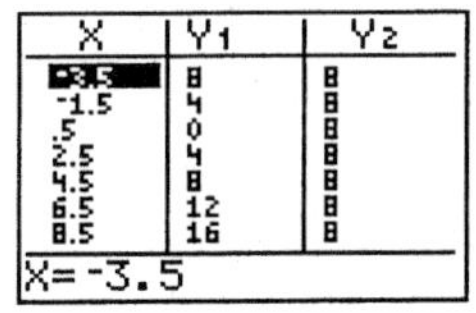

11. Both are solutions.

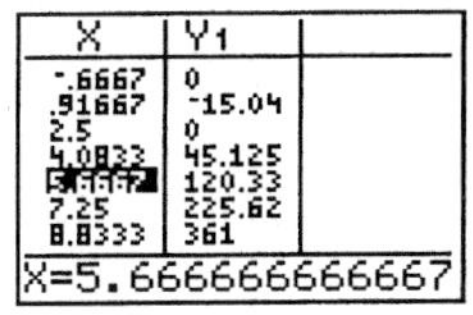

Practice

13.
$$
\begin{aligned}
4(x-1) &= 28 \\
4x - 4 &= 28 \\
4x &= 32 \\
x &= 8
\end{aligned}
$$

15.
$$
\begin{aligned}
13(x-9) - 2 &= 7x - 5 \\
13x - 117 - 2 &= 7x - 5 \\
13x - 119 &= 7x - 5 \\
6x - 119 &= -5 \\
6x &= 114 \\
x &= 19
\end{aligned}
$$

17.
$$
\begin{aligned}
\frac{1}{3}x - 2 &= \frac{1}{2}x + 2 \\
6\left(\frac{1}{3}x - 2\right) &= 6\left(\frac{1}{2}x + 2\right) \\
2x - 12 &= 3x + 12 \\
-12 &= x + 12 \\
-24 &= x
\end{aligned}
$$

19. $6(x-2)=3(4+x)+3x$
$6x-12=12+3x+3x$
$6x-12=6x+12$
$-12=12$
No solution.

21. $6x-5=2(x-3)+1+4x$
$6x-5=2x-6+1+4x$
$6x-5=6x-5$
$0=0$
The solution set is all real numbers.

23. $2x+y=7$
$y=7-2x$

25. $4x-y=8$
$4x=y+8$
$4x-8=y$

27. $2x-3y=5$
$-3y=5-2x$
$y=\frac{5-2x}{-3}=\frac{2x-5}{3}$

29. $x+2y=6$
$2y=6-x$
$y=3-\frac{1}{2}x$

31. $V=\frac{1}{3}\pi r^2h$
$3V=\pi r^2h$
$\frac{3V}{\pi r^2}=h$

33. $|x+1|=10$
$x+1=10$ or $x+1=-10$
$x=9$ or $x=-11$

35. $|2x-1|+4=6$
$|2x-1|=2$
$2x-1=2$ or $2x-1=-2$
$2x=3$ or $2x=-1$
$x=\frac{3}{2}$ or $x=\frac{-1}{2}$

37. $|4x-1|=-2$ No solution since $|4x-1|\geq 0$.

39. $2|x-1|-2=0$
$2|x-1|=2$
$|x-1|=1$
$x-1=1$ or $x-1=-1$
$x=2$ or $x=0$

41. $|3x+2|=|2x-3|$
$3x+2=2x-3$ or $3x+2=-(2x-3)$
$3x=2x-5$ or $3x+2=-2x+3$
$x=-5$ or $5x=1$
$x=-5$ or $x=\frac{1}{5}$

43. $(x+1)(x-2)=0$
$x+1=0$ or $x-2=0$
$x=-1$ or $x=2$

45. $x^2-16=0$
$(x-4)(x+4)=0$
$x-4=0$ or $x+4=0$
$x=4$ or $x=-4$

47. $x^2+21x+20=0$
$(x+20)(x+1)=0$
$x+20=0$ or $x+1=0$
$x=-20$ or $x=-1$

49. $2x^3 - 10x^2 - 48x = 0$

$2x(x^2 - 5x - 24) = 0$

$2x(x-8)(x+3) = 0$

$2x = 0$ or $x - 8 = 0$ or $x + 3 = 0$

$x = 0$ or $x = 8$ or $x = -3$

51. $x(x-12) = -20$

$x^2 - 12x + 20 = 0$

$(x-10)(x-2) = 0$

$x - 10 = 0$ or $x - 2 = 0$

$x = 10$ or $x = 2$

53. $\frac{11}{x} + \frac{13}{x} = 12$ Multiply both sides by x.

$x\left(\frac{11}{x} + \frac{13}{x}\right) = 12x$

$11 + 13 = 12x$

$24 = 12x$

$2 = x$

55. $\frac{3}{p+6} - 2 = \frac{7}{p+6}$ Multiply both sides by p+6.

$3 - 2(p+6) = 7$

$3 - 2p - 12 = 7$

$-2p - 9 = 7$

$-2p = 16$

$p = -8$

57. $\frac{z^2}{z+1} + 2 = \frac{1}{z+1}$ Multiply both sides by z +1.

$z^2 + 2(z+1) = 1$

$z^2 + 2z + 2 = 1$

$z^2 + 2z + 1 = 0$

$(z+1)^2 = 0$

$z + 1 = 0$

$z = -1$

-1 is extraneous, so no solution.

59. $\frac{3}{y-2} + 1 = \frac{3}{y-2}$ Multiply both sides by y – 2

$3 + (y-2) = 3$

$y - 2 = 0$

$y = 2$

2 is extraneous, so no solution.

61. $\frac{2p}{3} - \frac{1}{p} = \frac{2p-1}{3}$ Multiply both sides by 3p.

$3p\left(\frac{2p}{3} - \frac{1}{p}\right) = 3p\left(\frac{2p-1}{3}\right)$

$2p^2 - 3 = 2p^2 - p$

$-3 = -p$

$3 = p$

Applications

63. Let x = length of the smaller piece. Then

$x + 3x = 20$

$4x = 20$

$x = 5$

He should cut off a 5 ft. piece.

65. Let x = miles driven. The rental cost for Alamo is 30 + .24x and the rental cost for Budget is 44 + .16x. Thus, x must satisfy

$30 + .24x = 44 + .16x$

$.08x = 14$

$x = 175$

The cost is the same for 175 miles driven.

67. Let h = height of the flagpole. Then h satisfies $\frac{h}{5} = \frac{30}{7}$

$7h = 150$

$h = \frac{150}{7} = 21\frac{3}{7}$

The flagpole is $21\frac{3}{7}$ ft. tall.

69. Let x = measure of a base angle. Then x

satisfies $x+x+4x=180$

$$6x=180$$

$$x=30$$

The angles are 30°, 30° and 120°.

71. Let x = amount to be invested. Then x satisfies $.13x=.11x+150$

$$.02x=150$$

$$x=7500$$

The teacher has \$7500 to invest.

Review

73. a. $\dfrac{x^{-2}y^{-4}}{(2x)^{-5}}=\dfrac{(2x)^5}{x^2y^4}=\dfrac{32x^5}{x^2y^4}=\dfrac{32x^3}{y^4}$

b. $\left(\dfrac{3x^4y^{-2}}{2^{-2}x^{-3}y^2}\right)^{-1}=\left(\dfrac{3\cdot 2^2x^4x^3}{y^2y^2}\right)^{-1}=$

$\left(\dfrac{12x^7}{y^4}\right)^{-1}=\dfrac{y^4}{12x^7}$

75. a. 3.14 E -3 = 3.14×10^{-3}

b. 4,000,000,000 = 4×10^9

c. 0.00000000000000000056 = 5.6×10^{-19}

Section 7.4 Graphing and Its Applications Revisited

Vocabulary

1. ordered pairs

3. standard viewing

5. y-intercept

7. function

Concepts

9.

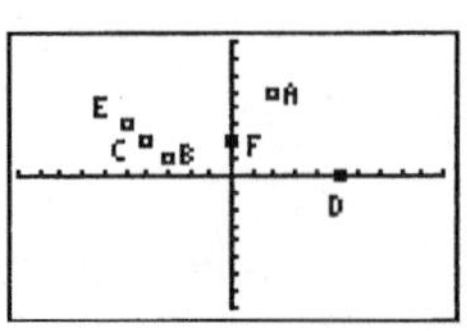

11. [0,150] by [-10,50]
a. (4,45) yes b. (5,150) no

13. [-20,20] by [-10,5]
a. (35, 10) no b. (-15,-5) yes

15. a. domain = {-3,-2,-1,0,1,2,3}
b. range = {4,-1,-4,-5}

17. $x+3y=6$
y-intercept: (0,2) x-intercept: (6,0)

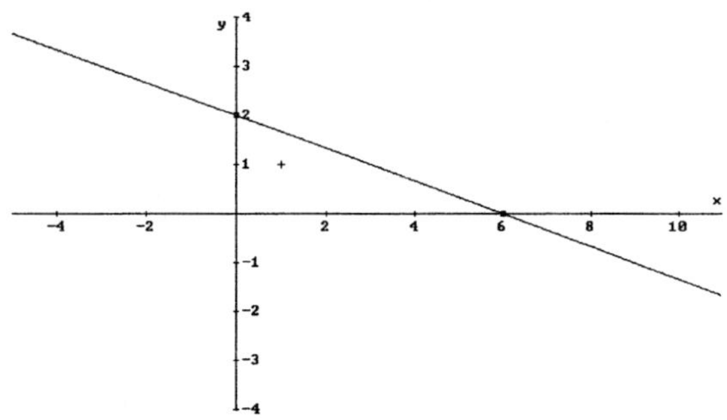

19. $2x-y=-2$

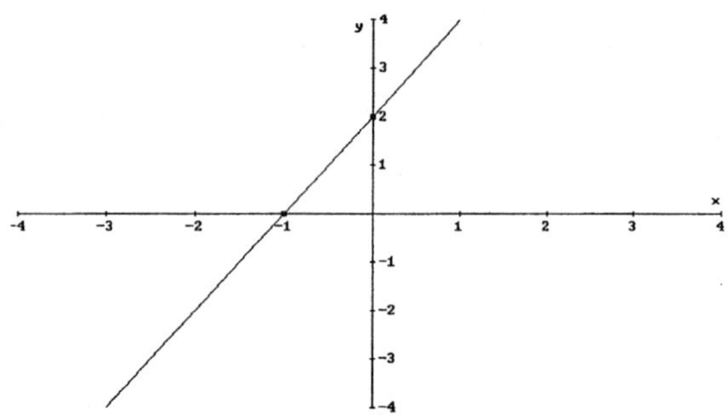

21. $4-x=5$
No y-intercept x-intercept: (-1,0)

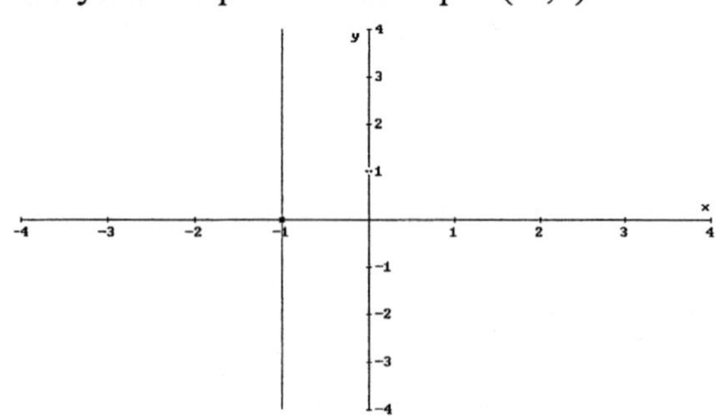

23. $y = 4$
y-intercept: (0,4) No x-intercept

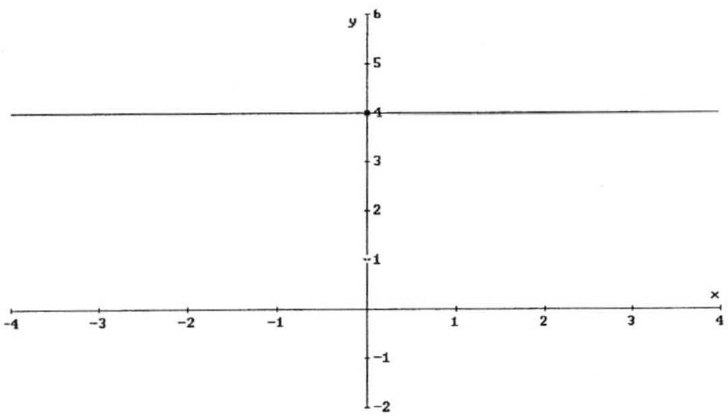

25. $f(x) = -3x + 40$

y-intercept: (0,40) x-intercept: $(\frac{40}{3},0)$

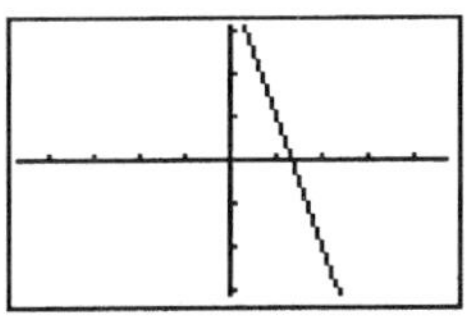

27. $-2x + y = 24$
y-intercept: (0,24) x-intercept: (-12,0)

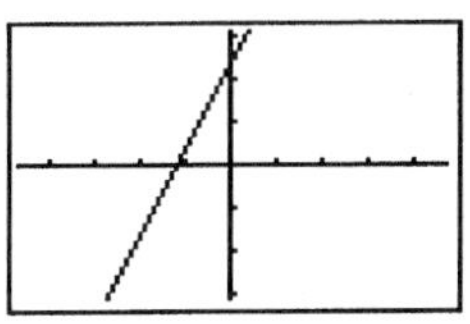

29. $5x - 8y = 80$
y-intercept: (0,-10) x-intercept: (16,0)

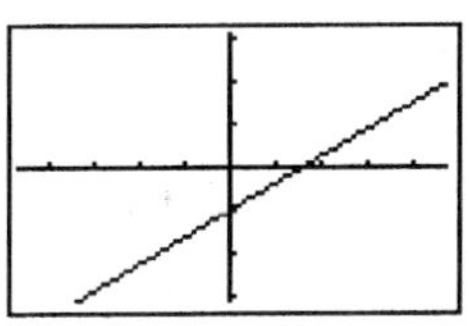

31. $5x + 9y = 55$

y-intercept: $(0, \frac{55}{9})$ x-intercept: (11,0)

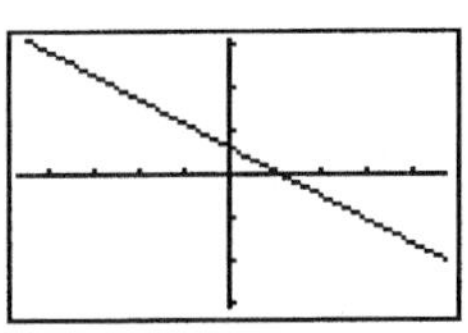

33. The graph is a function.

35. The graph is not a function since it has 2 or more points with the same first coordinate.

37. The graph is not a function since it has 2 or more points with the same first coordinate.

39. domain = {-8,9,7,-½,-4.5},
range = {5,2,-3,3, ⅔}

41. domain = {-6,-5.5,-5,-4.5,-4,-3.5,-3}
range = {31,25.25,20,15.25,11,7.25,4}

43. $2 + \frac{x}{3} = \frac{5}{6} - x$ The solution is -0.875.

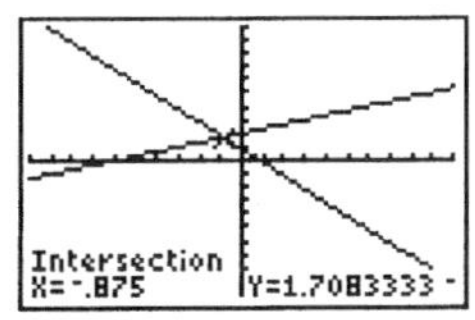

45. $2(x - 8) - 3 = 4x + 10 - 2x$ No solution.
Graphing in [-15,15] by [-25,25] gives

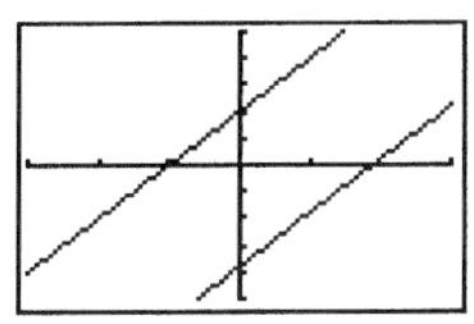

47. $x^2 = -2x + 8$ The solutions are 2 and – 4.
Graphing in [-10,10] by [-5,20] gives

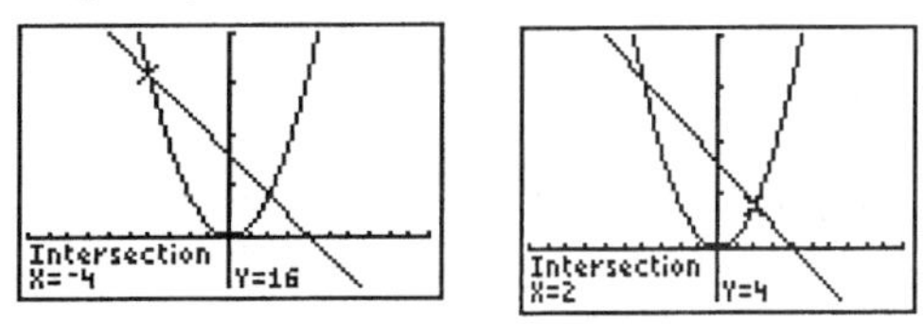

49. $\frac{x-3}{5} = \frac{8}{x}$ The solutions are -5 and 8.

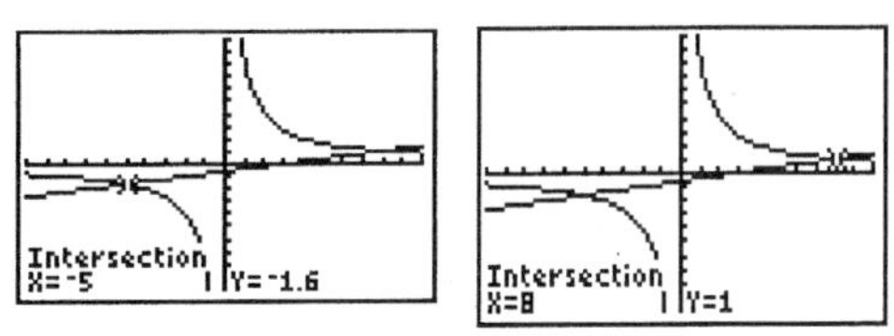

51. $|3x - 5| = 10$ The solutions are 5 and $\frac{-5}{3}$.

Graphing in [-10,10] by [-10, 15] gives

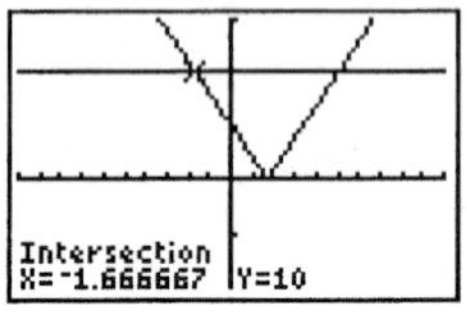

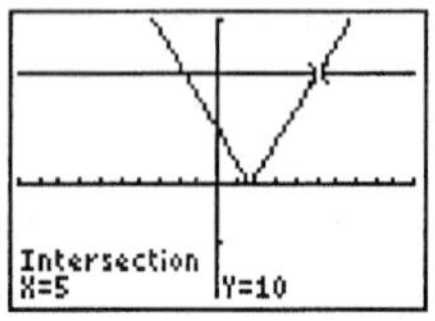

53. $-|2x - 8| = -12$ The solutions are -2 and 10. Graphing in [-10,15] by [-10,15] gives

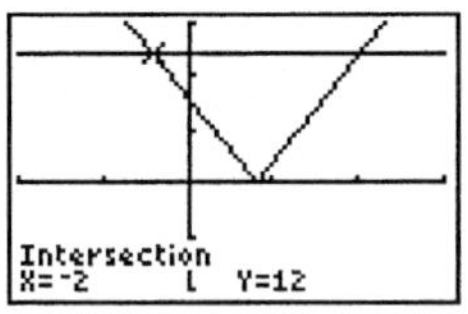

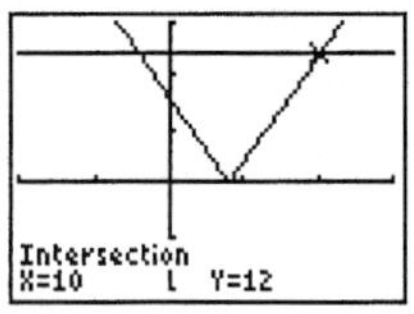

55. $|x + 2| + |x - 3| = 12$ The solutions are 6.5 and – 5.5. Graphing in [-10,10] by [-10,20],

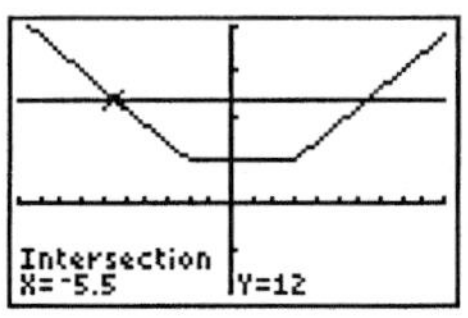

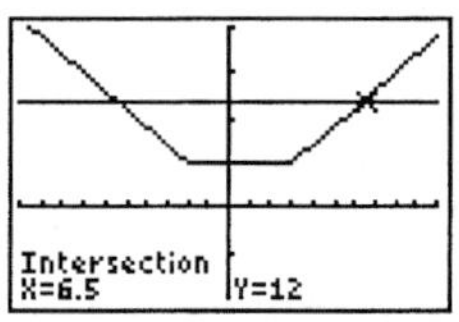

57. $5 = \sqrt{3x + 4}$ The solution is 7.

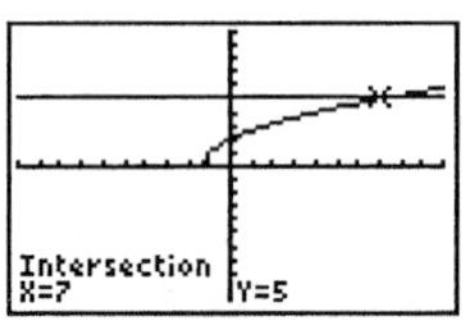

59. $2 + \frac{x}{3} - \frac{5}{6} + x = 0$ The solution is – 0.875.

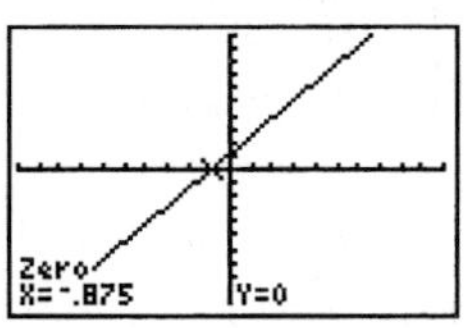

61. $\frac{x-3}{5} - \frac{8}{x} = 0$ The solutions are – 5 and 8.

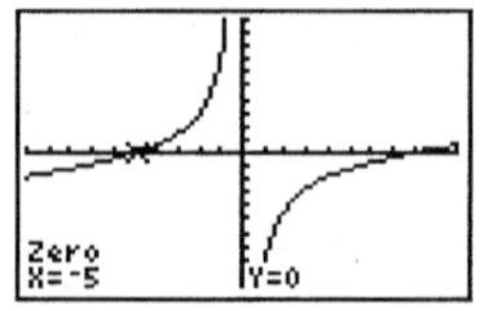

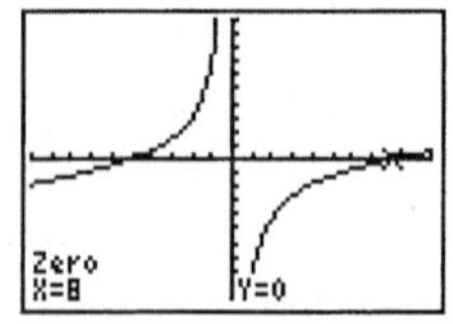

63. $|3x - 5| - 10 = 0$ The solutions are $\frac{-5}{3}$ and 5.

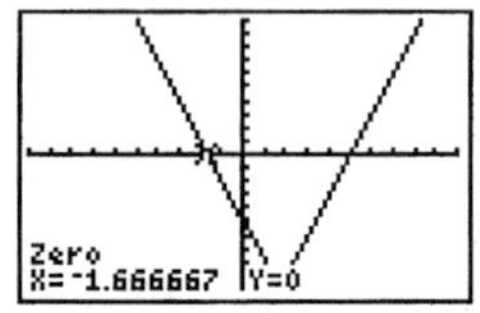

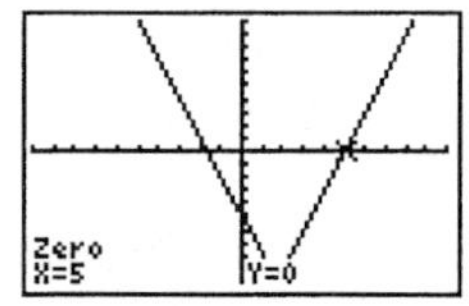

65. $-|2x - 8| + 12 = 0$ The solutions are -2, 10. Graphing in [-10,15] by [-10,10] gives

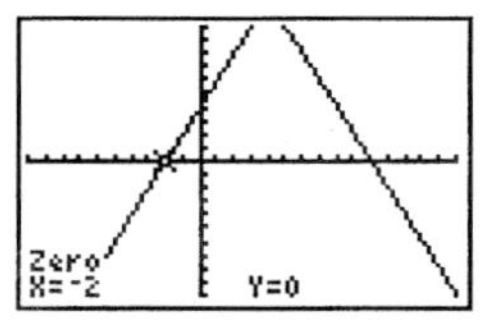

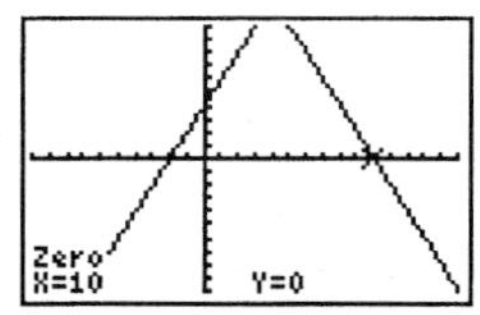

67. $x^2 + 2x - 8 = 0$ The solutions are - 4 and 2.

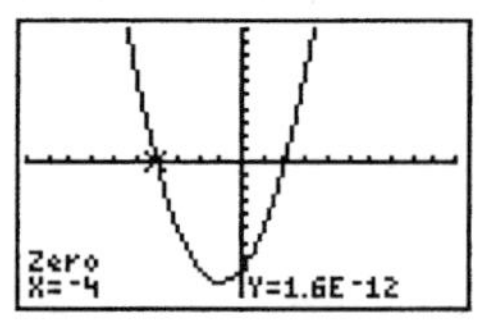

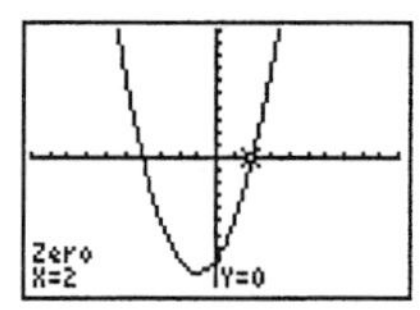

69. $-6x^2 - 7x + 20 = 0$ The solutions are – 2.5 and $\frac{4}{3}$.

Graphing in the window [-5,5] by [-10,10

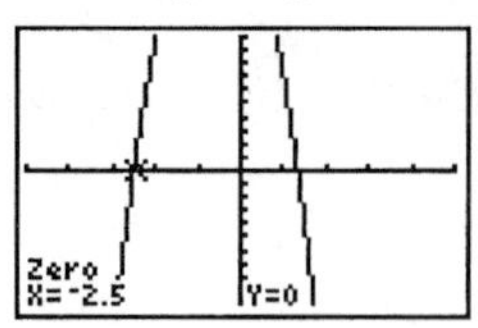

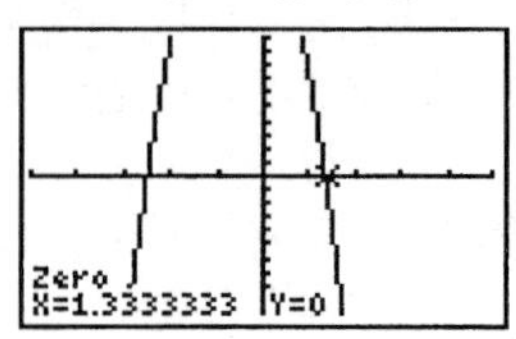

71. $x^4 - 1 = 0$ The solutions are –1 and 1. Graphing in [-5,5] by [-5,5],

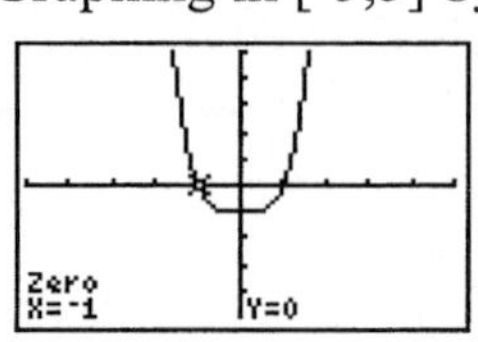

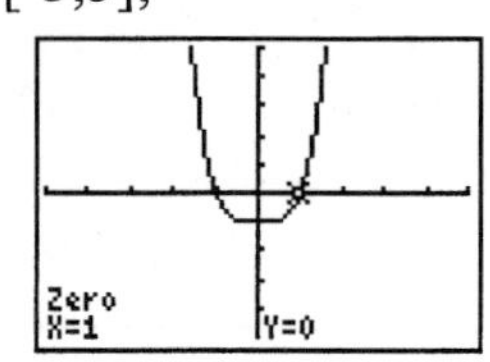

73. $14x^3 - 40x^2 + 6x^4 = 0$ The solutions are 0, -4 and $\frac{5}{3}$. Graphing in [-5,5] by [- 40,20],

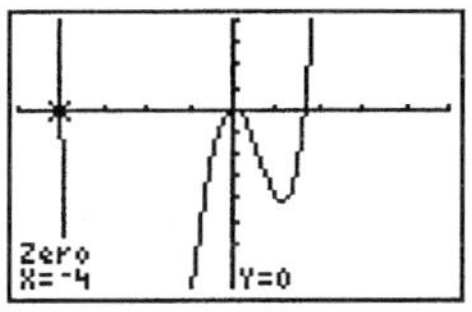

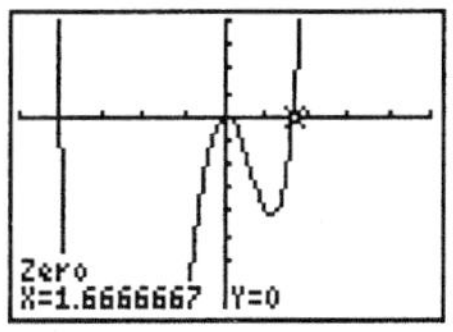

Applications

75. a. $y = 0.24x + 45$
    b. [0,150] by [0,100]
    c. $76.40 to travel 131 miles. (May vary with window.)

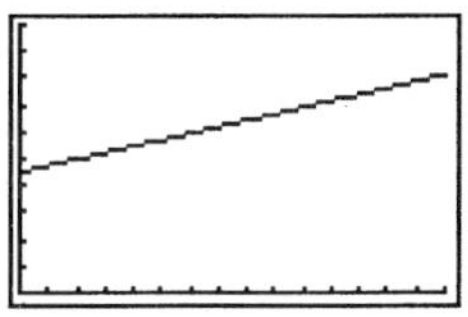

d. The exact cost is $76.44 to travel 131 mi.

Review

77. The crude oil reserves = 42 (2.61×10^{11}) = $109.62 \times 10^{11} = 1.0962 \times 10^{13}$ gallons.

79. Evaluate $\frac{\left(2x^{-4}y^2\right)^3}{4x^{-2}y^{-1}}$ for $x = \frac{1}{2}$, $y = \frac{1}{5}$.

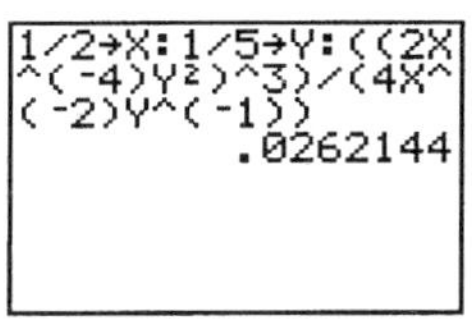

Answer: $\frac{2^{11}}{5^7}$

Section 7.5 Solving Simple Linear Inequalities

Vocabulary

1. inequality

3. interval

Concepts

5. The solution is -9.

7. a. y1 < y2 -2
   b. y1 ≤ y2 -2
   c. y1 > y2 -2
   d. y1 ≥ y2 -2

9. a. y1 < y2 -6
   b. y1 ≤ y2 -6
   c. y1 > y2 -6
   d. y1 ≥ y2 -6

Practice

11. $2x - 3 \le 5$ Interval: (-∞, 4]
    $2x \le 8$
    $x \le 4$
    4

13. $-3x - 7 > -1$ Interval: (-∞, -2)
    $-3x > 6$
    $x < -2$
    -2

15. $2x + 9 \le x + 8$ Interval: (-∞, -1]
    $x + 9 \le 8$
    $x \le -1$
    -1

17. $7x - 16 < 6x$ Interval: [-13,∞)
    $x - 16 < 0$
    $x < 16$
    -13

19. $7 - x \le 3x - 1$ Interval: $[2, \infty)$

$$7 \le 4x - 1$$
$$8 \le 4x$$
$$2 \le x$$

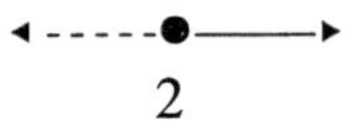

21. $3(x - 8) < 5x + 6$ Interval: $(-15, \infty)$

$$3x - 24 < 5x + 6$$
$$-24 < 2x + 6$$
$$-30 < 2x$$
$$-15 < x$$

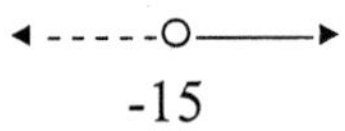

23. $8(5 - x) \le 10(8 - x)$ Interval: $(-\infty, 20]$

$$40 - 8x \le 80 - 10x$$
$$40 + 2x \le 80$$
$$2x \le 40$$
$$x \le 20$$

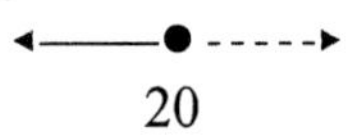

25. $\frac{5}{2}(7x - 15) + x \ge \frac{13}{2}x - \frac{3}{2}$ Interval: $[3, \infty)$

Multiply both sides by 2.

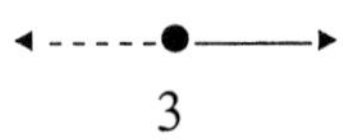

$$5(7x - 15) + 2x \ge 13x - 3$$
$$35x - 75 + 2x \ge 13x - 3$$
$$37x - 75 \ge 13x - 3$$
$$24x - 75 \ge -3$$
$$24x \ge 72$$
$$x \ge 3$$

27. $\frac{3x - 3}{2} < 2x + 2$ Interval: $(-7, \infty)$

Multiply both sides by 2.

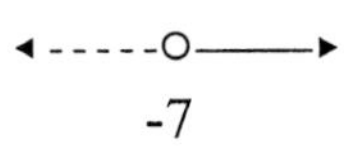

$$3x - 3 < 2(2x + 2)$$
$$3x - 3 < 4x + 4$$
$$-3 < x + 4$$
$$-7 < x$$

29. $\frac{2(x + 5)}{3} \le 3x - 6$ Interval: $[4, \infty)$

Multiply both sides by 3.

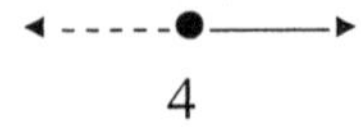

$$2(x + 5) \le 3(3x - 6)$$
$$2x + 10 \le 9x - 18$$
$$10 \le 7x - 18$$
$$28 \le 7x$$
$$4 \le x$$

Applications

31. Let x = her fourth test score. Then x satisfies

$$\frac{68 + 75 + 79 + x}{4} \ge 80$$
$$68 + 75 + 79 + x \ge 320$$
$$222 + x \ge 320$$
$$x \ge 98$$

She must make a 98% or better to earn a B.

33. Let x = economy rating of the third model.

Then $\frac{17 + 19 + x}{3} \ge 21$

$$17 + 19 + x \ge 63$$
$$36 + x \ge 63$$
$$x \ge 27$$

The third model must have an economy rating of 27 mpg or better.

35. Let x = number of servings on the seventh day. Then $\frac{5 + 4 + 3 + 6 + 7 + 4 + x}{7} \ge 5$

$$29 + x \ge 35$$
$$x \ge 6$$

You must eat at least 6 servings of fruit on the seventh day.

37. Let x = number of days the machine is rented Then the total cost of x days rental is $0.67x$ and so x satisfies $0.67x > 385$ or $x > 574.6$. Thus, after 574 days it will be cheaper to buy a washer.

39. Let x = length of a side of the square. Then $4x \geq 68$ so $x \geq 17$. A side can be no less than 17 cm.

Review

41. Domain = {1,-1,-2}, Range = {2,3,4}

43.

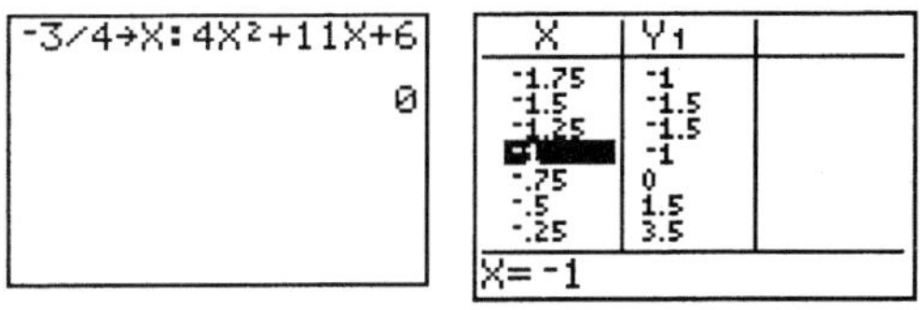

Section 7.6 Compound Inequalities

Vocabulary

1. compound

3. intersection

Concepts

5. a. y1 < y2 < y3: (1.5, 6.5)
   b. y2 < y1 or y2 > y3: (- ∞, 1.5) ∪ (6.5, ∞)

7. a. y1 < y2 < y3: (-12, -6)
   b. y2 < y1 or y2 > y3: (- ∞, -12) ∪ (-6, ∞)

9. a. iii b. iv c. ii d. i

Practice

11. x ≤5 and x > -2
    (- ∞, 5] ∩ (-2, ∞) = (-2, 5]
    -2 5

13. x < 10 and x ≤ 4
    (- ∞, 10) ∩ (- ∞, 4] = (- ∞, 4], 4

15. x ≤ 6 and x ≥ 12
    (- ∞, 6] ∩ [12, ∞) = { }

17. x > 5 or x < -2
    (5, ∞) ∪ (- ∞, -2), -2 5

19. x > 10 or x ≥ 4
    (10, ∞) ∪ [4, ∞) = [4, ∞), 4

21. x < 6 or x ≥ -2
    (- ∞, 6) ∪ [-2, ∞) = (-∞,∞)

23. $5x - 3 > 12$ and $7x - 23 < 5$

$$5x > 15 \text{ and } 7x < 28$$
$$x > 3 \text{ and } x < 4$$

Interval: (3,4) 3 4

25. $-9x \geq -x$ and $4 - 3x \leq x$

$$0 \geq 8x \text{ and } 4 \leq 4x$$
$$0 \geq x \text{ and } 1 \leq x$$

No solution, { }.

27. $7 - 4x \geq 8$ and $2 + 3x \leq 8$

$$-4x \geq 1 \text{ and } 3x \leq 6$$
$$x \leq -\frac{1}{4} \text{ and } x \leq 2$$

Interval: (- ∞, - ¼], -1/4

29. $-3(x+2) > 2(x+1)$ and $-4(x-1) < x+8$

$$-3x - 6 > 2x + 2 \text{ and } -4x + 4 < x + 8$$
$$-6 > 5x + 2 \text{ and } 4 < 5x + 8$$
$$-8 > 5x \text{ and } -4 < 5x$$
$$-\frac{8}{5} > x \text{ and } -\frac{4}{5} < x$$

No solution. { }

31.
$$-2 < -x + 3 < 5$$
$$-2 - 3 < -x + 3 - 3 < 5 - 3$$
$$-5 < -x < 2$$
$$5 > x > -2$$

Interval: (-2, 5), -2 5

33. $15 > 2x - 7 > 9$

$15 + 7 > 2x - 7 + 7 > 9 + 7$

$22 > 2x > 16$

$11 > x > 8$

Interval: (8, 11),

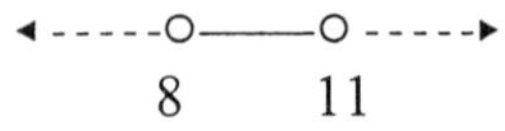

35. $0 \le \frac{4-x}{3} \le 2$

Multiply by 3:

$0 \le 4 - x \le 6$

$-4 + 0 \le -4 + 4 - x \le -4 + 6$

$-4 \le -x \le 2$

$4 \ge x \ge -2$

Interval: [-2, 4],

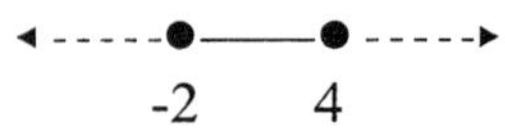

37. $3x + 2 < 8$ or $2x - 3 > 11$

$3x < 6$ or $2x > 14$

$x < 2$ or $x > 7$

Interval: $(-\infty, 2) \cup (7, \infty)$,

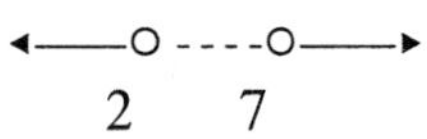

39. $-4(x + 2) \ge 12$ or $3x + 8 < 11$

$-4x - 8 \ge 12$ or $3x < 3$

$-4x \ge 20$ or $x < 1$

$x \le -5$ or $x < 1$

Interval: $(-\infty, 1)$,

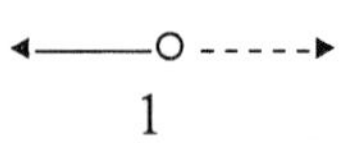

41. $4x < -24$ or $-3x > 18$

$x < -6$ or $x < -6$

Interval: $(-\infty, -6)$,

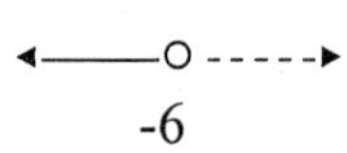

43. $-2 < 3x + 4 \le 7$ Interval: (-2, 1]
The graph of $y = 3x + 4$ is above the graph of $y = -2$ and on or below the graph of $y = 7$ for $-2 < x < 1$.

45. $0 \le 1 - \frac{2}{3}x < 5$ Interval: (-6, 1.5]

The graph of $y = 1 - \frac{2}{3}x$ is on or above the graph of $y = 0$ and below the graph of $y = 5$ for $-6 < x \le 1.5$.

47. $3x + 4 < -2$ or $3x + 4 \ge 7$
Intervals: $(-\infty, -2) \cup [1, \infty)$ The graph of $y = 3x + 4$ is below the graph of $y = -2$ for $x < -2$ and is on or above the graph of $y = 7$ for $x \ge 1$

49. $1 - \frac{2}{3}x \le 0$ or $1 - \frac{2}{3}x > 5$
Intervals: $[1.5, \infty) \cup (-\infty, -6)$ The graph of $y = 1 - ⅔x$ is on or below the x-axis for $x \ge 1.5$ and above the graph of $y = 5$ for $x < -6$.

Applications

51. Let x = the Celsius temperature. Then x will satisfy the inequality $1540 \le x \le 1650$.

53. Let x be the length in meters. Since 1 m. = 3.281 ft., then x m. = 3.281x ft. Thus, x satisfies $0 < 3.281x \le 25$ and so $0 < x < \frac{25}{3.281} = 7.62$. The python can grow up to 7.62 m. in length.

55. Let x = number of stereos. Then 3x = number of TV's, and $1 \cdot x + 2 \cdot 3x = 7x$ = total hours to assemble x stereos and 3x TV's. Thus, $640 \le 7x \le 810$, so $91.4 \le x \le 115.7$. The number of stereos can be between 92 and 115 inclusive.

57. Let x = length of the rectangle. Then the width is 3x – 8 and the perimeter P = 2x + 2(3x – 8) = 8x – 16. Thus, $8 < 8x - 16 < 16$

$24 < 8x < 32$

$3 < x < 4$

The length is between 3' and 4'.

Review

59. $\frac{2}{3}(x-5) \le 6-12x$ Interval: $(-\infty, \frac{14}{19}]$

$2(x-5) \le 3(6-12x)$

$2x-10 \le 18-36x$

$38x-10 \le 18$

$38x \le 28$

$x \le \frac{28}{38} = \frac{14}{19}$

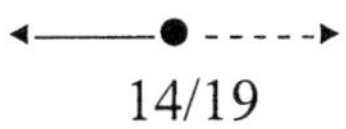

61. a. $5x^4 - 125y^2 = 5(x^4 - 25y^2) =$
$5[(x^2)^2 - (5y)^2] = 5(x^2 - 5y)(x^2 + 5y)$

b. $6x^2 - 17x + 12 = (3x - 4)(2x - 3)$

c. $4x^3 - 19x^2 - 5x = x(4x^2 - 19x - 5) =$
$x(4x + 1)(x - 5)$

Section 7.7 Absolute Value Inequalities

Vocabulary

1. absolute value

Concepts

3. a. $-|3x-2| + 4 = -3$: $\{3, -4/3\}$
   b. $-|3x-2| + 4 < -3$: $(-\infty, -4/3) \cup (3, \infty)$
   c. $-|3x-2| + 4 > -3$: $(-4/3, 3)$

5. a. $y1 = y2$: $\{-4, -1\}$
   b. $y1 < y2$: $(-4, -1)$
   c. $y1 > y2$: $(-\infty, -4) \cup (-1, \infty)$

7. a. $y1 = y2$: $\{\ \}$
   b. $y1 < y2$: $\mathbb{R}$
   c. $y1 > y2$: $\{\ \}$

9. a. $y1 = y2$: $\{8\}$
   b. $y1 < y2$: $(-\infty, 8) \cup (8, \infty)$
   c. $y1 > y2$: $\{\ \}$

Practice

11. $|3x| < 27$

$-27 < 3x < 27$

$-9 < x < 9$

Interval: (-9, 9)

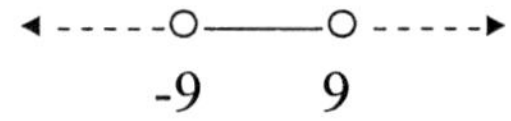

13. $|x-8| \le 12$

$-12 \le x-8 \le 12$

$-4 \le x \le 20$

Interval: [- 4, 20]

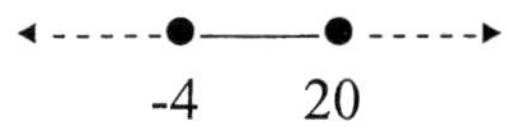

15. $|3x-2| < 10$

$-10 < 3x-2 < 10$

$-8 < 3x < 12$

$\frac{-8}{3} < x < 4$

Interval: (-8/3, 4)

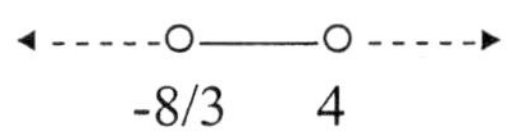

17. $|5x - 12| < -5$ No solution, since the absolute value of a real number is always zero or positive.

19. $|4-3x| \le 13$

$-13 \le 4-3x \le 13$

$-17 \le -3x \le 9$

$\frac{17}{3} \ge x \ge -3$

Interval: [-3, 17/3]

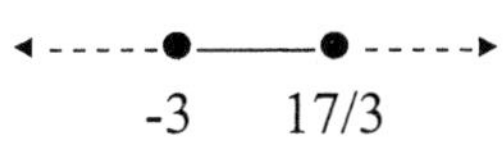

21. $|7x| > 7$

$7x > 7$ or $7x < -7$

$x > 1$ or $x < -1$

Intervals: $(-\infty, -1) \cup (1, \infty)$

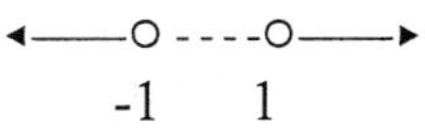

23. $|x+5| \ge 7$

$x+5 \ge 7$ or $x+5 \le -7$

$x \ge 2$ or $x \le -12$

Intervals: (- ∞, -12] ∪ [2, ∞) ◄—●---●—►
-12 2

25. $|2x-5|>25$

$2x-5>25$ or $2x-5<-25$

$2x>30$ or $2x<-20$

$x>15$ or $x<-10$

Intervals: (- ∞, -10) ∪ (15, ∞) ◄—○---○—►
-10 15

27. $|4x+3|>0$
$|4x+3| \geq 0$ always and $|4x+3| = 0$ only when $4x + 3 = 0$ or $x = -3/4$. Thus, the solution set is (- ∞, -3/4) ∪ (-3/4, ∞) and the number line is ◄——○——►
-3/4

29. $|8x-3|>0$
$|8x-3| \geq 0$ always and $|8x-3| = 0$ only when $8x - 3 = 0$ or $x = 3/8$. Thus, the solution set is (- ∞, 3/8) ∪ (3/8, ∞), and the number line is ◄——○——►.
3/8

31. $\left|\frac{x-2}{3}\right| \leq 4$

$-4 \leq \frac{x-2}{3} \leq 4$

$-12 \leq x-2 \leq 12$

$-10 \leq x \leq 14$

Interval: [-10, 14] ◄-----●——●-----►
-10 14

33. $|3x+1|+2<6$

$|3x+1|<4$

$-4<3x+1<4$

$-5<3x<3$

$\frac{-5}{3}<x<1$

Interval: (-5/3, 1) ◄-----○——○-----►
-5/3 1

35. $3|2x+5| \geq 9$

$|2x+5| \geq 3$

$2x+5 \geq 3$ or $2x+5 \leq -3$

$2x \geq -2$ or $2x \leq -8$

$x \geq -1$ or $x \leq -4$

Intervals: [-1, ∞) ∪ (- ∞, -4] ◄—●---●—►
-4 -1

37. $|5x-1| + 4 \leq 0$ or $|5x-1| \leq -4$ There is no solution since $|5x-1| \geq 0$ always. { }

39. $\left|\frac{1}{3}x+7\right|-5>6$

$\left|\frac{1}{3}x+7\right|>11$

$\frac{1}{3}x+7>11$ or $\frac{1}{3}x+7<-11$

$x+21>33$ or $x+21<-33$

$x>12$ or $x<-54$

Intervals: (12, ∞) ∪ (- ∞, -54) ◄—○---○—►
-54 12

41. $\left|\frac{1}{5}x-5\right|+4>4$

$\left|\frac{1}{5}x-5\right|>0$ so $\frac{1}{5}x-5 \neq 0$

$\frac{1}{5}x \neq 5$ or $x \neq 25$

Intervals: (- ∞, 25) ∪ (25, ∞) ◄——○——►
25

43. $\left|\frac{3}{5}x+\frac{7}{3}\right|<2$

$-2<\frac{3}{5}x+\frac{7}{3}<2$

Multiply all 3 parts by 15.

$-30<9x+35<30$

$-65<9x<-5$

$\frac{-65}{9}<x<\frac{-5}{9}$

Interval: (-65/9, -5/9) ◄-----o———o-----►
-65/9 -5/9

45. $\left|3\left(\frac{x+4}{4}\right)\right|>0$ This only fails to be true if $3\left(\frac{x+4}{4}\right)=0$, which occurs when x + 4 = 0 or x = - 4. The solution is (- ∞, - 4) ∪ (- 4,∞). The number line is ◄———o———►
- 4

47. $\left|\frac{1}{7}x+1\right|\le 0$ This is true only when $\frac{1}{7}x+1=0$, so that x = -7. The solution is {-7}.

49. $\left|\frac{x-5}{10}\right|<0$ No solution since absolute value is never less than 0.

51. x and y must have opposite signs.

53. For |x| < k, the solutions are <u>between</u> -k and k. For |x| > k, the solutions either lie to the right of k or to the left of -k.

55. If k > 0, |x| + k > 0 is equivalent to |x| > - k where - k is a negative number. Since the smallest value of |x| is 0, every value of |x| will be greater than a negative number.

Review

57. |3x – 5| + 3 = 2 or |3x – 5| = -1 No solution

59. The graph is a function since it passes the vertical line test. (To each value of x there corresponds a unique value of y.)

Chapter 7 Summary

1. a. $(-8)^2 = 64$ b. $-8^2 = -64$
c. $3^2 + 2^2 = 9 + 4 = 13$ d. $(3 + 2)^2 = 5^2 = 25$

2. a. $(-z^2)(z^3y^2) = -z^5y^2$ b. $(4x)x^2 = 4x^3$
c. $(x^2x^3)^3 = (x^5)^3 = x^{15}$ d. $(2x^2y)^2 = 4x^4y^2$

3. a. $\left(\frac{x^2y}{xy^2}\right)^2=\left(\frac{x}{y}\right)^2=\frac{x^2}{y^2}$

b. $\left(\frac{3x}{6x^2}\right)^3=\left(\frac{1}{2x}\right)^3=\frac{1}{(2x)^3}=\frac{1}{8x^3}$

c. $x^{-2}x^3 = x^{-2+3} = x^1 = x$

d. $y^4y^{-3} = y^{4+(-3)} = y^1 = y$

4. a. $\frac{x^3}{x^{-7}}=x^{3-(-7)}=x^{10}$

b. $(x^{-3}x^{-4})^{-2} = (x^{-7})^{-2} = x^{14}$

c. $\left(\frac{x^2}{x}\right)^{-5}=x^{-5}=\frac{1}{x^5}$

d. $\left(\frac{15x^4}{5y^3}\right)^{-2}=\left(\frac{5y^3}{15x^4}\right)^2=\left(\frac{y^3}{3x^4}\right)^2=\frac{y^6}{9x^8}$

5. a. $728 = 7.28 \times 10^2$ b. $0.00942 = 9.42 \times 10^{-3}$

c. $0.018 \times 10^{-2} = 1.8 \times 10^{-4}$
d. $600 \times 10^{-3} = 6 \times 10^{-1}$

6. a. $7.26 \times 10^5 = 726{,}000$
b. $3.91 \times 10^{-4} = 0.000391$
c. $\frac{(4800)(20{,}000)}{600{,}000}=\frac{(4.8\times10^3)(2\times10^4)}{6\times10^5}=$
$\frac{9.6}{6}\times10^{7-5}=1.6\times10^2=160$

7. 4E -7 = $4 \times 10^{-7} = 0.0000004$

8. $2.1721E3 = 2.1721 \times 10^3 = 2172.1$

9. a. $xy + 2y + 4x + 8 = (xy + 2y) + (4x + 8) =$
$y(x + 2) + 4(x + 2) = (x + 2)(y + 4)$

b. $ab + bc + 3a + 3c = (ab + bc) + (3a + 3c) =$
$b(a + c) + 3(a + c) = (a + c)(b + 3)$

c. $z^2 - 16 = z^2 - 4^2 = (z - 4)(z + 4)$

d. $2x^4 - 98 = 2(x^4 - 49) = 2[(x^2)^2 - 7^2] =$
$2(x^2 - 7)(x^2 + 7)$

e. $y^2 + 21y + 20 = y^2 + y + 20y + 20 = (y^2 + y)$
$+ (20y + 20) = y(y + 1) + 20(y + 1) =$
$(y + 1)(y + 20)$

f. $-y^2 + 5y + 24 = -1(y^2 - 5y - 24) =$
$-1(y - 8)(y + 3) = (8 - y)(y + 3)$

g. $15x^2 - 57x - 12 = 3(5x^2 - 19x - 4) =$
$3(5x + 1)(x - 4)$

h. $x^3 + 343 = x^3 + 7^3 = (x + 7)(x^2 - 7x + 7^2) =$
$(x + 7)(x^2 - 7x + 49)$

10. a. $2(x-1)+5=\frac{1}{2}(2x-3)$

$$2x-2+5=x-1.5$$
$$2x+3=x-1.5$$
$$x+3=-1.5$$
$$x=-1.5-3=-4.5$$

b. $\frac{x-4}{5}+1=\frac{2x}{3}$ Multiply both sides by 15.

$$3(x-4)+15\cdot 1=5(2x)$$
$$3x-12+15=10x$$
$$3x+3=10x$$
$$3=7x$$
$$\frac{3}{7}=x$$

c. $(x+2)(x-5)=(x-4)(x-1)$

$$x^2-3x-10=x^2-5x+4$$
$$-3x-10=-5x+4$$
$$2x-10=4$$
$$2x=14$$
$$x=7$$

d. $2x^2-x-1=0$

$$(2x+1)(x-1)=0$$
$$2x+1=0 \text{ or } x-1=0$$
$$x=\frac{-1}{2} \text{ or } x=1$$

e. $6a^2-6=0$

$$6(a^2-1)=0$$
$$6(a-1)(a+1)=0$$
$$a-1=0 \text{ or } a+1=0$$
$$a=1 \text{ or } a=-1$$

f. $x^3+5x^2+6x=0$

$$x(x^2+5x+6)=0$$
$$x(x+2)(x+3)=0$$
$$x=0 \text{ or } x+2=0 \text{ or } x+3=0$$
$$x=0 \text{ or } x=-2 \text{ or } x=-3$$

g. $|3x+1|=10$

$$3x+1=10 \text{ or } 3x+1=-10$$
$$3x=9 \text{ or } 3x=-11$$
$$x=3 \text{ or } x=\frac{-11}{3}$$

h. $\left|\frac{3}{2}x-4\right|=-9$ No solution.

$\frac{4}{x}+\frac{5}{x}=8$ Multiply both sides by x.

i. $4+5=8x$

$$9=8x$$

$$\frac{9}{8} = x$$

j. $\frac{x}{x+2} + \frac{1}{x-1} = 1$ LCD = (x + 2)(x – 1).

Multiply both sides by the LCD.

$$(x-1)x + (x+2)\cdot 1 = (x+2)(x-1)$$
$$x^2 - x + x + 2 = x^2 + x - 2$$
$$2 = x - 2$$
$$4 = x$$

11. a. $P = 2l + 2w$

$$P - 2l = 2w$$
$$\frac{P - 2l}{2} = w$$

b. 4x + y = 2 so that y = 2 – 4x.

c. x + 2y = 6 so that 2y = 6 – x and so y = 3 – ½ x.

12. a. 2x + 5y = 10

y-intercept: (0, 2) x-intercept: (5,0)

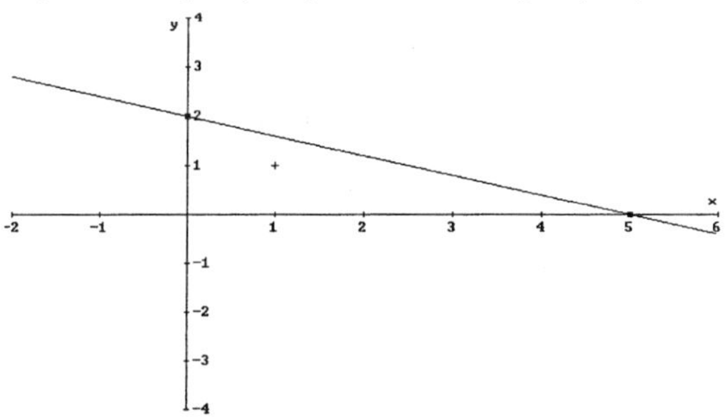

b. 3x – 4y = 12

y-intercept: (0, -3) x-intercept: (4, 0)

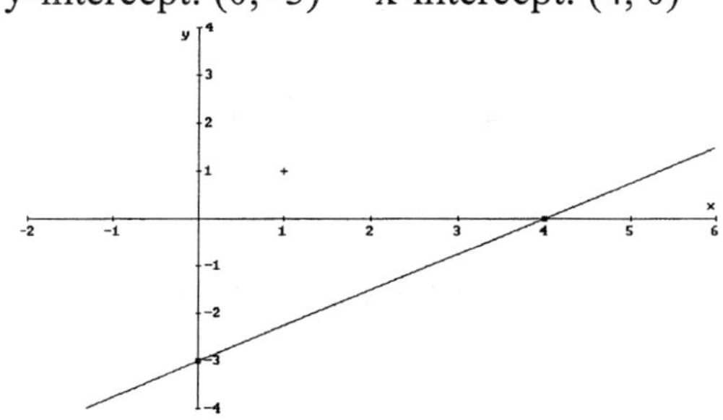

c. 3x – y = 6

y-intercept: (0, -6) x-intercept: (2, 0)

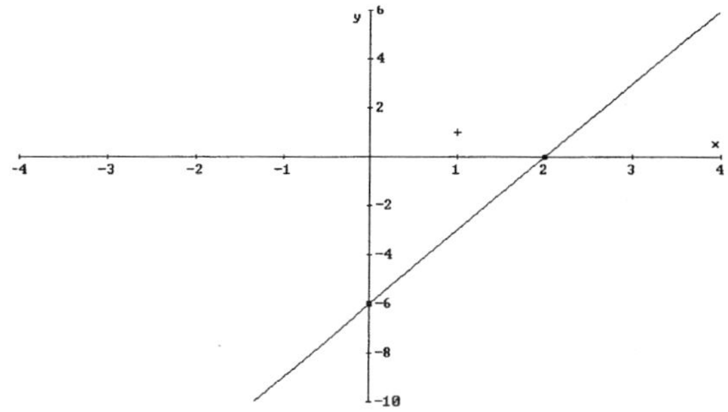

d. 5x + y = 7

y-intercept: (0, 7) x-intercept: (7/5, 0)

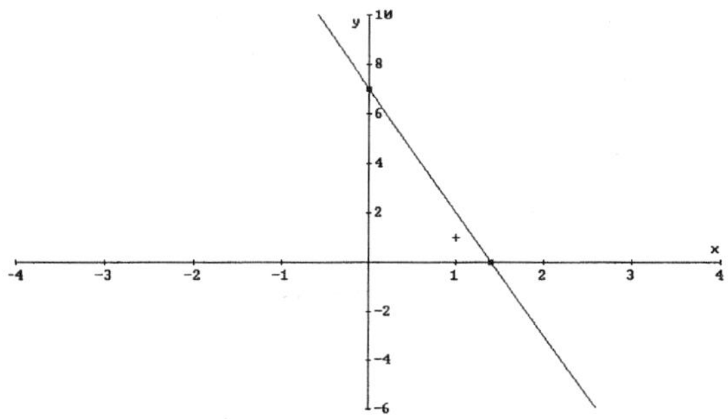

13. a. 4(x – 1) = 8 {3}

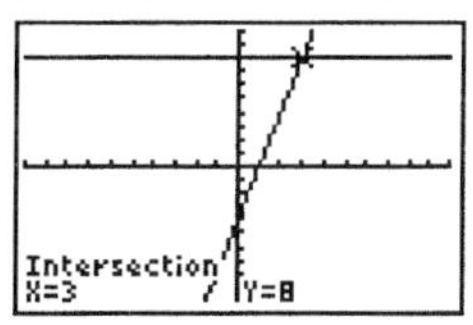

b. 3(x + 7) = -6 {-9}

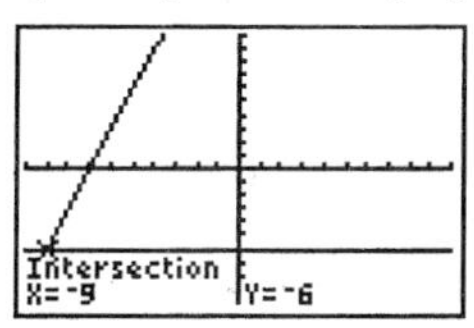

c. |3x + 1| = 10 {3, -11/3}

Graphing in [-5,5] by [-5,15] gives

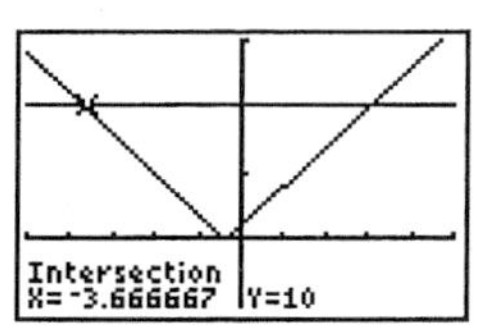

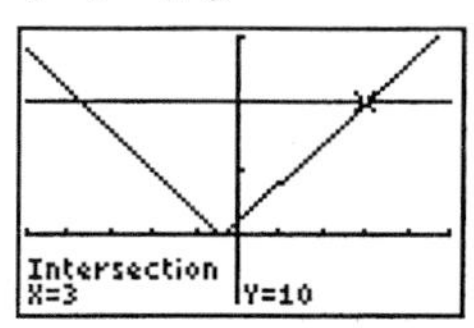

d. $\left|\frac{3}{2}x - 4\right| = 9$ {26/3, -10/3}

Graphing in [-10,10] by [-5,15] gives

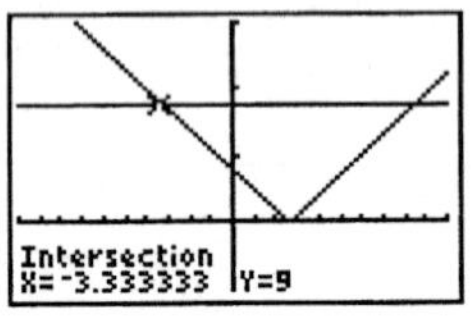

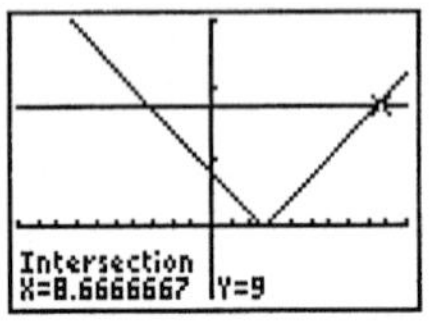

e. $x^2 - 11x + 30 = 0$ {5,6}

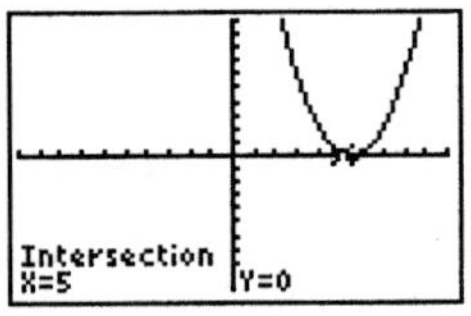

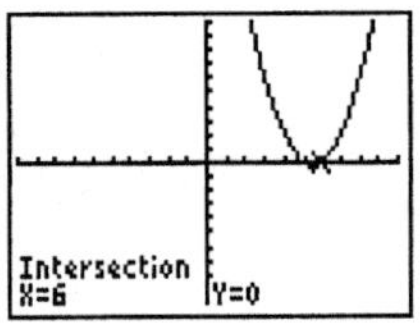

f. $x^2 - 5x - 24 = 0$ {-3,8}

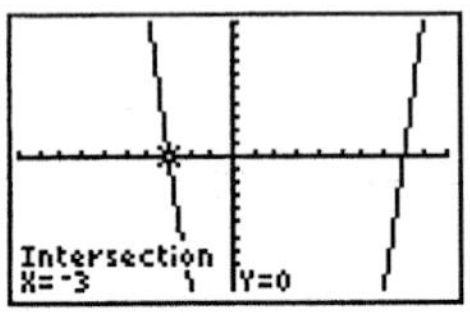

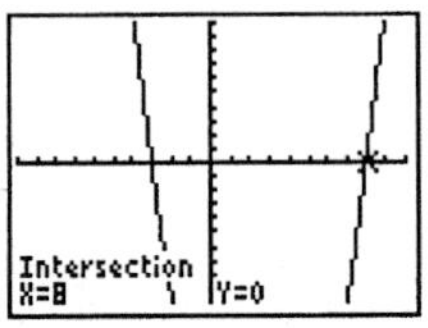

14. a. Function b. Function

15. a. $3x - 4 < 5$ Interval: (- ∞, 3)

$3x < 9$

$x < 3$

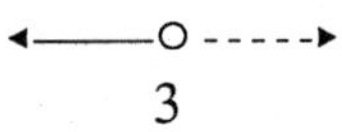

b. $5 - 2x < 6$ Interval: (- ½ , ∞)

$-2x < 1$

$x > -\frac{1}{2}$

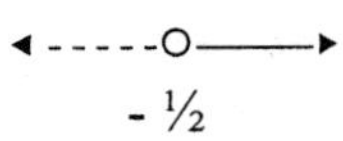

c. $2(x - 3) \le 7$ Interval: (- ∞, 13/2]

$2x - 6 \le 7$

$2x \le 13$

$x \le \frac{13}{2}$

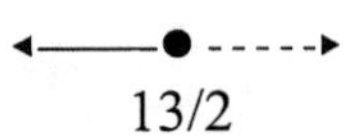

d. $-3(x - 1) + 5 \ge 6$ Interval: (- ∞, 2/3]

$-3(x - 1) \ge 1$

$-3x + 3 \ge 1$

$-3x \ge -2$

$x \le \frac{2}{3}$

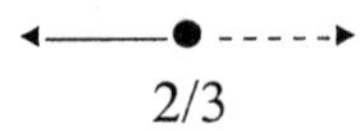

16. a. y1 = y2: {-1}
b. y1 < y2: (-1, ∞)
c. y1 > y2: (- ∞, -1)

17. a. y1 = y2: {8}
b. y1 < y2: (- ∞, 8)
c. y1 > y2: (8, ∞)

18. a. $3 < 3x + 4 < 10$ Interval: (-1/3, 2)

$-1 < 3x < 6$

$\frac{-1}{3} < x < 2$

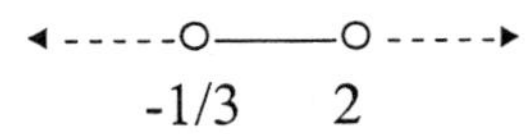

b. $4 > 3x + 2 > -3$ Interval: (-5/3, 2/3)

$2 > 3x > -5$

$\frac{2}{3} > x > \frac{-5}{3}$

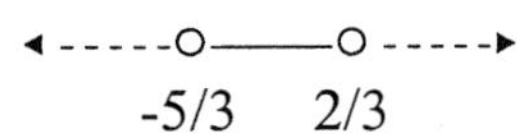

c. $6x + 1 < 5$ or $3(x - 9) > -2$

$6x < 4$ or $3x - 27 > -2$

$x < \frac{4}{6}$ or $3x > 25$

$x < \frac{2}{3}$ or $x > \frac{25}{3}$

Intervals: (- ∞, 2/3) ∪ (25/3, ∞)

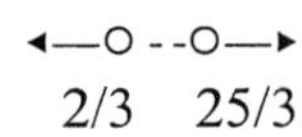

d. $5 - \frac{1}{2}x \geq -1$ or $6 - x < -1$

$-\frac{1}{2}x \geq -6$ or $-x < -7$

$x \leq 12$ or $x > 7$

Interval: (- ∞, ∞)

19. a. y1 < y2 < y3: (2.5, 7)
b. y2 < y1 or y2 > y3: (- ∞, 2.5) ∪ (7, ∞)

20. a. y1 < y2 < y3: (-1,7)
b. y2 < y1 or y2 > y3: (7, ∞) ∪ (- ∞, -1)

21. a. $|x + 3| \leq 4$ Interval: [-7, 1]

$-4 \leq x + 3 \leq 4$

$-7 \leq x \leq 1$

b. $|2x + 3| < 11$ Interval: (-7, 4)

$-11 < 2x + 3 < 11$

$-14 < 2x < 8$

$-7 < x < 4$

c. $|2x - 4| > 22$

$2x - 4 > 22$ or $2x - 4 < -22$

$2x > 26$ or $2x < -18$

$x > 13$ or $x < -9$

Intervals: (13, ∞) ∪ (- ∞, -9) 

d. $|3x + 1| \geq 5$

$3x + 1 \geq 5$ or $3x + 1 \leq -5$

$3x \geq 4$ or $3x \leq -6$

$x \geq \frac{4}{3}$ or $x \leq -2$

Intervals: [4/3, ∞) ∪ (- ∞, -2] 

e. |4x – 1|+7 < 2 or |4x – 1| < -5
No solution.

f. $\left|\frac{1}{2}x + 3\right| - 8 > -11$ or $\left|\frac{1}{2}x + 3\right| > -3$

The solution is all real numbers.

22. a. y1 = y2: { }
b. y1 < y2: ℝ
c. y1 > y2: { }

23. a. y1 = y2: { }
b. y1 < y2: ℝ
c. y1 > y2: { }

24. Let x = length of the shorter piece. Then 2x + 1 = length of the longer piece and

$x + (2x + 1) = 22$

$x + 2x + 1 = 22$

$3x + 1 = 22$

$3x = 21$

$x = 7$

The pieces are 7' and 15' long.

25. Let x = width. Then 2x = length. There is 624 feet of fencing so that x + 2x + x = 624. Thus, 4x = 624 or x = 156. The pasture will be 156' by 312'.

26. The possible profit is expressed by p < 42 – 27 or p < 15.

27. Let x be the number of discs the student can buy. Then the total cost to the student is given by 8.50x + 175, and so x satisfies

$8.50x + 175 \leq 330$

$8.50x \leq 155$

$x \leq 18.2 \ (18\frac{4}{17})$

The student can buy at most 18 discs.

Chapter 7 Test

1. $\dfrac{-12y^3(xz)^2}{18x^3(y^2z)^2} = \dfrac{-12y^3x^2z^2}{18x^3y^4z^2} = \dfrac{(6y^3x^2z^2)(-2)}{(6y^3x^2z^2)(3xy)} =$
$\dfrac{-2}{3xy}$

2. $\dfrac{x^2y^{-2}}{x^3z^2} = \dfrac{x^2}{x^3y^2z^2} = \dfrac{1}{xy^2z^2}$

3. $a^3 - 125 = a^3 - 5^3 = (a-5)(a^2 + 5a + 25)$

4. $2x^2 - 32 = 2(x^2 - 16) = 2(x^2 - 4^2) =$
$2(x-4)(x+4)$

5. $6x^2 + x - 2 = (3x+2)(2x-1)$

6. $9(x+4) + 4 = 4(x-5)$
$9x + 36 + 4 = 4x - 20$
$9x + 40 = 4x - 20$
$5x + 40 = -20$
$5x = -60$
$x = -12$

7. $\dfrac{y-1}{5} + 2 = \dfrac{2y-3}{3}$ Multiply both sides by 15.
$3(y-1) + 15 \cdot 2 = 5(2y-3)$
$3y - 3 + 30 = 10y - 15$
$3y + 27 = 10y - 15$
$27 = 7y - 15$
$42 = 7y$
$6 = y$

8. $A = p + prt$
$A - p = prt$
$\dfrac{A-p}{pr} = t$

9. $3x - 7y = 8$
$-7y = 8 - 3x$
$y = \dfrac{8-3x}{-7} = \dfrac{3x-8}{7}$

10. Let x = length of the shortest piece. Then 2x and 6x are the lengths of the other pieces.
Then $x + 2x + 6x = 20$
$9x = 20$
$x = \dfrac{20}{9}$ so that $6x = \dfrac{120}{9} = \dfrac{40}{3}$
The longest piece is 40/3 feet.

11. Let x = width of the rectangle. Then x + 5 = length, and since the perimeter is 26,
$2x + 2(x+5) = 26$
$2x + 2x + 10 = 26$
$4x = 16$
$x = 4,\ x + 5 = 9$
The area is 4·9= 36 cm.^{2}

12. $|2x+3| = 11$
$2x + 3 = 11$ or $2x + 3 = -11$
$2x = 8$ or $2x = -14$
$x = 4$ or $x = -7$

13. $\dfrac{3}{5} + \dfrac{7}{x+2} = 2$ LCD = 5(x + 2)
Multiply both sides by the LCD.
$3(x+2) + 5 \cdot 7 = 2 \cdot 5(x+2)$
$3x + 6 + 35 = 10x + 20$
$3x + 41 = 10x + 20$
$41 = 7x + 20$
$21 = 7x$
$3 = x$

14. $3x - y = 6$

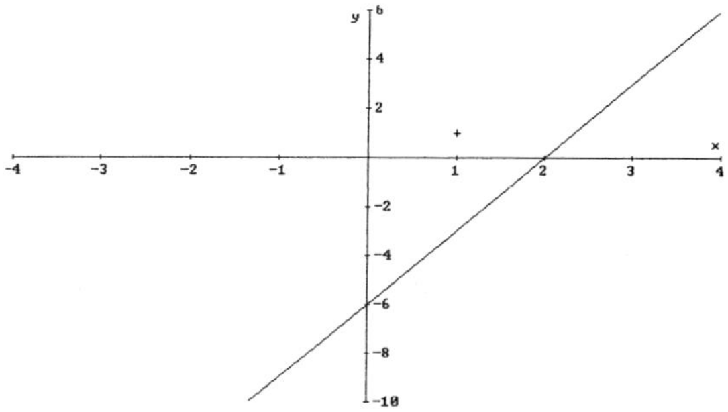

15. $y = 5$

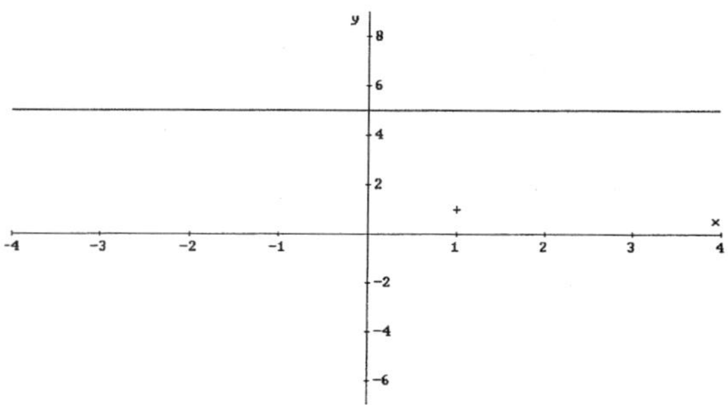

16. a. y1 = y2: $\{-1\}$
 b. y1 < y2: $(-1, \infty)$
 c. y1 > y2: $(-\infty, -1)$

17. a. y1 = y2: $\{2,6\}$
 b. y1 < y2: $(2,6)$
 c. y1 > y2: $(-\infty, 2) \cup (6, \infty)$

18. The graph is not a function since both (1,1) and (1, -1) are on the graph.

19. $\frac{1}{4}x - 2 \le -2$ Interval: $(-\infty, 0]$

$\frac{1}{4}x \le 0$

$x \le 0$

0

20. $-2 \le 4 - x < 5$ Interval: $(-1, 6]$

$-6 \le -x < 1$

$6 \ge x > -1$

Number line: 

-1 6

21. $5 - 2x > 6$ or $x + 2 > 1$ All reals. $(-\infty, \infty)$

$-2x > 1$ or $x > -1$

$x < \frac{-1}{2}$ or $x > -1$

22. $4x + 1 < x$ and $2x - 5 < 1$ Interval: $(-\infty, -1/3)$

$3x + 1 < 0$ and $2x < 6$

$3x < -1$ or $x < 3$

$x < \frac{-1}{3}$ and $x < 3$

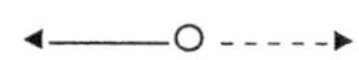

-1/3

23. $|2x - 5| < 4$ Interval: $(½, 9/2)$

$-4 < 2x - 5 < 4$

$1 < 2x < 9$

$\frac{1}{2} < x < \frac{9}{2}$

½ 9/2

24. $\left|\frac{1}{3}x - 1\right| \ge 5$

$\frac{1}{3}x - 1 \ge 5$ or $\frac{1}{3}x - 1 \le -5$

$\frac{1}{3}x \ge 6$ or $\frac{1}{3}x \le -4$

$x \ge 18$ or $x \le -12$

Intervals: $[18, \infty) \cup (-\infty, -12]$ 

-12 18

25. $\left|\frac{1}{2}x + 3\right| + 4 \le 2$ or $\left|\frac{1}{2}x + 3\right| \le -2$

No solution. { }

Chapter 8 Functions

Section 8.1 Relations and Functions

Vocabulary

1. relation
3. range
5. ordered pairs
7. y

Concepts

9. a. f(2) = 2
   b. f(-2) = -2
   c. f(0) = 0
   d. f(5) = 5

11. {(2,5),(3,-1),(-4,7)}
    The relation is a function.
    domain = {2,3,-4} range = {5,-1,7}

13. {(-1,2),(-2,3),(-1,-2)}
    The relation is not a function.
    domain = {-1,-2} range = {-2,2,3}

15. y = 2x + 3 determines y as a function of x, since for each real number x, 2x + 3 has exactly one value.

17. $y = 3x^2$ determines y as a function of x, since for each real number x, $3x^2$ has exactly one value.

19. $y^2 = x + 1$ does not determine y as a function of x. If x = 0, for example, y = ± 1.

21. y = |x| determines y as a function of x, since for each real number x, |x| has exactly one value.

Note: #23 and #25 are graphed in ZInteger.

23. $y = -x^2 + 5$
    domain = ℝ
    range = (-∞, 5]

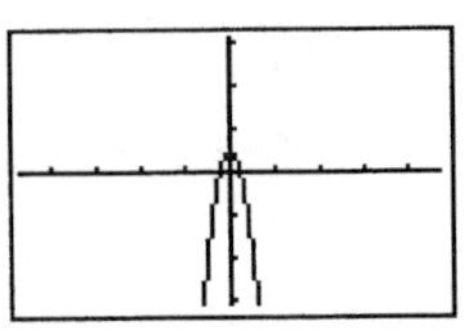

25. y = |x – 15| – 10
    domain = ℝ
    range = [-10,∞)

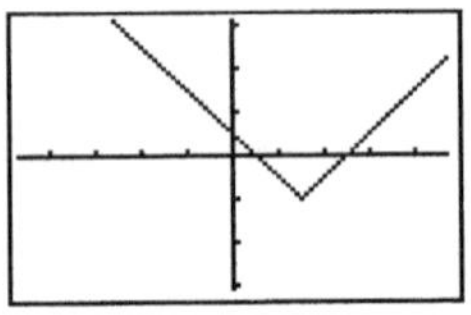

27. Not a function, fails the vertical line test.

29. The relation is a function.

31. The relation is a function.

33. Not a function, fails the vertical line test.

35. Not a function, since for x = 4, y = ± 1.

37. The relation is a function, since for each value of x, there is exactly one value of y.

39. f(x) = 3x + 2
    a. f(2) = 3·2 + 2 = 8
    b. f(-3) = 3(-3) + 2 = -9 + 2 = -7
    c. f(a) = 3a + 2
    d. f(a + 2) = 3(a +2) + 2 = 3a + 6 + 2 = 3a + 8

41. $f(x) = x^2 - 2x$
    a. $f(2) = 2^2 - 2(2) = 4 - 4 = 0$
    b. $f(-3) = (-3)^2 - 2(-3) = 9 - (-6) = 15$
    c. $f(a) = a^2 - 2a$
    d. $f(a + 2) = (a + 2)^2 - 2(a + 2) = a^2 + 4a + 4 - 2a - 4 = a^2 + 2a$

43. $f(x) = \dfrac{3}{x-4}$

    a. $f(2) = \dfrac{3}{2-4} = \dfrac{3}{-2} = -\dfrac{3}{2}$

    b. $f(-3) = \dfrac{3}{-3-4} = \dfrac{3}{-7} = -\dfrac{3}{7}$

    c. $f(a) = \dfrac{3}{a-4}$

    d. $f(a + 2) = \dfrac{3}{(a+2)-4} = \dfrac{3}{a-2}$

45. $f(-4) = -3$, since $(-4,-3)$ is on the graph.

47. $f(-3) = 2$, since $(-3,2)$ is on the graph.

49. $f(3) = 5$, since $(3,5)$ is on the graph.

51. $f(-4) = -6$, since $Y1 = -6$ when $x = -4$.

53. $f(-3) = 4$, since $Y1 = 4$ when $x = -3$.

55. No, because 4 is not in the pictured domain of f.

In 57 – 63, $f(x) = 2x + 1$

57. $f(3) + f(2) = (2\cdot3 + 1) + (2\cdot2 + 1) = 7 + 5 = 12$

59. $f(b) - f(a) = (2b + 1) - (2a + 1) =$
$2b + 1 - 2a - 1 = 2b - 2a$

61. $f(b) - 1 = (2b + 1) - 1 = 2b + 1 - 1 = 2b$

63. $f(0) - f(-\frac{1}{2}) = (2\cdot0 + 1) - [2(-\frac{1}{2}) + 1] =$
$1 - (-1 + 1) = 1 - 0 = 1$

65. $y = 3x^2 + 2$, not linear

67. $x = 3y - 4$, linear, $y = \frac{1}{3}x + \frac{4}{3}$

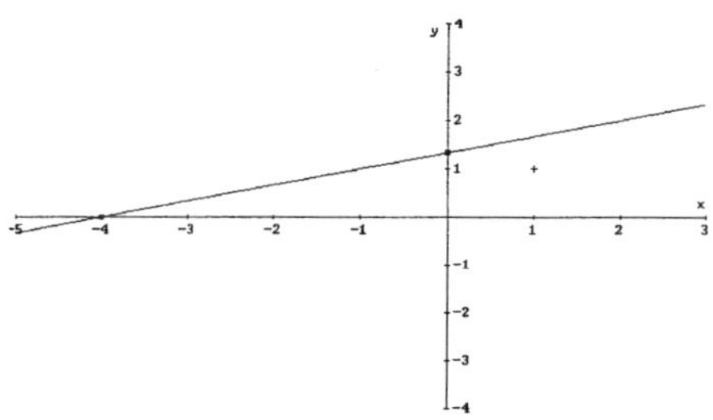

69. $4x - 2y = 8$, linear, $y = 2x - 4$

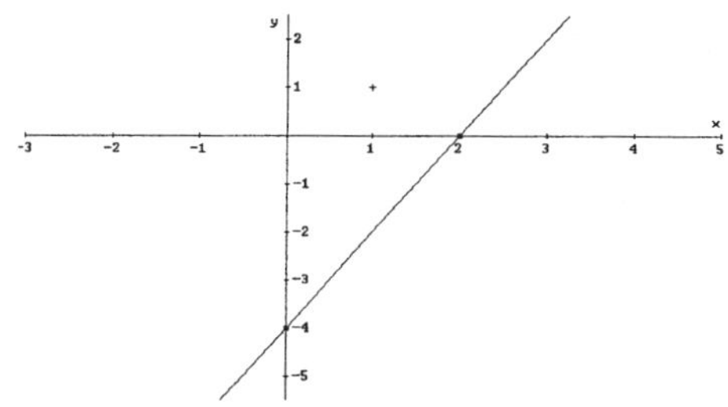

Review

71. The number 5 is a solution of $2x + 4 \ge 9$ since $2(5) + 4 = 14$ and $14 \ge 9$.

73. For $x = -2$ and $y = 5$, $3x + 2y = 3(-2) + 2(5) = -6 + 10 = 4$.

75. The number -7 is a solution of $x^2 - 49 = 0$ since $(-7)^2 - 49 = 49 - 49 = 0$.

Section 8.2 Functions and Graphs

Vocabulary

1. increasing

3. left; right

Concepts

5. The number of plane deaths was increasing in these yearly intervals: 1980-81, 1981-82, 1984-85, 1986-87, 1987-88, 1990-91.

7. The maximum number of plane deaths recorded during 1980-1993 was 285.

9. The function is increasing on $(-\infty, \infty)$.

11. The function is increasing on $[0, \infty)$

13. The function is increasing on $(-\infty,0]$.

15. The function is increasing on $[-4,4]$.

17. a. increasing on [-8,-6] and on [3,6].
b. decreasing on [2,3]
c. constant on [-6,2]

19. A turning point of a function is a point where the graph changes direction.

21. $y = x^2 - 2x + 5$, relative minimum at (1,4)

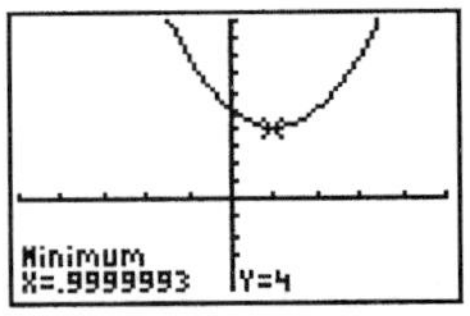

23. $y = 4 - |x + 2|$, relative maximum at (-2,4).

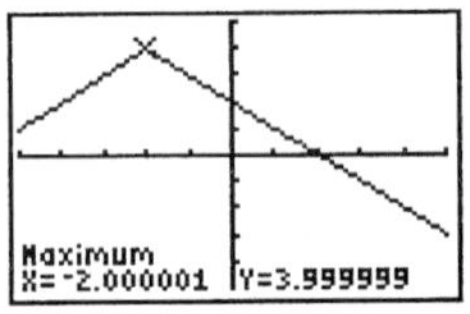

25. The graph is a function, since it passes the vertical line test.

27. The turning point is (4,1) and it is a relative maximum.

29. f(-5) = 5 since (-5,5) is on the graph.

31. Various answers are possible. Note that the rates of increase and decrease are constant.

33. $y = -.01x^2 + .64x + 9.76$

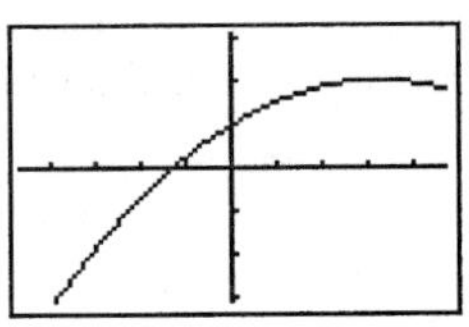

a. The x-values represent how far he has traveled horizontally.
b. The y-values represent his height above the ground.
c. He has traveled approximately 77 ft.
d. He will have traveled 62 ft.
e. He reaches a maximum height of 20 ft.
f. He will have traveled 32 ft. horizontally when he reaches his maximum height.
g. Before he is shot out, his head is 9.76 ft. above the ground.

35. $y = -x^2 + 35x - 150$
a. \$17.50 per ticket
b. \$156.25 maximum profit

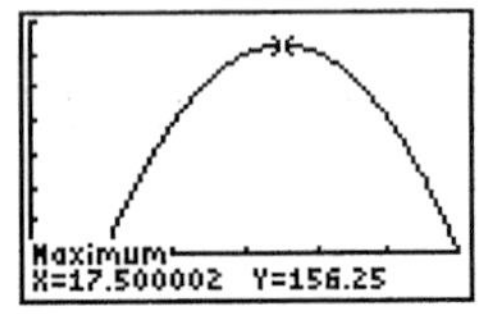

c. \$136 profit at \$13 per ticket.
d. A price of \$15 per ticket gives a \$150 profit.

Review

37. Algebraically
x-intercepts: Set y = 0, solve for x.
y-intercept: Set x = 0, solve for y.

Graphically
x-intercepts: Use 'root' or 'zero'
y-intercept: Use 'value' or 'eval'

39. $y = |x + 2| - 3$

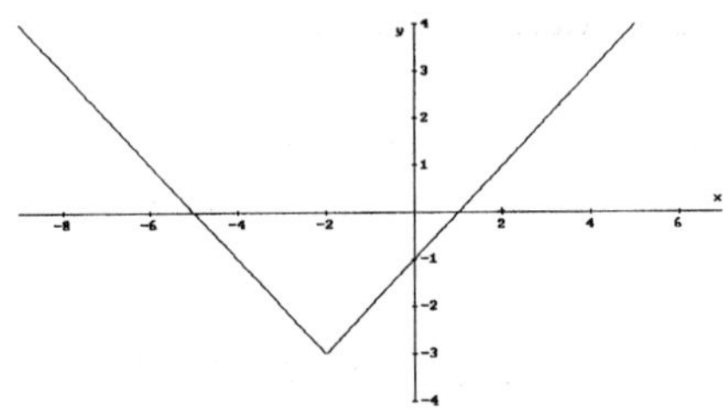

Section 8.3 Slope of a Non-vertical Line

Vocabulary

1. y; x

3. positive

5. zero

7. equal

Concepts

9. $m = \frac{\Delta y}{\Delta x} = \frac{1.5-1}{1-0} = .5$

11. $m = \frac{\Delta y}{\Delta x} = \frac{2-0}{1-2} = -2$

13. negative slope

15. no defined value for its slope

17. zero slope

Practice

19. (-5,-8), (3,8) $m = \frac{\Delta y}{\Delta x} = \frac{-8-8}{-5-3} = \frac{-16}{-8} = 2$

21. (0,-8), (-5,0) $m = \frac{\Delta y}{\Delta x} = \frac{-8-0}{0-(-5)} = \frac{-8}{5}$

23. (-7,-5), (-7,-2) $m = \frac{\Delta y}{\Delta x} = \frac{-5-(-2)}{-7-(-7)} = \frac{-3}{0}$
There is no defined value for the slope.

25. The graph of Y1 has the greater absolute value for its slope since it is the steeper graph.

27. The graph of Y2 has the greater absolute value for its slope since it is the steeper graph.

29. $3x + 2y = 12$
$2y = -3x + 12$
$y = \frac{-3}{2}x + 6$
The slope is $m = \frac{-3}{2}$ and the y-intercept is the point (0,6).

31. $3x = 4y - 2$
$-4y = -3x - 2$
$y = \frac{3}{4}x + \frac{1}{2}$
The slope is $m = \frac{3}{4}$ and the y-intercept is the point (0, ½).

33. $x = \frac{3-y}{4}$
$4x = 3 - y$
$y = -4x + 3$
The slope m = -4 and the y-intercept is the point (0,3).

35. f(x) = 3x – 5 has slope m = 3. A parallel line will have the same slope of 3.

37. 4x – y = 2 or y = 4x – 2 has slope m = 4. A perpendicular line will have slope $m = -\frac{1}{4}$.

39. f(x) = 3x – 4 and $f(x) = -\frac{1}{3}x + 2$. The product of the slopes is -1 and so they are perpendicular lines.

41. 2x – y = 7 and 2x + y = 8 or y = 2x – 7 and y = -2x + 8. The lines are neither parallel (2≠-2) nor perpendicular (2(-2)≠-1).

43. The x-axis has equation y = 0 and slope 0.

45. A calculator will graph only functions and a vertical line is not a function, but a horizontal line is a function.

47. slope = $\frac{1}{5}$, y-intercept = (0,4). The line has equation $y = \frac{1}{5}x + 4$.

49. Parallel to a line of slope 5 and y-intercept (0,-2). The line has equation y = 5x + (-2) = 5x – 2.

Applications

51. The slope m = rate of growth in sales per year $= \frac{110{,}000 - 50{,}000}{2} = \frac{60{,}000}{2} = 30{,}000$ dollars per year.

53. y = 5 + 0.75x
 a. The slope is m = .75 = cost per ride.
 b. y = 7 + 0.50x
 c. No, the slope could never be negative since this would mean a negative cost per ride.

55. a. y = 250x + 3000 , where y = value and x = number of years since purchased.
 b. The slope m = increase in value per year.
 c. If the slope were negative, it would be losing value each year.

Section 8.4 Equations of Lines

Vocabulary

1. slope-intercept; slope; y-intercept

3. standard

5. x = k; undefined

Concepts

7.

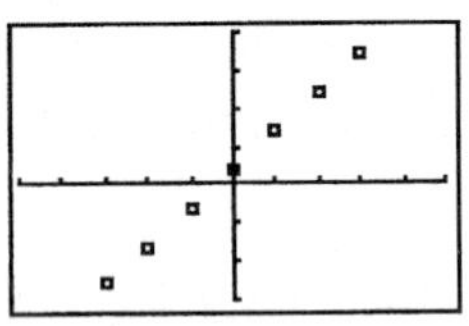

 a. The graph indicates a linear relationship.
 b. The slope is $m = \frac{\Delta y}{\Delta x} = \frac{7-2}{1-0} = 5$.
 c. The y-intercept is (0,2).
 d. The line is y = 5x + 2.

9. a. y = x + 3 b. y = -x + 5
 c. y = 5x – 4 d. y = x – 3
 e. y = 5x + 4 f. y = x + 5

Practice

11. m = 5, P(0,7) : y – 7 = 5(x – 0)
 standard form: -5x + y = 7

13. m = -3, P(2,0): y – 0 = -3(x – 2)
 standard form: 3x + y = 6

15. $m = \frac{3}{2}$, P(2,5): $y - 5 = \frac{3}{2}(x-2)$
 standard form: $2(y-5) = 3(x-2)$
 $2y - 10 = 3x - 6$
 $-3x + 2y = 4$

17. P(0,0), Q(4,4) $m = \frac{\Delta y}{\Delta x} = \frac{4-0}{4-0} = 1$
 Point-slope form: y – 0 = 1(x – 0)
 Slope-intercept form: y = x

19. P(3,4), Q(0,-3) $m = \frac{\Delta y}{\Delta x} = \frac{4-(-3)}{3-0} = \frac{7}{3}$
 Point-slope form: $y - (-3) = \frac{7}{3}(x-0)$
 Slope-intercept form: $y = \frac{7}{3}x - 3$.

21. P(-2,4), Q(3,-5) $m = \frac{\Delta y}{\Delta x} = \frac{4-(-5)}{-2-3} = \frac{9}{-5}$
 Point-slope form: $y - 4 = -\frac{9}{5}(x+2)$
 Slope-intercept form: $y = -\frac{9}{5}(x+2) + 4$
 $y = -\frac{9}{5}x - \frac{18}{5} + \frac{20}{5}$
 $y = -\frac{9}{5}x + \frac{2}{5}$

23. $y + 1 = x$ or $y = 1 \cdot x - 1$
slope $m = 1$ and y-intercept $b = -1$

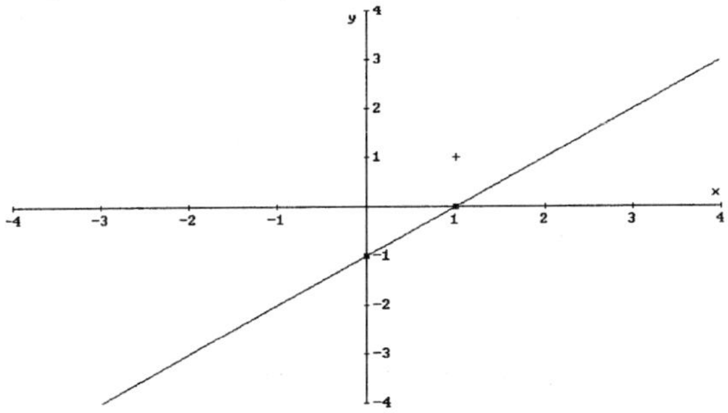

25. $y = \frac{2}{3}x + \frac{9}{2}$ slope $m = \frac{2}{3}$, y-intercept $b = \frac{9}{2}$

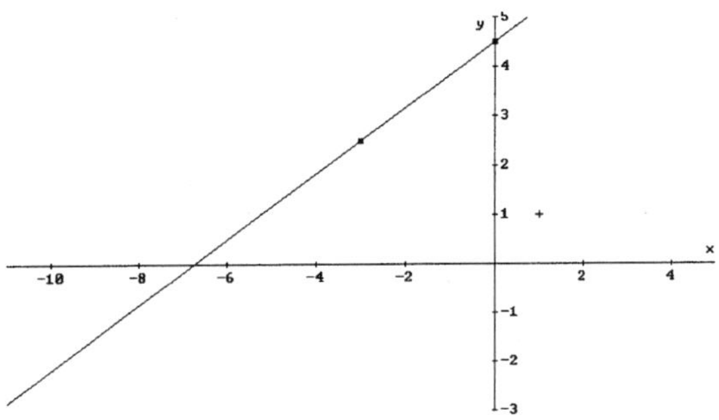

27. $3(y - 4) = -2(x - 3)$

$3y - 12 = -2x + 6$

$3y = -2x + 18$

$y = \frac{-2}{3}x + 6$

slope $m = -2/3$ and y-intercept $b = 6$

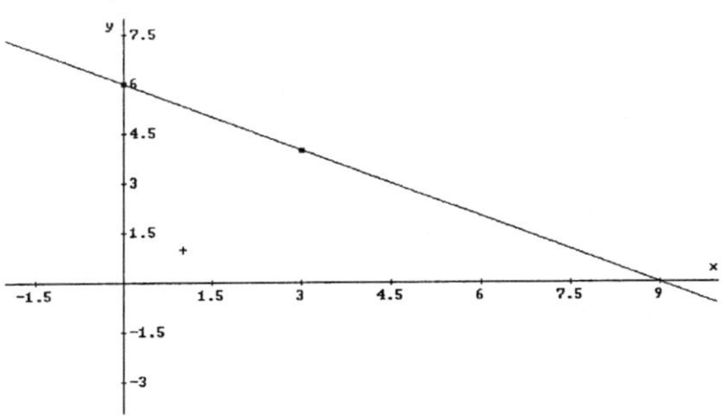

29. $P(0,0)$, L_1: $y = 4x - 7$, so $m_1 = 4$. A line L parallel to L_1 has the same slope, so $m = 4$. The equation of L is $y - 0 = 4(x - 0)$ or $y = 4x$.

31. $P(2,5)$, L_1: $4x - y = 7$. Put L_1 in slope-intercept form: $y = 4x - 7$, so $m_1 = 4$. A line L parallel to L_1 will have the same slope $m = 4$. The equation of L is $y - 5 = 4(x - 2)$ or $y = 4x - 3$.

33. $P(4,-2)$, L_1: $x = \frac{5}{4}y - 2$. Put L_1 in slope-intercept form: $4x = 5y - 8$

$4x + 8 = 5y$

$\frac{4}{5}x + \frac{8}{5} = y$

Thus, $m_1 = 4/5$. A line parallel to L_1 will have the same slope $m = 4/5$. The equation of L is $y - (-2) = \frac{4}{5}(x - 4)$ or $y = \frac{4}{5}x - \frac{26}{5}$.

35. $P(0,0)$, L_1: $y = 4x - 7$, so $m_1 = 4$. A line L perpendicular to L_1 will have slope $m = -1/m_1 = -1/4$. The equation of L is

$y - 0 = -\frac{1}{4}(x - 0)$ or $y = -\frac{1}{4}x$.

37. $P(2,5)$, L_1: $4x - y = 7$. From Exercise 31, the slope of L_1 is $m_1 = 4$. A line L perpendicular to L_1 will have slope $m = -1/m_1 = -1/4$. The equation of L is

$y - 5 = -\frac{1}{4}(x - 2)$ or $y = -\frac{1}{4}x + \frac{11}{2}$.

39. $P(4,-2)$, L_1: $x = \frac{5}{4}y - 2$. From Exercise 33, the slope of L_1 is $m_1 = 4/5$. A line L perpendicular to L_1 will have slope $m = -1/m_1 = -5/4$. The equation of L is

$y - (-2) = -\frac{5}{4}(x - 4)$ or $y = -\frac{5}{4}x + 3$.

41. The midpoint between (2,4) and (-6,10) is $\left(\frac{2+(-6)}{2}, \frac{4+10}{2}\right) = (-2,7)$. Since y = 3 is a horizontal line, the desired line perpendicular to it will be a vertical line through (-2,7). Its equation is x = -2.

43. From Exercise 42, the midpoint between (2,-4) and (8,12) is (5,4). Since x = 3 is a vertical line, the desired line parallel to it will be a vertical line through (5,4). Its equation is x = 5.

45.
$$Ax + By = C$$
$$By = -Ax + C$$
$$y = -\frac{A}{B}x + \frac{C}{B}$$

This line has slope $m = -\frac{A}{B}$ and y-intercept $b = \frac{C}{B}$.

Applications

47. From the graph, $m = \frac{24{,}300 - 1900}{0 - 7} =$ -\$3200/ year. The y-intercept is (0, 24300) so the depreciation equation is y = -3200x + 24,300.

49. The line will pass through (0,475000) and (10, 2· 475000). Thus, $m = \frac{475{,}000}{10 - 0} =$ \$47,500/yr. The y-intercept is (0, 475000) so the appreciation equation is y = 47,500x + 475,000.

51. From the ad, the line will pass through (0,3900) and (3,1890). Thus, $m = \frac{3900 - 1890}{0 - 3} = \frac{2010}{-3} = -\$670/\text{yr}.$ The y-intercept is (0,3900) so the depreciation equation is y = -670x + 3900.

53. The line passes through (0, 1050) with slope m = -\$120/yr. The depreciation equation is y = -120x + 1050. After x = 8 years, it has a salvage value y = -120(8) + 1050 = \$90.

55. The line passes through (2,450) with slope m = \$40/yr. Thus, the appreciation equation is y – 450 = 40(x – 2) or y = 40x + 370. In x = 13 years, it will be worth y = 40(13) + 370 = \$890.

57. Let b = original purchase price. The line passes through (3,47700) and (0,b) with slope m = \$3500/yr. Thus, 3500 = $\frac{\Delta y}{\Delta x} = \frac{b - 47700}{0 - 3}$. Thus, -3(3500) = b – 47,700 so that b = 47,700 – 10,500 = \$37,200.

59. The hourly labor rate is $\frac{69}{1.5} = \$46/\text{hr}.$ For a 5 hour job, labor would cost 5(46) = \$230.

61. The cost y = \$60 + mx where x = number of feet and m = cost/ft. Since 250 ft. of gutter costs \$450,
$$250x + 60 = 435$$
$$250x = 375$$
$$x = 1.50$$
The cost per foot is \$1.50. To have 300 ft. installed will cost 300(1.50) + 60 = \$510.

Review

63. a.
$$2x^2 + x - 3 = 0$$
$$(2x+3)(x-1) = 0$$
$$2x + 3 = 0 \text{ or } x - 1 = 0$$
$$x = \frac{-3}{2} \text{ or } x = 1$$

b. 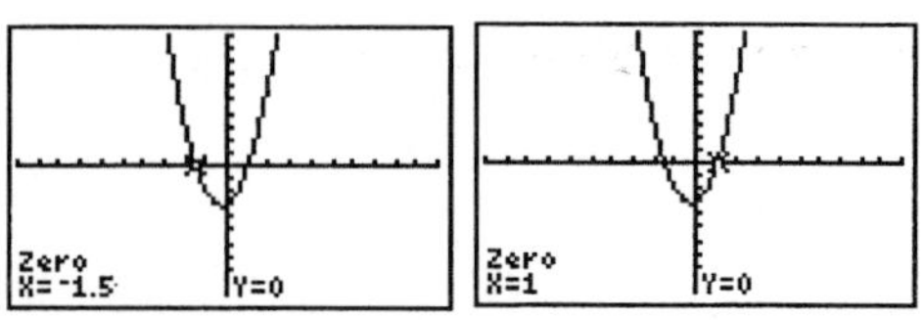

65. a. $4x^2 - 9 = 0$

$(2x-3)(2x+3) = 0$

$x = \frac{3}{2}$ or $x = -\frac{3}{2}$

b.

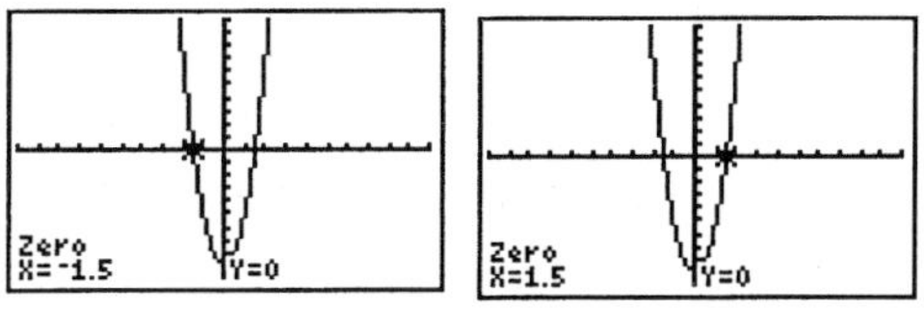

67. Let x = number of ounces of copper to be added. Then the ratio $\frac{\text{gold}}{\text{alloy}} = \frac{1}{4} = \frac{20}{60+x}$, so that 60 + x = 4(20) and x = 80 – 60 = 20. Thus, 20 ounces of copper must be added.

Section 8.5 Variation

Vocabulary

1. direct; increases

3. inverse; decreases

5. joint

Concepts

7. direct variation

9. inverse variation

11. direct variation

13. inverse variation

15. inverse variation, k = 8

17. combined variation, k = 3

19. joint variation, k = 3

Practice

21. $A = kp^2$

23. $v = \frac{k}{r^3}$

25. $B = kmn$

27. $P = \frac{ka^2}{j^3}$

29. $F = \frac{km_1m_2}{d^2}$

31. L varies jointly with m and n.

33. E varies jointly with a and b^2.

35. X varies directly with x^2 and inversely with y^2.

37. R varies directly with L and inversely with d^2.

Applications

39. $A = \pi r^2$, $A = \pi(6^2) = 36\pi$ in.^{2}

41. $d = kg$ $\quad 288 = 12k$ Thus, $d = 24k$.

$24 = k$

On 18 gal., d = 24(18) = 432 miles.

43. $d = \frac{k}{a}$ $\quad 10 = \frac{k}{25}$ Thus, $d = \frac{250}{a}$. For 10

$k = 250$

cows, d = 250/10 = 25 days.

45. $V = \frac{k}{p}$ $\quad 20 = \frac{k}{6}$ Thus, $V = \frac{120}{p}$.

$k = 120$

When p = 10, V = 120/10 = 12 in.^{3}

47. $F = \frac{k}{l}$ $\quad 256 = \frac{k}{2}$ Thus, $F = \frac{512}{l}$.

$k = 512$

When l = 6 ft., F = 512/6 = 85 ⅓ times/sec.

49. $V = klwh$ Replacing l by $2l$, w by 3w and h by 2h, $V = k(2l)(3w)(2h) = 12klwh$ and volume is increased by a factor of 12.

51. $O = 23.5\ h\ r^2$ For h = 20' and r = 15/2 = 7.5', $O = 23.5(20)(7.5)^2 = 26{,}437.5$ gal.

53. $V = kA$ 6 = 2k, so that k = 3. The resistance is 3 ohms.

55. $D = \frac{k}{wd^3}$ $1.1 = \frac{k}{4(4^3)}$ $D = \frac{281.6}{wd^3}$ For
$k = 281.6$
this beam, deflection $D = \frac{281.6}{2(8^3)} = 0.275$ in.

57. $P = \frac{kT}{V}$ $1 = \frac{k(273)}{1}$ Thus, $P = \frac{T}{273V}$.
$k = \frac{1}{273}$
If V = 2 m.^{3}, and P = 1 atmosphere, then
$1 = \frac{T}{2 \cdot 273}$, so that T = 546°K.

59. One similarity is the constant of variation. One difference is the relationship among the related variables.

61. No, this is not inverse variation. The amount of change ≠ k ÷ purchase price.

Review

63. $f(x) = x^2 + 3$ Domain: ℝ Range: $[3, \infty)$

65. $f(x) = |x + 3|$ Domain: ℝ Range: $[0, \infty)$

Chapter 8 Summary

1. a. $y = 4x - 1$ Function
   Domain: ℝ Range: ℝ
   b. $x = 3y - 10$ Function
   Domain: ℝ Range: ℝ
   c. $y = 3x^2 + 1$ Function
   Domain: ℝ Range: $[1, \infty)$
   d. $y = \frac{4}{2 - x}$ Function
   Domain: $\{x | x \neq 2\}$ Range: $\{y | y \neq 0\}$
   e. $x = \frac{y + 3}{2}$ Function
   Domain: ℝ Range: ℝ
   f. $y^2 = 4x$ Not a function
   Domain: $[0, \infty)$ Range: ℝ

2. a. Function b. Not a function
   c. Not a function d. Not a function

3. $f(x) = 3x + 2$, $g(x) = x^2 - 4$
   a. $f(-3) = 3(-3) + 2 = -7$
   b. $g(8) = 8^2 - 4 = 60$
   c. $g(-2) = (-2)^2 - 4 = 0$
   d. $f(5) = 3(5) + 2 = 17$

4. a. $f(-2) = -3$ b. $f(4) = 3$

5. a. $y = 3x + 2$, linear
   b. $y = (x+5)/4$, linear
   c. $4x - 3y = 12$ linear
   d. $y = x^2 - 25$ non-linear

6. a. increasing on $[0, \infty)$
   b. decreasing on $[0, \infty)$
   c. decreasing on $(-\infty, 5]$
   d. increasing on $[0, 3]$; decreasing on $[3, \infty)$
   e. decreasing on $(-\infty, 0]$, $[2, \infty)$; increasing on $[0, 2]$
   f. increasing on $(-\infty, -4]$, $[1, \infty)$; decreasing on $[-2, 1]$; constant on $[-4, -2]$

7. a. $y = -x^2 + 36$ Relative maximum at (0, 36)

   b. $y = |x + 3| - 5$ Relative minimum at (-3,-5)
   c. $y = -2x^3 + 8x^2 - 6x$ Relative minimum at (0.45, -1.26); Relative maximum at (2.22, 4.23).

   d. $y = (x - 3)(2x + 5)(x - 1)$ Relative minimum at (2.11, -9.11); Relative maximum at (-1.11, 24.1)

8. a. Graph $y = 128x - 16x^2$ in [0,10] by [0,260]

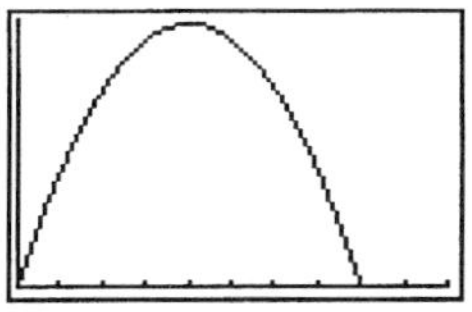

b. The relative maximum in (4, 256).
c. The ball is 256 ft. high at x = 4 seconds.
d. The ball is at a height of 245.76 ft. going up and coming down.
e. The ball is on the ground when y = 0.

9. a. $y = x(80 - 2x)$
b. The graph in [0,50] by [0, 1000] is

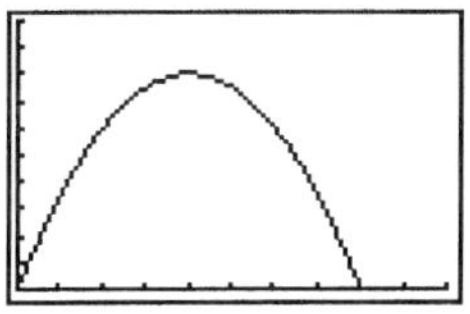

c. The relative maximum is (20, 800) so the maximum area he can enclose is 800 ft.^{2}
d. The width will be 20 ft. (See part c.)

10. a. P(2,5) and Q(5,8)
$$m = \frac{\Delta y}{\Delta x} = \frac{5-8}{2-5} = \frac{-3}{-3} = 1$$
b. P(-3,-2) and Q(6,12)
$$m = \frac{\Delta y}{\Delta x} = \frac{-2-12}{-3-6} = \frac{-14}{-9} = \frac{14}{9}$$
c. P(-3,4) and Q(-5,-6)
$$m = \frac{\Delta y}{\Delta x} = \frac{4-(-6)}{-3-(-5)} = \frac{10}{2} = 5$$
d. P(5,-4) and Q(-6,-9)
$$m = \frac{\Delta y}{\Delta x} = \frac{-4-(-9)}{5-(-6)} = \frac{5}{11}$$

11. a. $2x - 3y = 18$ or $y = \frac{2}{3}x - 6$ so $m = \frac{2}{3}$.
b. $2x + y = 8$ or $y = -2x + 8$ so $m = -2$.
c. $-2(x - 3) = 10$ or $x = -2$. No slope.
d. $3x + 1 = 7$ or $x = 2$. No slope.

12. a. (0, -5), m = 2/3 : $y = \frac{2}{3}x - 5$ using slope-intercept form.
b. (0, 2), m = -1/2: $y = -\frac{1}{2}x + 2$ using slope-intercept form.

13. a. positive slope b. negative slope
c. zero slope d. No defined slope

14. a. perpendicular, $(m_1 m_2 = -1)$
b. parallel, $(m_1 = m_2)$
c. neither, $(m_2 = -m_1)$
d. perpendicular, $(m_1 m_2 = -1)$

15. a. P(-8,5), m = 3
$y - 5 = 3[x - (-8)]$ or $3x - y = -29$ is the equation of the line in standard form.

b. Points (-2,4) and (6,-9). First determine m. $m = \frac{\Delta y}{\Delta x} = \frac{4-(-9)}{-2-6} = \frac{-13}{8}$. Using point-slope form, $y - 4 = -\frac{13}{8}(x + 2)$. Thus, $8(y - 4) = -13(x + 2)$ or $8y - 32 = -13x - 26$ so that $13x + 8y = 6$ is the equation of the line in standard form.

c. P(-3, -5), parallel to L_1: $3x - 2y = 7$. Put L_1 in slope-intercept form: $y = \frac{3}{2}x + \frac{7}{2}$. Thus, the slope of L_1 is $m_1 = 3/2$. A line L parallel to L_1 will have the same slope m = 3/2. The equation of the desired line L is $y - (-5) = \frac{3}{2}(x - (-3))$ or $2(y + 5) = 3(x + 3)$ so that $2y + 10 = 3x + 9$ and $3x - 2y = 1$ is the equation of the line in standard form.

d. P(-3, -5), perpendicular to L_1: $3x - 2y = 7$. From part c. $m_1 = 3/2$, and the slope of a line L perpendicular to L_1 will be $m = -2/3$. The equation of L is given by
$$y - (-5) = -\frac{2}{3}(x - (-3)) \text{ or } 3(y + 5) = -2(x + 3)$$
so that $3y + 15 = -2x - 6$ and $2x + 3y = -21$ is the equation in standard form.

16. a. From the graph, the y-intercept is (0,-2) and the slope $m = \frac{\Delta y}{\Delta x} = \frac{2}{3}$. Using slope-intercept form the equation is $y = \frac{2}{3}x - 2$.
b. The line is vertical passing through (3,0) and has equation x = 3.

17. a. The rate of depreciation is $\frac{2500-850}{1998-2001} =$ – \$550/year.
b. y = -550x + 2500 is the linear model, where x = years since 1998 and y = value.
c. Good viewing window : [0,10] by [0,3000]

18. The line passes through (1940, 15) and (1998,200). The rate of appreciation is the slope $m = \frac{200-15}{1998-1940} = \frac{185}{58} \approx \$3.19/\text{year}$

19. x = ky $\begin{array}{c} 12 = k \cdot 2 \\ 6 = k \end{array}$ and x = 6y. When y = 12 x = 6(12) = 72.

20. $x = \frac{k}{y}$ $\quad 24 = \frac{k}{3}$ $\quad k = 72$ Thus, $x = \frac{72}{y}$. When x = 12, 12y = 72 and so y = 6.

21. x = kyz 24 = k(3)(4), so k = 2.

22. $x = \frac{kt}{y}$ $\quad 2 = \frac{k \cdot 8}{64}$

$k = \frac{128}{8} = 16$

23. $BMI = \frac{kw}{h^2}$ $\quad 26.95 = \frac{k(182)}{69^2}$, so $k \approx 705$
Thus, for a person weighing 132 lb. and 55" tall, $BMI = \frac{705(132)}{55^2} = 30.76 \approx 31$.

24. V = khA Replacing h by 3h and A by 2A, V = k(3h)(2A) = 6khA. The volume is increased by a factor of 6.

Chapter 8 Test

1. y = |x + 1| determines y as a function of x, since each value of x results in exactly one value of y..

2. From the graph, the domain is (- ∞, 2] and the range is (- ∞,3].

3. $f(x) = x^2 + 4x$, so $f(-3) = (-3)^2 + 4(-3) = -3$

4. a. increasing on (- ∞, -3]
b. decreasing on [2, 4]
c. constant on [-3, 2]

5. $y = -16x^2 + 32x + 10$
a. [0,3] by [0,30] is an appropriate window.
b. The maximum height is 26 feet.
c. It hits the ground after 2.27 sec. (y = 0)
d. The turning point is (1,26).

6. Put $x = \frac{3y-8}{2}$ in slope-intercept form.
2x = 3y – 8 or 3y = 2x + 8, so $y = \frac{2}{3}x + \frac{8}{3}$, and the slope m = 2/3.

7. P(-5, -6) and Q(5, -2)
$m = \frac{\Delta y}{\Delta x} = \frac{-2-(-6)}{5-(-5)} = \frac{4}{10} = \frac{2}{5}$

8. P(0,3) and $m = \frac{4}{5}$. Using slope-intercept form, the equation of the line is given by $y = \frac{4}{5}x + 3$.

9. The slope of $y = \frac{2}{3}x + 4$ is $\frac{2}{3}$ and the slope of $y = \frac{3}{2}x - 4$ is $\frac{3}{2}$. Since the slopes are

only reciprocals, the lines are neither perpendicular or parallel.

10. A line perpendicular to the y-axis will be a horizontal line. The equation of the horizontal line through (3, -1) is y = -1.

11. Given (4,1) and (-7,6). The slope is given by $m = \frac{1-6}{4-(-7)} = \frac{-5}{11}$. Using point-slope form with the point (4, 1), $y - 1 = \frac{-5}{11}(x - 4)$.

12. y(x + 3) + 4 = x(y – 2) simplifies to 3y + 4 = -2x so that $y = -\frac{2}{3}x - \frac{4}{3}$ so that y is a linear function of x.

13. $y = \frac{kw}{z^2}$

14. $P = kC^2$, where C = current.

15. Assume $P = 5C^2$
   a. The function is increasing on $[0, \infty)$.
   b. At (3, 45) on the graph, 3 amps of current produces a power loss of 45 watts due to heat.

16. x = ky   30 = k(4), so k = 7.5 and x = 7.5y.
   If y = 9, the x = 7.5(9) = 67.5.

17. $v = \frac{k}{t}$ $55 = \frac{k}{20}$ Thus, $v = \frac{1100}{t}$.
   $k = 55(20) = 1100$
   If v = 75, 75t = 1100 or t = 44/3.

Cumulative Review: Chapters 7 - 8

Vocabulary/Concepts

1. scientific notation

2. a. $\frac{1}{x^{-3}} = x^3$ b. $\frac{x^8}{x^{-2}} = x^{8-(-2)} = x^{10}$
   c. $x^{-5} \cdot x^0 = \frac{1}{x^5} \cdot 1 = \frac{1}{x^5}$
   d. $\frac{x^{-2}}{x^5} = x^{-2-5} = x^{-7} = \frac{1}{x^7}$

3. a. $1{,}320{,}000 = 1.32 \times 10^6$
   b. $2{,}000{,}000 = 2 \times 10^6$
   c. $0.0013 = 1.3 \times 10^{-3}$
   d. $0.00412 = 4.12 \times 10^{-3}$

4. a. $1.3 \times 10^{-2} = 0.013$
   b. $6.1 \times 10^5 = 610{,}000$
   c. $0.051 \times 10^{-2} = 0.00051$
   d. $3.82 \times 10^6 = 3{,}820{,}000$

5. solution
6. function

7. a. Not a function since for x = 4, y = ± 2.
   b. Function, since for each value of x there is exactly one value of y.

8. intersection
9. union

10. As a compound inequality, $|x + 2| \le 4$ becomes $-4 \le x + 2 \le 4$
11. a. Function   b. Not a function.
   c. Function   d. Function
   e. Function   f. Not a function

12. $f(x) = x^2 - 3x + 2$
   a. $f(-1) = (-1)^2 - 3(-1) + 2 = 1 + 3 + 2 = 6$
   b. $f(3) = 3^2 - 3(3) + 2 = 9 - 9 + 2 = 2$
   c. $f(-3) = (-3)^2 - 3(-3) + 2 = 9 + 9 + 2 = 20$
   d. $f(8) = 8^2 - 3(8) + 2 = 64 - 24 + 2 = 42$

13. x = 3y – 4, x + 4 = 3y and so $y = \frac{1}{3}x + \frac{4}{3}$.

14. y; x

15. y = a constant
16. x = a constant

17. 0

18. undefined

19. slope-intercept; slope; y-intercept

20. point-slope; slope; (x_1, y_1)

Practice

21. a. $\frac{(2x^{-4}y^2)^3}{4x^{-2}y^{-1}} = \frac{8x^{-12}y^6}{4x^{-2}y^{-1}} = 2x^{-12-(-2)}y^{6-(-1)} =$

$2x^{-10}y^7 = \frac{2y^7}{x^{10}}$

b. 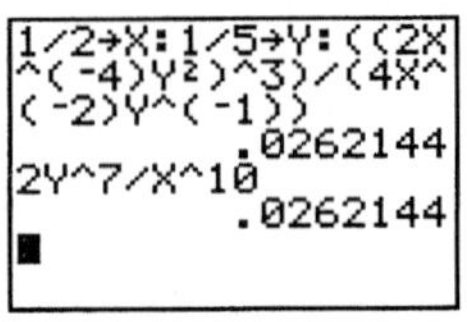

22. $\frac{8(x-5)}{3} = 2(x-4)$

$8(x-5) = 6(x-4)$

$8x - 40 = 6x - 24$

$2x = 16$

Thus, x = 8.

23. $2x^2 - 7x = 4$

$2x^2 - 7x - 4 = 0$

$(2x+1)(x-4) = 0$

$2x + 1 = 0$ or $x - 4 = 0$

$x = -\frac{1}{2}$ or $x = 4$

24. $2|x+1| - 2 = 0$

$2|x+1| = 2$

$|x+1| = 1$

$x + 1 = 1$ or $x + 1 = -1$

$x = 0$ or $x = -2$

25. 3 + |2x – 1| = 2 or |2x – 1| = -1. No solution, since absolute value is never negative.

26. $\frac{x}{x-2} - \frac{5}{x-2} = 3$

Multiply both sides by x – 2.

$x - 5 = 3(x-2)$

$x - 5 = 3x - 6$

$1 = 2x$

$\frac{1}{2} = x$

27. $-2 \le \frac{5-3x}{2} \le 2$ Interval: [1/3, 3]

$-4 \le 5 - 3x \le 4$

$-9 \le -3x \le -1$

$3 \ge x \ge \frac{1}{3}$

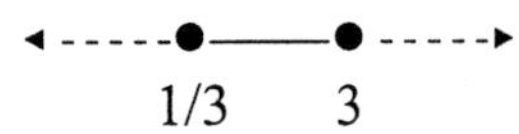

28. $3|2x+5| \ge 9$

$|2x+5| \ge 3$

$2x + 5 \ge 3$ or $2x + 5 \le -3$

$2x \ge -2$ or $2x \le -8$

$x \ge -1$ or $x \le -4$

Intervals: [-1, ∞) ∪ (- ∞, -4]

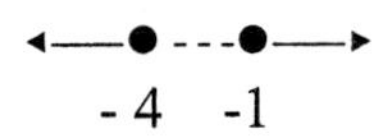

29. f(x) = 3x – 1. f(2) – f(3) = (3·2 – 1) – (3·3 – 1) = 5 – 8 = -3

30. a. increasing on [-3, -1]
b. decreasing on [-6, -3] and [4, 6]
c. constant on [-1, 4]

31. 2x + 3y = 6
y-intercept: (0,2) x-intercept: (3,0)

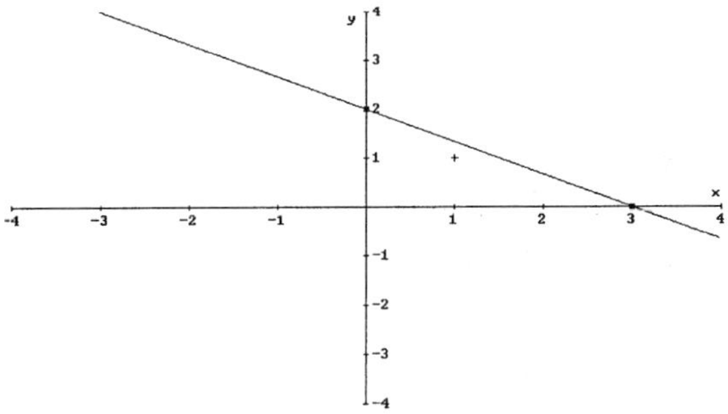

32. Domain = {1,2,-1,-3}, Range = {-1,5,-5}

33. From the table, $m = \frac{\Delta y}{\Delta x} = \frac{-3-(-4)}{2-0} = \frac{1}{2}$

34. $x = \frac{3}{4} - \frac{1}{4}y$ slope m = -4; y-intercept b = 3

$4x = 3 - y$

$y = -4x + 3$

35. The slope of x – 2y = -18 is ½ and the slope of 2x + y = -6 is - 2. Since the slopes are negative reciprocals, the lines are perpendicular lines.

36. P(0,5) and slope = 2/3. Using slope-intercept form, $y = \frac{2}{3}x + 5$. Then 3y = 2x + 15 or 2x – 3y = -15 in standard form.

37. P(5, 1) and slope 3. Using point-slope form, y – 1 = 3(x – 5). Then y – 1= 3x – 15 or 3x – y = 14 in standard form.

38. A line with slope 0 is a horizontal line. The one through (6,2) has equation y = 2.

39. Points (3,2) and (-1,4). First find the slope. $m = \frac{\Delta y}{\Delta x} = \frac{4-2}{-1-3} = \frac{2}{-4} = -\frac{1}{2}$. Using point-slope form the equation is $y - 2 = -\frac{1}{2}(x-3)$.

Now 2(y – 2) = – (x – 3) or 2y – 4 = -x + 3 and x + 2y = 7 is in standard form.

40. P(-6,3), L_1: y + 3x = -12
Put L_1 in slope-intercept form: y = -3x – 12. Thus, m_1 = -3. A line L perpendicular to L_1 will have slope m = $-1/m_1$ = 1/3. The equation of L is $y - 3 = \frac{1}{3}(x+6)$. Now 3(y – 3) = x + 6 or 3y – 9 = x + 6 and x – 3y = -15 is the equation in standard form.

41. $-3x + 15 = \sqrt{x-5}$ Solution set: {5}

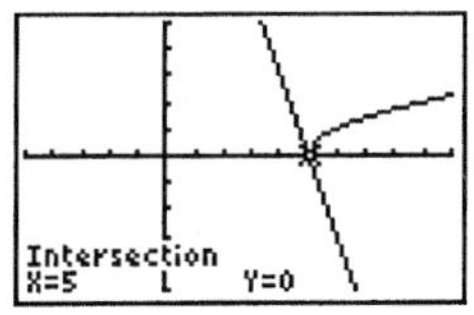

42. $\sqrt{2x+5} - 4 = 0$ Solution set: {5.5}

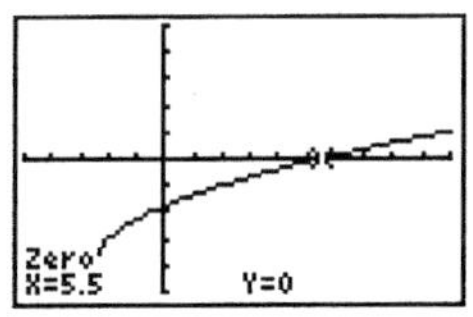

43. $x^3 - x^2 - 12x = 0$ Solution set: {0, 4, -3}

44. $y = 3(x-1)^2 + 6$ relative minimum: (1,6)

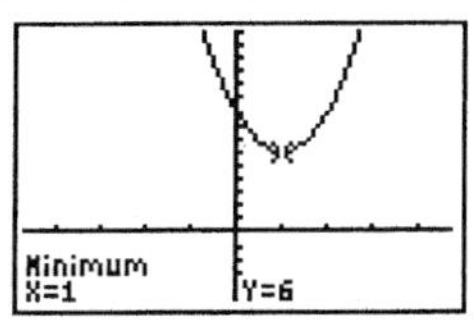

45. Purchased at $300, increases at $15/year.
   a. y = 15x + 300
   b. The slope is the yearly rate of increase.
   c. y = $600

46. Fixed charges of $45, hourly rate of $20/hour
   a. y = 20x + 45
   b. 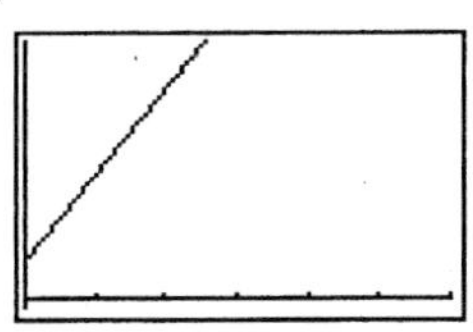 In [0,30] by [-10,300]

c. For a repair taking 5 hours, the cost is \$145, found by using 'value', or by letting $x = 5$ in the equation in part a.

d. No negative x values were included in the window since working a negative number of hours is not possible.

e. For a repair taking 20 hours, the cost is \$445, found using 'value' or by letting $x = 20$ in the equation in part a.

Applications

47. Let x = number of miles per day you can drive. Since rental is \$45/day plus \$0.24 per mile and you have \$100 to spend, then

$$.24x + 45 = 100$$
$$.24x = 55$$
$$x = \frac{55}{.24} \approx 229.2$$

You could drive approximately 229 miles.

48. Let x = her grade on the fourth test. Then

$$\frac{78 + 80 + 85 + x}{4} \geq 70$$
$$\frac{243 + x}{4} \geq 70$$

Then $243 + x \geq 280$ or $x \geq 37$. She only needs to make 37 or higher to keep her C.

49. $A = \pi r^2$, so for $r = 9$ mm., $A = \pi(9^2) = 81\pi$ mm.^{2}

50. $V = \frac{k}{a}$ $8000 = \frac{k}{5}$ Thus, $V = \frac{40000}{a}$.

$$k = 40000$$

When $a = 10$, $V = \$4000$.

Chapter 9 Rational Exponents and Radicals

Section 9.1 Introduction to Roots and Radicals

Vocabulary

1. principal

3. exact; approximate; irrational

Concepts

5. b. and d. are undefined for real numbers, since even roots of negative numbers are non-real.

7. a. A positive number has 1 real cube root.
   b. A negative number has 1 real cube root.
   c. A negative number has no real square roots

Practice

9. $\sqrt{144} = \sqrt{12^2} = 12$

11. $\sqrt[3]{-8} = \sqrt[3]{(-2)^3} = -2$

13. $\sqrt[3]{-27x^6} = \sqrt[3]{(-3x^2)^3} = -3x^2$

15. $-\sqrt{\frac{4}{25}} = -\sqrt{\left(\frac{2}{5}\right)^2} = -\frac{2}{5}$

17. $-\sqrt[4]{625} = -\sqrt[4]{5^4} = -5$

19. $\sqrt{-49}$ is not defined for real numbers.

21. $\sqrt[6]{-729}$ is not defined for real numbers.

23. $-\sqrt{1} = -\sqrt{1^2} = -1$

25. $\sqrt[3]{512} = \sqrt[3]{8^3} = 8$

27. $-\sqrt{-4}$ is not defined for real numbers.

29. $\sqrt[3]{343a^6b^3} = \sqrt[3]{(7a^2b)^3} = 7a^2b$

31. $\sqrt[5]{-\frac{243}{32}} = \sqrt[5]{\left(-\frac{3}{2}\right)^5} = -\frac{3}{2}$

33. $\sqrt{-25}$ is not defined for real numbers.

35. $\sqrt[6]{64} = \sqrt[6]{2^6} = 2$

37. $\sqrt{(-9)^2} = \sqrt{81} = \sqrt{9^2} = 9$

39. $\sqrt[3]{\frac{125}{216}} = \sqrt[3]{\left(\frac{5}{6}\right)^3} = \frac{5}{6}$

41. $\sqrt[3]{\frac{27}{1000}a^6b^6} = \sqrt[3]{\left(\frac{3}{10}a^2b^2\right)^3} = \frac{3}{10}a^2b^2$

43. $\sqrt[6]{729} = \sqrt[6]{3^6} = 3$

45. $\sqrt{\frac{49}{81}} = \sqrt{\left(\frac{7}{9}\right)^2} = \frac{7}{9}$

47. $\sqrt[3]{0.001} = \sqrt[3]{(0.1)^3} = 0.1$

49. $-\sqrt[4]{\frac{81}{256}} = -\sqrt[4]{\left(\frac{3}{4}\right)^4} = -\frac{3}{4}$

51. $\sqrt[4]{16y^4} = \sqrt[4]{(2y)^4} = |2y| = 2|y|$

53. $\sqrt[6]{64x^6} = \sqrt[6]{(2x)^6} = |2x| = 2|x|$

54. $\sqrt[8]{x^{24}} = \sqrt[8]{(x^3)^8} = |x^3| = |x|^3$

57. $\sqrt{x^2 + 10x + 25} = \sqrt{(x+5)^2} = |x+5|$

59. $\sqrt[4]{\frac{1}{81}x^8} = \sqrt[4]{\left(\frac{1}{3}x^2\right)^4} = \frac{1}{3}x^2$

61. $\sqrt[3]{-x^6} = \sqrt[3]{(-x^2)^3} = -x^2$

63. $\sqrt{(a+6)^2} = |a+6|$

65. $\sqrt[5]{-32x^5} = \sqrt[5]{(-2x)^5} = -2x$

67. $\sqrt{(-8c)^2} = |-8c| = 8|c|$

69. a. $\sqrt[6]{4096}$ b. rational

71. a. $5\sqrt[3]{9}$ b. irrational

73. a. $5\sqrt[3]{27-9}$ b. irrational

75. a. $\frac{8}{\sqrt{49}+1}$ b. rational

77. >; $\sqrt{25-9} = \sqrt{16} = 4,\ \sqrt{25} - \sqrt{9} = 5 - 3 = 2$

79. >; $\sqrt{36} + \sqrt{16} = 6 + 4 = 10,\ \sqrt{52} \approx 7.2$

81. <; $\sqrt{100} = 10,\ 4\sqrt{25} = 4 \cdot 5 = 20$

Applications

83. a. The longer the skid, the greater the speed.
 b. A car skidding 400 ft. was traveling 64.8 miles per hour.

Review

85. $(-5x^2y)^3 = (-5)^3(x^2)^3y^3 = -125x^6y^3$

87. $-(2a^2b^3c)^4 = -2^4(a^2)^4(b^3)^4c^4 = -16a^8b^{12}c^4$

89. $(x+2y)^2 = x^2 + 2(2y)x + (2y)^2 =$
 $x^2 + 4xy + 4y^2$

Section 9.2 Rational Exponents and Simplifying Radical Expressions

Vocabulary

1. base; factor

3. rational

Concepts

5. $7^{\frac{1}{3}} = \sqrt[3]{7}$

7. $3x^{\frac{1}{4}} = 3\sqrt[4]{x}$

9. $(3x)^{\frac{1}{4}} = \sqrt[4]{3x}$

11. $\sqrt[4]{3a} = (3a)^{\frac{1}{4}}$

13. $\sqrt[6]{\frac{1}{7}abc} = \left(\frac{1}{7}abc\right)^{\frac{1}{6}}$

15. $\sqrt[3]{a^2 - b^2} = (a^2 - b^2)^{\frac{1}{3}}$

Practice

17. $4^{\frac{1}{2}} = \sqrt{4} = 2$

19. $8^{\frac{1}{3}} = \sqrt[3]{8} = 2$

21. $\left(\frac{1}{8}\right)^{\frac{1}{3}} = \sqrt[3]{\frac{1}{8}} = \frac{1}{2}$

23. $-16^{\frac{1}{4}} = -\sqrt[4]{16} = -2$

25. $0^{\frac{1}{3}} = \sqrt[3]{0} = 0$

27. $(81)^{\frac{3}{4}} = \left(\sqrt[4]{81}\right)^3 = 3^3 = 27$

29. $\left(\frac{1}{8}\right)^{\frac{2}{3}} = \left(\sqrt[3]{\frac{1}{8}}\right)^2 = \left(\frac{1}{2}\right)^2 = \frac{1}{4}$

31. $4^{-\frac{1}{2}} = \frac{1}{4^{\frac{1}{2}}} = \frac{1}{\sqrt{4}} = \frac{1}{2}$

33. $4^{-\frac{3}{2}} = \frac{1}{4^{\frac{3}{2}}} = \frac{1}{\left(\sqrt{4}\right)^3} = \frac{1}{2^3} = \frac{1}{8}$

35. $\left(\frac{27}{8}\right)^{-\frac{4}{3}} = \frac{1}{\left(\frac{27}{8}\right)^{\frac{4}{3}}} = \frac{1}{\left(\sqrt[3]{\frac{27}{8}}\right)^4} = \frac{1}{\left(\frac{3}{2}\right)^4} = \frac{1}{\frac{81}{16}} =$
$\frac{16}{81}$

37. $\left(25x^4\right)^{\frac{3}{2}} = \left(\sqrt{25x^4}\right)^3 = (5x^2)^3 = 125x^6$

39. $\left(\frac{8x^3}{27}\right)^{\frac{2}{3}} = \left(\sqrt[3]{\frac{8x^3}{27}}\right)^2 = \left(\frac{2x}{3}\right)^2 = \frac{4x^2}{9}$

41. $(16x^2)^{-\frac{3}{2}} = \frac{1}{(16x^2)^{\frac{3}{2}}} = \frac{1}{\left(\sqrt{16x^2}\right)^3} = \frac{1}{(4x)^3} =$
$\frac{1}{64x^3}$

43. $\left(-27y^3\right)^{-\frac{2}{3}} = \frac{1}{\left(-27y^3\right)^{\frac{2}{3}}} = \frac{1}{\left(\sqrt[3]{-27y^3}\right)^2} =$
$\frac{1}{(-3y)^2} = \frac{1}{9y^2}$

45. $\left(-32p^5\right)^{-\frac{2}{5}} = \frac{1}{\left(-32p^5\right)^{\frac{2}{5}}} = \frac{1}{\left(\sqrt[5]{-32p^5}\right)^2} =$
$\frac{1}{(-2p)^2} = \frac{1}{4p^2}$

47. $\left(-\frac{8x^3}{27}\right)^{-\frac{1}{3}} = \left(-\frac{27}{8x^3}\right)^{\frac{1}{3}} = \left(\sqrt[3]{-\frac{27}{8x^3}}\right) = -\frac{3}{2x}$

49. $\sqrt{6}\cdot\sqrt{6} = \sqrt{6^2} = 6$

51. $\sqrt[3]{5x^2}\cdot\sqrt[3]{25x} = \sqrt[3]{125x^3} = 5x$

53. $\frac{\sqrt{180ab^4}}{\sqrt{5ab^2}} = \sqrt{\frac{180ab^4}{5ab^2}} = \sqrt{36b^2} = 6b$

55. $\frac{\sqrt[3]{189a^4}}{\sqrt[3]{7a}} = \sqrt[3]{\frac{189a^4}{7a}} = \sqrt[3]{27a^3} = 3a$

57. $\sqrt{20} = \sqrt{4\cdot 5} = \sqrt{4}\sqrt{5} = 2\sqrt{5}$

59. $-\sqrt{200} = -\sqrt{100\cdot 2} = -\sqrt{100}\sqrt{2} = -10\sqrt{2}$

61. $\sqrt[3]{80} = \sqrt[3]{8\cdot 10} = \sqrt[3]{8}\sqrt[3]{10} = 2\sqrt[3]{10}$

63. $\sqrt[3]{-81} = \sqrt[3]{-27\cdot 3} = \sqrt[3]{-27}\sqrt[3]{3} = -3\sqrt[3]{3}$

65. $\sqrt[5]{96} = \sqrt[5]{32}\sqrt[5]{3} = 2\sqrt[5]{3}$

67. $\sqrt{\dfrac{7}{9}} = \dfrac{\sqrt{7}}{\sqrt{9}} = \dfrac{\sqrt{7}}{3}$

69. $\sqrt[5]{\dfrac{3}{32}} = \dfrac{\sqrt[5]{3}}{\sqrt[5]{32}} = \dfrac{\sqrt[5]{3}}{2}$

71. $\sqrt{175a^2b^3} = \sqrt{25a^2b^2(7b)} = \sqrt{25a^2b^2}\sqrt{7b} =$
$5ab\sqrt{7b}$

73. $\sqrt[4]{32x^{12}y^4} = \sqrt[4]{(16x^{12}y^4)2} = \sqrt[4]{2^2(x^3)^4y^4}\sqrt[4]{2} =$
$2x^3y\sqrt[4]{2}$

75. $\sqrt{\dfrac{z^2}{16x^2}} = \dfrac{\sqrt{z^2}}{\sqrt{16x^2}} = \dfrac{z}{4x}$

77. $\sqrt[4]{\dfrac{5x}{16z^4}} = \dfrac{\sqrt[4]{5x}}{\sqrt[4]{2^4z^4}} = \dfrac{\sqrt[4]{5x}}{2z}$

Review

79. $\left(3x^2y\right)^{-2} = \dfrac{1}{(3x^2y)^2} = \dfrac{1}{9x^4y^2}$

81. $-\left(\dfrac{2xy}{4x^2y^3}\right)^{-3} = -\left(\dfrac{1}{2xy^2}\right)^{-3} = -(2xy^2)^3 =$
$-(8x^3y^6) = -8x^3y^6$

83. a. For x = -1 and y = 4, $\left(\dfrac{3x}{6y}\right)^{-2} = \left(\dfrac{3(-1)}{6(4)}\right)^{-2} =$
$\left(-\dfrac{1}{8}\right)^{-2} = (-8)^2 = 64$

b. 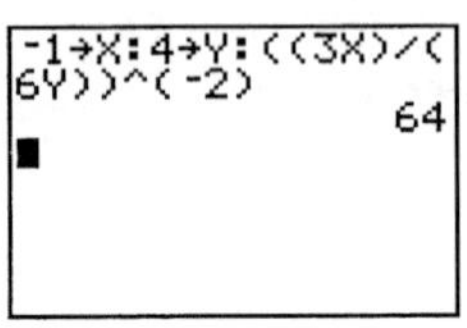

Section 9.3 Arithmetic Operations with Radicals

Vocabulary

1. index; radicand; distributive

3. rationalizing

Concepts

5. T

7. F $\sqrt{3}$ and $-3\sqrt{5}$ have different radicands and cannot be combined.

9. T The denominator will be
$\sqrt[3]{5}\sqrt[3]{25} = \sqrt[3]{125} = 5$

Practice

11. $4\sqrt{2x} + 6\sqrt{2x} = (4+6)\sqrt{2x} = 10\sqrt{2x}$

13. $8\sqrt[5]{7a^2} - 7\sqrt[5]{7a^2} = (8-7)\sqrt[5]{7a^2} = \sqrt[5]{7a^2}$

15. $\sqrt{2} - \sqrt{8} = \sqrt{2} - 2\sqrt{2} = (1-2)\sqrt{2} = -\sqrt{2}$

17. $3\sqrt{24} + \sqrt{54} = 3\cdot 2\sqrt{6} + 3\sqrt{6} = (6+3)\sqrt{6} =$
$9\sqrt{6}$

19. $\sqrt[3]{32} - \sqrt[3]{108} = 2\sqrt[3]{4} - 3\sqrt[3]{4} = (2-3)\sqrt[3]{4} = -\sqrt[3]{4}$

21. $\sqrt[4]{32} + \sqrt[4]{162} = 2\sqrt[4]{2} + 3\sqrt[4]{2} = (2+3)\sqrt[4]{2} = 5\sqrt[4]{2}$

23. $23\sqrt[4]{768} + \sqrt[4]{48} = 23\sqrt[4]{256\cdot 3} + \sqrt[4]{16\cdot 3} =$
$23\cdot 4\sqrt[4]{3} + 2\sqrt[4]{3} = 92\sqrt[4]{3} + 2\sqrt[4]{3} = 94\sqrt[4]{3}$

25. $\sqrt{80} - \sqrt{128} + \sqrt{288} = \sqrt{16\cdot 5} - \sqrt{64\cdot 2} +$
$\sqrt{144\cdot 2} = 4\sqrt{5} - 8\sqrt{2} + 12\sqrt{2} =$
$4\sqrt{5} + (-8+12)\sqrt{2} = 4\sqrt{5} + 4\sqrt{2}$

27. $\sqrt{25yz^2} + \sqrt{9yz^2} = 5z\sqrt{y} + 3z\sqrt{y} =$
$(5z + 3z)\sqrt{y} = 8z\sqrt{y}$

29. $3\sqrt{2x} - \sqrt{8x} = 3\sqrt{2x} - \sqrt{4 \cdot 2x} =$
$3\sqrt{2x} - 2\sqrt{2x} = (3-2)\sqrt{2x} = \sqrt{2x}$

31. $3\sqrt[4]{x^4y} - 2\sqrt[4]{x^4y} = 3x\sqrt[4]{y} - 2x\sqrt[4]{y} =$
$(3x - 2x)\sqrt[4]{y} = x\sqrt[4]{y}$

33. $\sqrt{8y^7} + \sqrt{32y^7} - \sqrt{2y^7} = \sqrt{4(y^3)^2 2y} +$
$\sqrt{16(y^3)^2 2y} - \sqrt{(y^3)^2 2y} = 2y^3\sqrt{2y} + 4y^3\sqrt{2y}$
$-y^3\sqrt{2y} = (2y^3 + 4y^3 - y^3)\sqrt{2y} = 5y^3\sqrt{2y}$

35. $\sqrt[3]{xy^4} + \sqrt[3]{8xy^4} - \sqrt[3]{27xy^4} = y\sqrt[3]{xy} + 2y\sqrt[3]{xy}$
$-3y\sqrt[3]{xy} = (y + 2y - 3y)\sqrt[3]{xy} = 0\sqrt[3]{xy} = 0$

37. $\sqrt{4x^2 + 12x + 9} + \sqrt{9x^2 + 6x + 1} = \sqrt{(2x+3)^2}$
$+\sqrt{(3x+1)^2} = (2x+3) + (3x+1) = 5x + 4$

39. $\sqrt{3}\sqrt{27} = \sqrt{3 \cdot 27} = \sqrt{81} = 9$

41. $\left(2\sqrt[3]{16}\right)\left(-\sqrt[3]{4}\right) = -2\sqrt[3]{16 \cdot 4} = -2\sqrt[3]{64} =$
$-2(4) = -8$

43. $\sqrt{8x}\sqrt{2x^3y} = \sqrt{16x^4y} = \sqrt{(4x^2)^2 y} = 4x^2\sqrt{y}$

45. $\sqrt{15rs^2}\sqrt{10r} = \sqrt{150r^2s^2} = \sqrt{5^2r^2s^2 6} = 5rs\sqrt{6}$

47. $\sqrt[3]{3xy^2}\sqrt[3]{9x^3} = \sqrt[3]{27x^4y^2} = 3x\sqrt[3]{xy^2}$

49. $\sqrt{y^2(x+y)}\sqrt{(x+y)^3} = \sqrt{y^2(x+y)^4} =$
$y(x+y)^2$

51. $2\sqrt{7}(3\sqrt{7} - 1) = 2\sqrt{7}(3\sqrt{7}) - 2\sqrt{7} =$
$2 \cdot 3\sqrt{7}\sqrt{7} - 2\sqrt{7} = 6 \cdot 7 - 2\sqrt{7} = 42 - 2\sqrt{7}$

53. $-\sqrt{3}(\sqrt{7} - \sqrt{5}) = -\sqrt{3}\sqrt{7} + \sqrt{3}\sqrt{5} =$
$-\sqrt{21} + \sqrt{15}$

55. $3\sqrt{7t}(2\sqrt{7t} + 3\sqrt{3t^2}) = 3\sqrt{7t}(2\sqrt{7t}) +$
$3\sqrt{7t}(3\sqrt{3t^2}) = 6\sqrt{(7t)^2} + 9\sqrt{21t^3} =$
$6(7t) + 9t\sqrt{21t} = 42t + 9t\sqrt{21t}$

57. $(2\sqrt{3} + 1)(\sqrt{3} - 1) = 2(\sqrt{3})^2 + \sqrt{3} - 2\sqrt{3} - 1 =$
$2(3) - \sqrt{3} - 1 = 5 - \sqrt{3}$

59. $(7\sqrt{y} + 2)(3\sqrt{y} - 5) = 21(\sqrt{y})^2 + 6\sqrt{y}$
$-35\sqrt{y} - 10 = 21y - 29\sqrt{y} - 10$

61. $(\sqrt{3m} + \sqrt{2n})(\sqrt{3m} + \sqrt{2n}) =$
$(\sqrt{3m})^2 + 2(\sqrt{3m})(\sqrt{2n}) + (\sqrt{2n})^2 =$
$3m + 2\sqrt{6mn} + 2n$

63. $(5\sqrt{p} - \sqrt{3q})(\sqrt{p} + 2\sqrt{3q}) =$
$5(\sqrt{p})^2 - \sqrt{3q}\sqrt{p} + 10\sqrt{3q}\sqrt{p} - 2(\sqrt{3q})^2 =$
$5p + 9\sqrt{3qp} - 2(3q) = 5p + 9\sqrt{3qp} - 6q$

65. $\left(2\sqrt{3t} + 5\right)^2 = (2\sqrt{3t})^2 + 2(2\sqrt{3t})5 + 5^2 =$
$4(3t) + 20\sqrt{3t} + 25 = 12t + 20\sqrt{3t} + 25$

67. $3(\sqrt{5x} - \sqrt{3})^2 = 3[(\sqrt{5x})^2 - 2\sqrt{5x}\sqrt{3} + (\sqrt{3})^2]$
$= 3(5x - 2\sqrt{15x} + 3) = 15x - 6\sqrt{15x} + 9$

69. $\sqrt{\frac{5}{3}} = \frac{\sqrt{5}}{\sqrt{3}} \cdot \frac{\sqrt{3}}{\sqrt{3}} = \frac{\sqrt{15}}{3}$

71. $\frac{\sqrt{3}}{\sqrt{50}}\cdot\frac{\sqrt{2}}{\sqrt{2}}=\frac{\sqrt{6}}{\sqrt{100}}=\frac{\sqrt{6}}{10}$

73. $\frac{2}{\sqrt[3]{6}}\cdot\frac{\sqrt[3]{6^2}}{\sqrt[3]{6^2}}=\frac{2\sqrt[3]{36}}{\sqrt[3]{6^3}}=\frac{2\sqrt[3]{36}}{6}=\frac{\sqrt[3]{36}}{3}$

75. $\frac{\sqrt[3]{9}}{\sqrt[3]{54}}=\frac{\sqrt[3]{9}}{\sqrt[3]{9}\sqrt[3]{6}}=\frac{1}{\sqrt[3]{6}}\cdot\frac{\sqrt[3]{36}}{\sqrt[3]{36}}=\frac{\sqrt[3]{36}}{\sqrt[3]{216}}=\frac{\sqrt[3]{36}}{6}$

77. $\frac{\sqrt{9xy}}{\sqrt{3x^2y}}=\sqrt{\frac{9xy}{3x^2y}}=\sqrt{\frac{3}{x}}=\frac{\sqrt{3}}{\sqrt{x}}\cdot\frac{\sqrt{x}}{\sqrt{x}}=\frac{\sqrt{3x}}{x}$

79. $\frac{\sqrt{5ab^2c}}{\sqrt{10abc}}=\sqrt{\frac{5ab^2c}{10abc}}=\sqrt{\frac{b}{2}}=\frac{\sqrt{b}}{\sqrt{2}}\cdot\frac{\sqrt{2}}{\sqrt{2}}=\frac{\sqrt{2b}}{2}$

81. $\frac{1}{\sqrt[5]{2}}\cdot\frac{\sqrt[5]{2^4}}{\sqrt[5]{2^4}}=\frac{\sqrt[5]{16}}{\sqrt[5]{32}}=\frac{\sqrt[5]{16}}{2}$

83. $\frac{4}{\sqrt[4]{32}}\cdot\frac{\sqrt[4]{8}}{\sqrt[4]{8}}=\frac{4\sqrt[4]{8}}{\sqrt[4]{256}}=\frac{4\sqrt[4]{8}}{4}=\sqrt[4]{8}$

85. $\frac{3}{\sqrt{3}-1}\cdot\frac{\sqrt{3}+1}{\sqrt{3}+1}=\frac{3\left(\sqrt{3}+1\right)}{3-1}=\frac{3(\sqrt{3}+1)}{2}$

87. $\frac{\sqrt{2}-1}{\sqrt{2}+1}\cdot\frac{\sqrt{2}-1}{\sqrt{2}-1}=\frac{2-2\sqrt{2}+1}{2-1}=3-2\sqrt{2}$

89. $\frac{\sqrt{3}+\sqrt{2}}{\sqrt{3}-\sqrt{2}}\cdot\frac{\sqrt{3}+\sqrt{2}}{\sqrt{3}+\sqrt{2}}=\frac{3+\sqrt{6}+\sqrt{6}+2}{3-2}=$
$5+2\sqrt{6}$

91. $\frac{3}{\sqrt{x}-2}\cdot\frac{\sqrt{x}+2}{\sqrt{x}+2}=\frac{3(\sqrt{x}+2)}{x-4}$

93. $\frac{\sqrt{x}+\sqrt{y}}{\sqrt{x}-\sqrt{y}}\cdot\frac{\sqrt{x}+\sqrt{y}}{\sqrt{x}+\sqrt{y}}=\frac{x+2\sqrt{x}\sqrt{y}+y}{x-y}=$
$\frac{x+2\sqrt{xy}+y}{x-y}$

Review

95. a. $x^2+5x=6$

$x^2+5x-6=0$

$(x+6)(x-1)=0$

$x=-6$ or $x=1$

b. 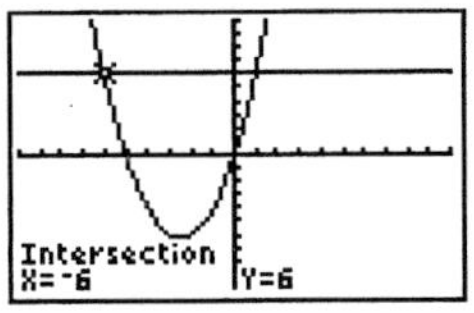

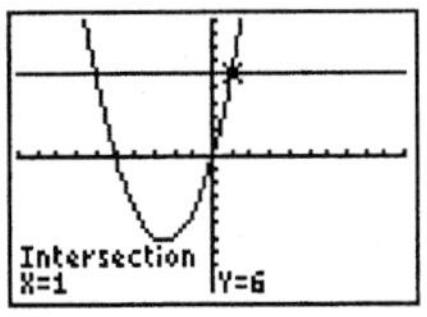

97. $2x-1>9$

$2x>10$

$x>5$

a. ◄-----o———► (5)   b. $(5,\infty)$

99. $x+1<0$ or $2x-6>0$

$x<-1$ or $x>3$

a. ◄———o-----o———► (-1, 3)   b. $(-\infty,-1)\cup(3,\infty)$

Section 9.4 Radical Equations and Their Applications

Vocabulary

1. radical; radicand
3. right
5. legs; hypotenuse

Concepts

7. a. iii   b. iv   c. v

d. ii e. i f. vi

Practice

9. $\sqrt{6x+1}+2=7$

$\sqrt{6x+1}=5$

$6x+1=5^2=25$

$6x=24$

$x=4$

11. $\sqrt[3]{7n-1}=3$

$7n-1=3^3=27$

$7n=28$

$n=4$

13. $\sqrt[4]{10p+1}=\sqrt[4]{11p-7}$ Raise each side to the fourth power. $10p+1=11p-7$

$8=p$

15. $r-9=(2r-3)^{\frac{1}{2}}$

$(r-9)^2=2r-3$

$r^2-18r+81=2r-3$

$r^2-20r+84=0$

$(r-14)(r-6)=0$

$r=14$ or $r=6$, but 6 is extraneous.

17. $(-5x+24)^{\frac{1}{2}}=6-x$

$-5x+24=(6-x)^2$

$-5x+24=36-12x+x^2$

$0=x^2-7x+12$

$0=(x-4)(x-3)$

$x=3$ or $x=4$

19. $\sqrt{x}\sqrt{x+16}=15$

$x(x+16)=15^2=225$

$x^2+16x-225=0$

$(x+25)(x-9)=0$

$x=-25$ or $x=9$, but -25 is extraneous.

21. $(x^3-7)^{\frac{1}{3}}=x-1$

$x^3-7=(x-1)^3$

$x^3-7=x^3-3x^2+3x-1$

$-7=-3x^2+3x-1$

$3x^2-3x-6=0$

$x^2-x-2=0$

$(x-2)(x+1)=0$

$x=2$ or $x=-1$

23. $\sqrt{2y+1}=1-2\sqrt{y}$ Square both sides.

$2y+1=(1-2\sqrt{y})^2$

$2y+1=1-4\sqrt{y}+4y$

$-2y=-4\sqrt{y}$ (square again)

$4y^2=16y$

$4(y^2-4y)$

$4y(y-4)=0$

$y=0$ or $y=4$, but 4 is extraneous.

25. $\sqrt{y+7}+3=\sqrt{y+4}$

$(\sqrt{y+7}+3)^2=y+4$

$(y+7)+2\cdot 3\sqrt{y+7}+9=y+4$

$y+16+6\sqrt{y+7}=y+4$

$6\sqrt{y+7}=-12$

$\sqrt{y+7}=-2$

There is no solution since $\sqrt{y+7}\ge 0$ always

27. $\sqrt{6x+1} - 3\sqrt{x} = -1$

$\sqrt{6x+1} = 3\sqrt{x} - 1$

$6x+1 = (3\sqrt{x} - 1)^2$

$6x+1 = 9x - 2 \cdot 3\sqrt{x} = 1$

$-3x = -6\sqrt{x}$

$x = 2\sqrt{x}$

$x^2 = 4x$

$x^2 - 4x = 0$

$x(x-4) = 0$

$x = 0$ or $x = 4$, but 0 is extraneous.

29. $\sqrt{x-1} + \sqrt{x+2} = 3$

$\sqrt{x-1} = 3 - \sqrt{x+2}$ Square both sides.

$x - 1 = (3 - \sqrt{x+2})^2$

$x - 1 = 9 - 2 \cdot 3\sqrt{x+2} + (x+2)$

$x - 1 = 11 + x - 6\sqrt{x+2}$

$-12 = -6\sqrt{x+2}$

$2 = \sqrt{x+2}$

$4 = x + 2$

$2 = x$

31. $\sqrt{\sqrt{x} + \sqrt{x+8}} = 2$ Square both sides.

$\sqrt{x} + \sqrt{x+8} = 4$

$\sqrt{x+8} = 4 - \sqrt{x}$ Square again.

$x + 8 = (4 - \sqrt{x})^2$

$x + 8 = 16 - 8\sqrt{x} + x$

$-8 = -8\sqrt{x}$

$1 = \sqrt{x}$

$1 = x$

33. $\dfrac{6}{\sqrt{x+5}} = \sqrt{x}$ Square both sides.

$\dfrac{36}{x+5} = x$

$36 = x(x+5) = x^2 + 5x$

$0 = x^2 + 5x - 36$

$0 = (x+9)(x-4)$

$x = -9$ or $x = 4$, but -9 is extraneous.

35. $\sqrt{x} + \sqrt{3x+2} + 4 = \sqrt{2x-5} - \sqrt{5x-1}$

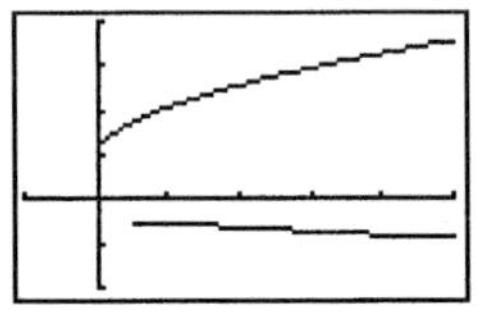

There is no solution.

37. $\sqrt{x} + \sqrt{4x} - 4 = \sqrt{6x-8} - \sqrt{2x-4}$

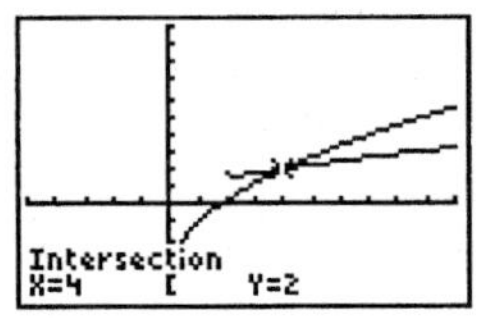

The solution is $x = 4$.

39. Both equations are solved by raising to an appropriate power, and then dividing. One equation is raised to the 2nd power, while the other is raised to the 3rd power.

41. $a = 10$ cm., $c = 26$ cm.

$10^2 + b^2 = 26^2$

$100 + b^2 = 676$

$b^2 = 576$

$b = 24$ cm.

43. $a = 14$ in., $c = 50$ in.

$14^2 + b^2 = 50^2$

$196 + b^2 = 2500$

$b^2 = 2304$

$b = 48$ in.

45. P(-5,-2), Q(7,3)

$d(PQ) = \sqrt{[7-(-5)]^2 + [3-(-2)]^2} =$

$\sqrt{12^2 + 5^2} = \sqrt{144+25} = \sqrt{169} = 13$

47. P(10,4), Q(2,-2)

$d(PQ) = \sqrt{(10-2)^2 + [4-(-2)]^2} = \sqrt{8^2 + 6^2}$

$= \sqrt{64+36} = \sqrt{100} = 10$

49. Q(2,-3), P(4,-8)

$d(PQ) = \sqrt{(2-4)^2 + [-3-(-8)]^2} =$

$\sqrt{(-2)^2 + 5^2} = \sqrt{4+25} = \sqrt{29}$

51. Let s = length of a side. Then $s^2 = 18$ so that $s = \sqrt{18} = 3\sqrt{2}$. The points B and C have coordinates (0,a) and (a,0) since they are the same distance from the origin (the diagonals of a square bisect each other) and on the y and x-axis respectively.

$$d(BC) = \sqrt{(0-a)^2 + (a-0)^2} = \sqrt{2a^2}$$

Then $3\sqrt{2} = s = \sqrt{2a^2}$

$9(2) = 2a^2$

$9 = a^2$

$3 = a$

Thus, B(0,3), C(3,0), A(-3,0) and D(0,-3).

53. The length d is the hypotenuse of a right triangle with legs 5 ft. and 12 ft. Thus, $d^2 = 5^2 + 12^2 = 169$, so that d = 13 ft.

55. The distance d is the hypotenuse of a right triangle with both legs 90 ft. Thus, $d^2 = 90^2 + 90^2 = 8100 + 8100 = 16{,}200$ and so $d = \sqrt{16{,}200} \approx 127$ ft.

57. The distance d is the hypotenuse of a right triangle whose legs are 30 ft. and 90 ft. Thus, $d^2 = 90^2 + 100^2 = 8100 + 10{,}000 = 18{,}100$ and so $d = \sqrt{18{,}100} \approx 135$ ft.

59. For a 12" x 24" x 17" box, the diagonal length $d = \sqrt{12^2 + 24^2 + 17^2} = \sqrt{1009} \approx 31.8$ inches. Thus, a 32" racquet will not fit.

61. For a 21" x 21" x 21" box, the diagonal length $d = \sqrt{3 \cdot 21^2} = \sqrt{1323} \approx 36.4"$. Thus, a 36" bone would fit.

63. The length AD can be thought of as the hypotenuse of a right triangle with legs 105 + 60 = 165 yd. (south to north) and 52 yd. (west to east). Thus $AD = \sqrt{165^2 + 52^2} = \sqrt{29{,}929} = 173$ yd.

65. The distance d from each post to the weight is the hypotenuse of a right triangle with legs 1 ft. and 20 ft. Thus, $d = \sqrt{1^2 + 20^2} = \sqrt{401}$. The total length of the stretched wire is $2\sqrt{401} \approx 40.05$. The amount of stretch is $40.05 - 40 = 0.05$ ft.

67. If A = 24, then $24 = 6\sqrt[3]{V^2}$

$4 = \sqrt[3]{V^2}$

$4^3 = 64 = V^2$

$8 = V$

The volume is 8 $cm.^3$

Review

69. function

71. relation only (fails the vertical line test)

Section 9.5 Radical Functions

Vocabulary

1. function

3. x; domain

Concepts

5. f(-1) = **1**

7. f(3) ≈ 3.606, to the nearest thousandth.

9. a.

11. b.

Practice

For #13 – #23,
$f(x)=\sqrt{2x-3},\ \ g(x)=\sqrt{3x-3}+2$

13. $f(2)=\sqrt{2\cdot 2-3}=\sqrt{4-3}=\sqrt{1}=1$

15. $g(10)=\sqrt{3\cdot 10-3}+2=\sqrt{27}+2=3\sqrt{3}+2$

17. $g(19)=\sqrt{3\cdot 19-3}+2=\sqrt{54}+2=3\sqrt{6}+2$

19. $g(-71)=\sqrt{3(-71)-3}+2=\sqrt{-216}+2$ which is undefined for real numbers.

21. $f(-1)=\sqrt{2(-1)-3}=\sqrt{-5}$ which is undefined for real numbers.

23. $f(14)=\sqrt{2\cdot 14-3}=\sqrt{25}=5$

25. $f(x)=\sqrt{x+2}$ domain: [-2,∞) range: [0,∞)

27. $f(x)=\sqrt{x}-5$ domain: [0,∞) range: [-5,∞)

29. $f(x)=2\sqrt{x+3}$ domain: [-3,∞) range: [0,∞)

31. $f(x)=\sqrt{5x-3}$ The domain needs 5x – 3 ≥0 or x ≥3/5. domain: $\left[\frac{3}{5},\infty\right)$ range: [0,∞)

33. $f(x)=\sqrt{3x-8}-2$ The domain needs 3x – 8 ≥0 or $x\geq\frac{8}{3}$. domain: $\left[\frac{8}{3},\infty\right)$ range: [-2,∞)

35. $f(x)=\sqrt{6x-5}-4$ The domain needs 6x – 5 ≥0 or x ≥5/6. domain: $\left[\frac{5}{6},\infty\right)$ range: [- 4,∞)

37. a.

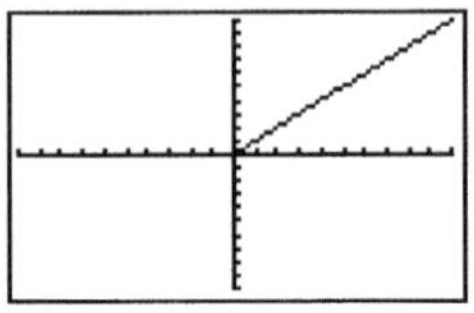

b.

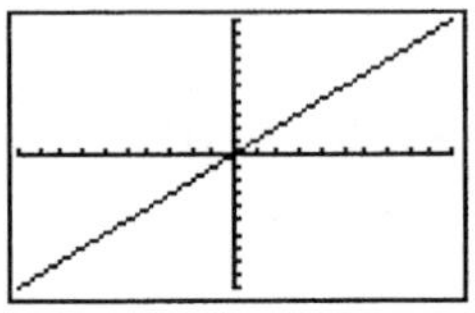

c. The graphs differ because $y=\sqrt{x}\sqrt{x}$ is defined only for x ≥0.

Applications

39. Given $d=1.4\sqrt{h}$. If d = 25, then

$$25=1.4\sqrt{h}$$
$$\sqrt{h}=\frac{25}{1.4}$$
$$h=\left(\frac{25}{1.4}\right)^2\approx 319$$

The tower should be 319 ft. tall.

41. Given $l=\sqrt{f^2+h^2}$. If l = 10' and h = 6', then $10=\sqrt{f^2+6^2}$

$$100=f^2+36$$
$$64=f^2$$
$$8=f$$

He should nail it 8' from the base of the wall.

43. Given $s=\sqrt{23x}$ and $d=\sqrt{312-2x^2}$. The equilibrium price x is where s = d. Thus,

$\sqrt{23x} = \sqrt{312 - 2x^2}$

$23x = 312 - 2x^2$

$2x^2 + 23x - 312 = 0$

$(x - 8)(2x + 39) = 0$

$x = 8 \text{ or } x = \dfrac{-39}{2}$

The equilibrium price is $8.

Review

45. $y = 3x^2 + 4x + 5$

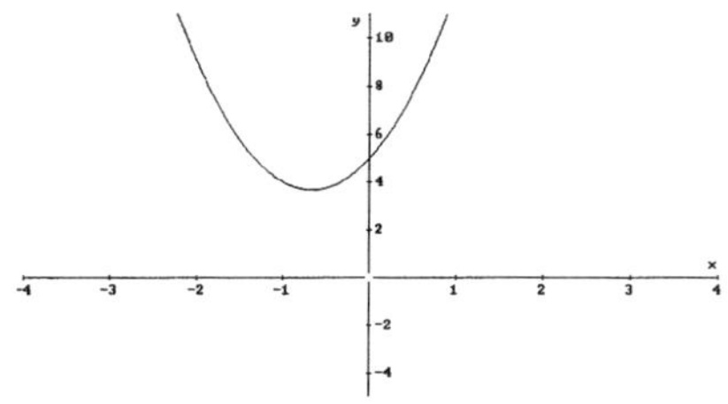

47. $y = \sqrt{x + 4}$

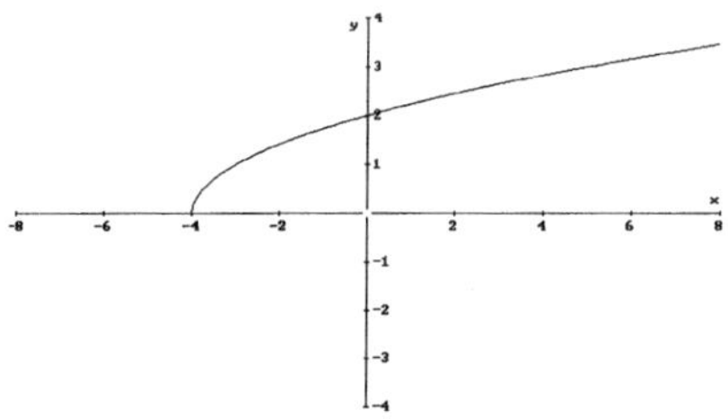

Section 9.6 $\sqrt{-1}$ and Complex Numbers

Vocabulary

1. **i**

3. real; imaginary

Concepts

5. $i^1 = \mathbf{i}$, $i^2 = \mathbf{-1}$, $i^3 = \mathbf{-i}$, $i^4 = \mathbf{1}$

7. $\sqrt{-36} = \sqrt{36}\sqrt{-1} = 6i$

9. $-\sqrt{-200} = -\sqrt{200}\sqrt{-1} = -10\sqrt{2}\, i$

11. $\sqrt{-18} = \sqrt{9 \cdot 2}\sqrt{-1} = 3\sqrt{2}\, i$

13. $-\sqrt{-32} = -\sqrt{16 \cdot 2}\sqrt{-1} = -4\sqrt{2}\, i$

15. $i^{19} = i^3 = \mathbf{-i}$, since 19 divided by 4 leaves a remainder of 3.

17. $i^{22} = i^2 = \mathbf{-1}$, since 22 divided by 4 leaves a remainder of 2.

19. T $\sqrt{4} + \sqrt{25}\, i = 2 + 5i = 2 - (-5i)$

21. T $-4i^2 + 7i^3 = (-4)(-1) + 7(-i) = 4 - 7i$

23. F $\sqrt{-9} - i = 3i - i = 2i$

Practice

25. $(5 + 3i) - (6 - 9i) = 5 + 3i - 6 + 9i = (5 - 6) + (3i + 9i) = -1 + 12i$

27. $(8 + 3i) + (-7 - 2i) = 8 + 3i - 7 - 2i = (8 - 7) + (3i - 2i) = 1 + i$

29. $(-9 + i) - 5i + (2 + 7i) = -9 + i - 5i + 2 + 7i = (-9 + 2) + (i - 5i + 7i) = -7 + 3i$

31. $(8 + 7i) - (-7 - \sqrt{-64}\ ) + (3 - i) = 8 + 7i + 7 + 8i + 3 - i = (8 + 7 + 3) + (7i + 8i - i) = 18 + 14i$

33. $(2 + 2\sqrt{2}\, i) + (-3 - \sqrt{2}\, i) =$

$2 + 2\sqrt{2}\, i - 3 - \sqrt{2}\, i = (2 - 3) + (2\sqrt{2}\, i - \sqrt{2}\, i)$

$= -1 + \sqrt{2}\, i$

35. $-4i(3 + 4i) = -12i - 16i^2 = -12i - 16(-1) = -12i + 16 = 16 - 12i$

37. $(3-2i)(4-3i) = 12-8i-9i+6i^2 = 12-17i + 6(-1) = 12-17i-6 = 6-17i$

39. $(5+\sqrt{3}\,i)(2-\sqrt{3}\,i) = 10+2\sqrt{3}\,i-5\sqrt{3}\,i-3i^2$
$10-3\sqrt{3}\,i-3(-1) = 13-3\sqrt{3}\,i$

41. $(-1+\sqrt{-4})(2+\sqrt{-9}) = (-1+2i)(2+3i) =$
$-2+4i-3i+6i^2 = -2+i+6(-1) = -8+i$

43. $(3-2i)^2 = 3^2-2(3)(2i)+(2i)^2 = 9-12i+4i^2 = 9-12i+4(-1) = 5-12i$

45. $i(-3-2i)(1-2i) = i(-3-2i+6i+4i^2) = i(-3+4i+4i^2) = -3i+4i^2+4i^3 = -3i+4(-1)+4(-i) = -4-7i$

47. $(3+2i)(3-2i)(1+i) = [3^2-(2i)^2](1+i) = (9-4i^2)(1+i) = [9-4(-1)](1+i) = 13(1+i) = 13+13i$

49. $\dfrac{3}{2i} = \dfrac{3}{2i}\cdot\dfrac{i}{i} = \dfrac{3i}{2i^2} = \dfrac{3i}{2(-1)} = -\dfrac{3}{2}i$

51. $\dfrac{5i^3}{2\sqrt{-4}} = \dfrac{5i^3}{2(2i)} = \dfrac{5i^3}{4i} = \dfrac{5i^2}{4} = \dfrac{5(-1)}{4} = -\dfrac{5}{4}$

53. $\dfrac{-4}{6i^7} = \dfrac{-4}{6i^7}\cdot\dfrac{i}{i} = \dfrac{-4i}{6i^8} = \dfrac{-4i}{6(1)} = -\dfrac{2}{3}i$

55. $\dfrac{26}{3-2i} = \dfrac{26}{3-2i}\cdot\dfrac{3+2i}{3+2i} = \dfrac{26(3+2i)}{3^2-(2i)^2} =$
$\dfrac{26(3+2i)}{9-4i^2} = \dfrac{26(3+2i)}{9-(-4)} = \dfrac{26(3+2i)}{13} =$
$2(3+2i) = 6+4i$

57. $\dfrac{2i}{5+3i} = \dfrac{2i}{5+3i}\cdot\dfrac{5-3i}{5-3i} = \dfrac{2i(5-3i)}{5^2-(3i)^2} =$
$\dfrac{2i(5-3i)}{25-(-9)} = \dfrac{2i(5-3i)}{34} = \dfrac{i(5-3i)}{17} = \dfrac{5i-3i^2}{17} =$
$\dfrac{5i-3(-1)}{17} = \dfrac{3+5i}{17} = \dfrac{3}{17}+\dfrac{5}{17}i$

59. $\dfrac{4}{3+\sqrt{-1}} = \dfrac{4}{3+i}\cdot\dfrac{3-i}{3-i} = \dfrac{4(3-i)}{3^2-i^2} = \dfrac{4(3-i)}{9-(-1)} =$
$\dfrac{4(3-i)}{10} = \dfrac{2(3-i)}{5} = \dfrac{6-2i}{5} = \dfrac{6}{5}-\dfrac{2}{5}i$

61. $\dfrac{-4i}{2-6i} = \dfrac{-4i}{2-6i}\cdot\dfrac{2+6i}{2+6i} = \dfrac{-4i(2+6i)}{2^2-(6i)^2} =$
$\dfrac{-8i-24i^2}{4-(-36)} = \dfrac{-8i-24(-1)}{40} = \dfrac{24-8i}{40} = \dfrac{8(3-i)}{40}$
$= \dfrac{3-i}{5} = \dfrac{3}{5}-\dfrac{1}{5}i$

63. $\dfrac{2+3i}{2-3i}\cdot\dfrac{2+3i}{2+3i} = \dfrac{2^2+2\cdot 2\cdot 3i+(3i)^2}{2^2-(3i)^2} =$
$\dfrac{4+12i+9i^2}{4-9i^2} = \dfrac{4+12i-9}{4-(-9)} = \dfrac{-5+12i}{13} =$
$\dfrac{-5}{13}+\dfrac{12}{13}i$

65. $\dfrac{\sqrt{3}+\sqrt{2}\,i}{\sqrt{3}-\sqrt{2}\,i}\cdot\dfrac{\sqrt{3}+\sqrt{2}\,i}{\sqrt{3}+\sqrt{2}\,i} = \dfrac{3+2\sqrt{3}\sqrt{2}\,i+2i^2}{3-2i^2} =$
$\dfrac{3+2\sqrt{6}\,i-2}{3-2(-1)} = \dfrac{1+2\sqrt{6}\,i}{5} = \dfrac{1}{5}+\dfrac{2\sqrt{6}}{5}i$

67. a. Let $x = 3-2i$. Then $x^2 = (3-2i)^2 = 9-12i+4i^2 = 9-12i-4 = 5-12i$. And $6x-13 = 6(3-2i)-13 = 18-12i-13 = 5-12i$.
Let $x = 3+2i$. Then $x^2 = (3+2i)^2 = 9+12i+4i^2 = 9+12i-4 = 5+12i$. And $6x-13 = 6(3+2i)-13 = 18+12i-13 = 5+12i$.

b. 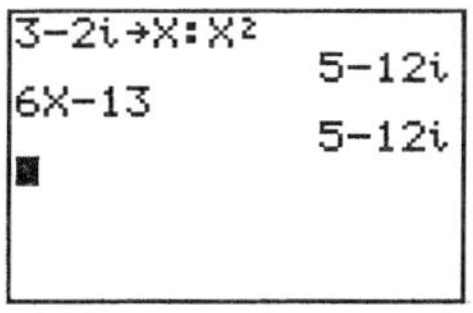

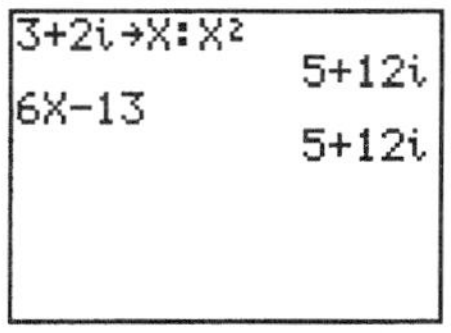

69. a. Let $x = 2 + i$. Then $(2 + i)^2 + (2 + i) + 1 =$ $4 + 4i + i^2 + 2 + i + 1 = 7 + 5i + (-1) =$ $6 + 5i \neq 0$
Let $x = 2 - i$. Then $(2 - i)^2 + (2 - i) + 1 =$ $4 - 4i + i^2 + 2 - i + 1 = 7 - 5i + (-1) =$ $6 - 5i \neq 0$.

b. 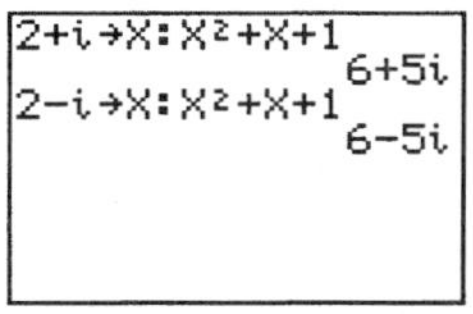

Review

71. $(3x^2 + 5x - 2) - (8x^2 - 2x + 4) = 3x^2 + 5x - 2 - 8x^2 + 2x - 4 = 3x^2 - 8x^2 + 5x + 2x - 2 - 4 = -5x^2 + 7x - 6$

73. $(3x + 2)(5x - 8) = (3x)(5x) + 2(5x) - 8(3x) + 2(-8) = 15x^2 + 10x - 24x - 16 = 15x^2 - 14x - 16$

75. $\dfrac{x^2 - 1}{x - 1} = \dfrac{(x+1)(x-1)}{x - 1} = x + 1$

Chapter 9 Summary

1. a. $\sqrt{49} = \sqrt{7^2} = 7$

b. $-\sqrt{121} = -\sqrt{11^2} = -11$

c. $-\sqrt{36} = -\sqrt{6^2} = -6$

d. $\sqrt{225} = \sqrt{15^2} = 15$

e. $\sqrt[3]{-27} = \sqrt[3]{(-3)^3} = -3$

f. $-\sqrt[3]{216} = -\sqrt[3]{6^3} = -6$

g. $\sqrt[4]{625} = \sqrt[4]{5^4} = 5$

h. $\sqrt[5]{-32} = \sqrt[5]{(-2)^5} = -2$

2. a. $\sqrt{25x^2} = \sqrt{(5x)^2} = 5x$

b. $\sqrt{x^2 + 4x + 4} = \sqrt{(x+2)^2} = x + 2$

c. $\sqrt[3]{27a^6b^2} = \sqrt[3]{(3a^2)^3b^2} = 3a^2\sqrt[3]{b^2}$

d. $\sqrt[4]{256a^8y^4} = \sqrt[4]{(4a^2y)^4} = 4a^2y$

3. a. $(-8)^{\frac{1}{3}} = \sqrt[3]{-8} = -2$

b. $-8^{\frac{2}{3}} = -\left(\sqrt[3]{8}\right)^2 = -2^2 = -4$

c. $\left(\dfrac{4}{9}\right)^{\frac{3}{2}} = \left(\sqrt{\left(\dfrac{2}{3}\right)^2}\right)^3 = \left(\dfrac{2}{3}\right)^3 = \dfrac{2^3}{3^3} = \dfrac{8}{27}$

d. $\dfrac{1}{25^{\frac{5}{2}}} = \dfrac{1}{\left(\sqrt{25}\right)^5} = \dfrac{1}{5^5} = \dfrac{1}{3125}$

e. $\left(\dfrac{4}{9}\right)^{-\frac{3}{2}} = \left(\sqrt{\dfrac{4}{9}}\right)^{-3} = \left(\dfrac{2}{3}\right)^{-3} = \dfrac{1}{\left(\dfrac{2}{3}\right)^3} = \dfrac{1}{\dfrac{8}{27}}$
$= \dfrac{27}{8}$

f. $\left(27x^3y\right)^{\frac{1}{3}} = \sqrt[3]{27x^3y} = \sqrt[3]{(3x)^3y} = 3x\sqrt[3]{y} =$
$3xy^{\frac{1}{3}}$

g. $\left(81x^4y^2\right)^{\frac{1}{4}} = \sqrt[4]{81x^4y^2} = \sqrt[4]{(3x)^4y^2} =$

$3x\sqrt[4]{y^2} = 3xy^{\frac{2}{4}} = 3xy^{\frac{1}{2}}$

h. $\left(25x^3y^4\right)^{\frac{3}{2}} = \left(\sqrt{25x^3y^4}\right)^3 = \left(5xy^2\sqrt{x}\right)^3 =$

$125x^3y^6x^{\frac{3}{2}} = 125x^{\frac{9}{2}}y^6$

i. $\left(8u^2v^3\right)^{-\frac{2}{3}} = \dfrac{1}{\left(8u^2v^3\right)^{\frac{2}{3}}} = \dfrac{1}{\left(\sqrt[3]{8u^2v^3}\right)^2} =$

$\dfrac{1}{\left(2v\sqrt[3]{u^2}\right)^2} = \dfrac{1}{4v^2(u^{\frac{2}{3}})^2} = \dfrac{1}{4v^2u^{\frac{4}{3}}}$

4. a. $\sqrt{240} = \sqrt{16\cdot 15} = \sqrt{16}\sqrt{15} = 4\sqrt{15}$

b. $\sqrt[3]{54} = \sqrt[3]{27\cdot 2} = \sqrt[3]{27}\sqrt[3]{2} = 3\sqrt[3]{2}$

c. $\sqrt[4]{32} = \sqrt[4]{16\cdot 2} = \sqrt[4]{16}\sqrt[4]{2} = 2\sqrt[4]{2}$

d. $\sqrt[5]{96} = \sqrt[5]{32\cdot 3} = \sqrt[5]{32}\sqrt[5]{3} = 2\sqrt[5]{3}$

e. $\sqrt[3]{16x^5y^4} = \sqrt[3]{8x^3y^3}\sqrt[3]{2x^2y} = 2xy\sqrt[3]{2x^2y}$

f. $\sqrt[3]{54x^7y^3} = \sqrt[3]{27x^6y^3}\sqrt[3]{2x} = 3x^2y\sqrt[3]{2x}$

g. $\dfrac{\sqrt{32x^3}}{\sqrt{2x}} = \sqrt{\dfrac{32x^3}{2x}} = \sqrt{16x^2} = 4x$

h. $\dfrac{\sqrt[3]{16x^5}}{\sqrt[3]{2x^2}} = \sqrt[3]{\dfrac{16x^5}{2x^2}} = \sqrt[3]{8x^3} = 2x$

i. $\sqrt[3]{\dfrac{2a^2b}{27x^3}} = \dfrac{\sqrt[3]{2a^2b}}{\sqrt[3]{27x^3}} = \dfrac{\sqrt[3]{2a^2b}}{3x}$

j. $\sqrt{\dfrac{17xy}{64a^4}} = \dfrac{\sqrt{17xy}}{\sqrt{64a^4}} = \dfrac{\sqrt{17xy}}{8a^2}$

5. a. $2\sqrt[3]{3} - \sqrt[3]{24} = 2\sqrt[3]{3} - \sqrt[3]{8\cdot 3} = 2\sqrt[3]{3} - 2\sqrt[3]{3} = 0$

b. $\sqrt[4]{32} + 2\sqrt[4]{162} = \sqrt[4]{16\cdot 2} + 2\sqrt[4]{81\cdot 2} =$

$2\sqrt[4]{2} + 2\cdot 3\sqrt[4]{2} = (2+6)\sqrt[4]{2} = 8\sqrt[4]{2}$

c. $2x\sqrt{8} + 2\sqrt{200x^2} + \sqrt{50x^2} =$

$2x\sqrt{4\cdot 2} + 2\sqrt{100x^2\cdot 2} + \sqrt{25x^2\cdot 2} =$

$2x\cdot 2\sqrt{2} + 2\cdot 10x\sqrt{2} + 5x\sqrt{2} =$

$(4x + 20x + 5x)\sqrt{2} = 29x\sqrt{2}$

d. $\sqrt[3]{54} - 3\sqrt[3]{16} + 4\sqrt[3]{128} =$

$\sqrt[3]{27\cdot 2} - 3\sqrt[3]{8\cdot 2} + 4\sqrt[3]{64\cdot 2} =$

$3\sqrt[3]{2} - 3\cdot 2\sqrt[3]{2} + 4\cdot 4\sqrt[3]{2} = (3 - 6 + 16)\sqrt[3]{2} =$

$13\sqrt[3]{2}$

6. a. $\sqrt[3]{3}\sqrt[3]{9} = \sqrt[3]{3\cdot 9} = \sqrt[3]{27} = 3$

b. $-\sqrt[3]{2x^2}\sqrt[3]{4x} = -\sqrt[3]{(2x^2)(4x)} = -\sqrt[3]{8x^3} =$

$-2x$

c. $\sqrt{2}(\sqrt{8} - 3) = \sqrt{2}\sqrt{8} - 3\sqrt{2} = \sqrt{16} - 3\sqrt{2} =$

$4 - 3\sqrt{2}$

d. $\sqrt{5}(\sqrt{2} - 1) = \sqrt{5}\sqrt{2} - \sqrt{5} = \sqrt{10} - \sqrt{5}$

e. $\left(\sqrt{2}+1\right)\left(\sqrt{2}-1\right) = \left(\sqrt{2}\right)^2 - 1^2 = 2 - 1 = 1$

f. $(2\sqrt{u} + 3)(3\sqrt{u} - 4) = 6u + 9\sqrt{u} - 8\sqrt{u} - 12$

$= 6u - 12 + \sqrt{u}$

7. a. $\dfrac{1}{\sqrt{3}}\cdot\dfrac{\sqrt{3}}{\sqrt{3}} = \dfrac{\sqrt{3}}{3}$

b. $\dfrac{\sqrt[3]{uv}}{\sqrt[3]{u^5v^7}}\cdot\dfrac{\sqrt[3]{uv^2}}{\sqrt[3]{uv^2}} = \dfrac{\sqrt[3]{u^2v^3}}{\sqrt[3]{u^6v^9}} = \dfrac{v\sqrt[3]{u^2}}{u^2v^3} = \dfrac{\sqrt[3]{u^2}}{u^2v^2}$

c. $\dfrac{2}{\sqrt{2}-1}\cdot\dfrac{\sqrt{2}+1}{\sqrt{2}+1}=\dfrac{2\left(\sqrt{2}+1\right)}{\left(\sqrt{2}\right)^2-1^2}=\dfrac{2\sqrt{2}+2}{2-1}=$

$2\sqrt{2}+2$

d. $\dfrac{2x-32}{\sqrt{x}+4}\cdot\dfrac{\sqrt{x}-4}{\sqrt{x}-4}=\dfrac{2(x-16)\left(\sqrt{x}-4\right)}{\left(\sqrt{x}\right)^2-4^2}=$

$\dfrac{2(x-16)\left(\sqrt{x}-4\right)}{x-16}=2\left(\sqrt{x}-4\right)=2\sqrt{x}-8$

8. a. $\sqrt{x+3}=\sqrt{2x-19}$ Square both sides.

$x+3=2x-19$

$22=x$

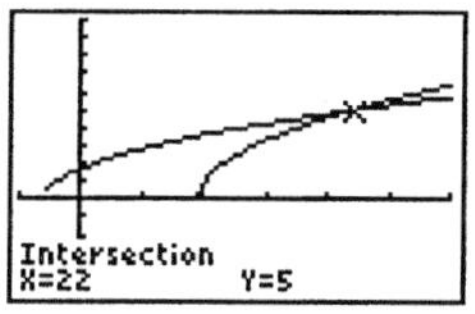

b. $x=\sqrt{12x-27}$ Square both sides.

$x^2=12x-27$

$x^2-12x+27=0$

$(x-9)(x-3)=0$

$x=9$ or $x=3$

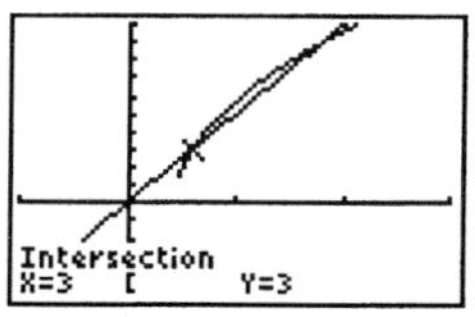

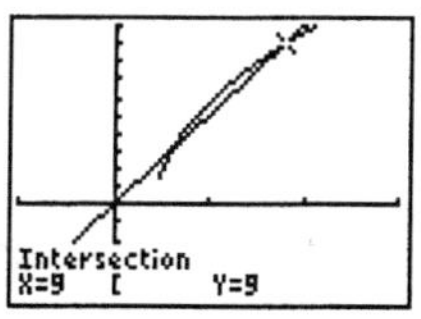

c. $\sqrt{x+1}+\sqrt{x}=2$

$\sqrt{x+1}=2-\sqrt{x}$ Square both sides.

$x+1=\left(2-\sqrt{x}\right)^2$

$x+1=4-4\sqrt{x}+x$

$-3=-4\sqrt{x}$ Square again.

$9=16x$

$x=\dfrac{9}{16}$

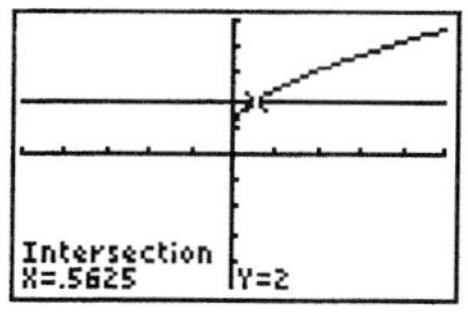

d. $\sqrt[3]{x^3+8}=x+2$ Cube both sides.

$x^3+8=(x+2)^3$

$x^3+8=x^3+6x^2+12x+8$

$0=6x^2+12x$

$0=6x(x+2)$

$x=0$ or $x=-2$

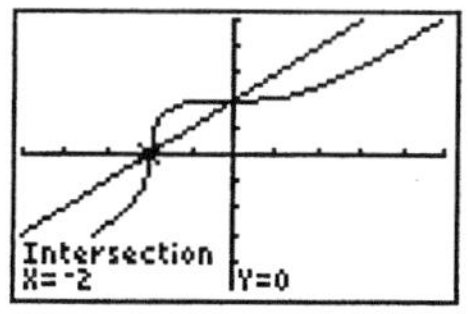

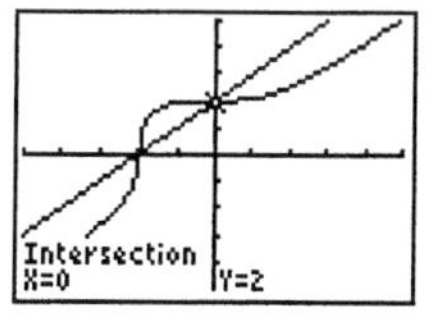

9. Notice that $\dfrac{d}{2}$ is one leg of a right triangle whose other leg is 117 yd. and whose hypotenuse is 125 yd. Thus,

$\left(\dfrac{d}{2}\right)^2+117^2=125^2$

$\dfrac{d^2}{4}+13{,}689=15{,}625$

$\dfrac{d^2}{4}=1936$

$d^2=7744$

$d=88$

The boat advances 88 yd. into the wind.

10. P(-2,-7), Q(4,1)

$d(PQ)=\sqrt{(-2-4)^2+(-7-1)^2}=$

$\sqrt{(-6)^2+(-8)^2}=\sqrt{36+64}=\sqrt{100}=10$

11. $f(x)=\sqrt{2x+4}-1$, $g(x)=\sqrt[3]{2x}$

a. $f(6)=\sqrt{2\cdot6+4}-1=\sqrt{16}-1=4-1=3$

b. $g(4)=\sqrt[3]{2\cdot4}=\sqrt[3]{8}=2$

c. $g(32) = \sqrt[3]{2 \cdot 32} = \sqrt[3]{64} = 4$

d. $f(3) = \sqrt{2 \cdot 3 + 4} - 1 = \sqrt{10} - 1 \approx 2.16$

e. $g(8) = \sqrt[3]{2 \cdot 8} = \sqrt[3]{16} \approx 2.52$

12. a. $f(x) = \sqrt{x - 5} - 4$
domain = $[5,\infty)$ range = $[-4,\infty)$

b. $f(x) = \sqrt{2x + 4}$
domain = $[-2,\infty)$ range = $[0,\infty)$

c. $f(x) = \sqrt[3]{x + 2}$
domain = $\mathbb{R}$ range = $\mathbb{R}$

d. $f(x) = \sqrt[3]{x + 4} - 3$
domain = $\mathbb{R}$ range = $\mathbb{R}$

13. Given $d = 1.4\sqrt{h}$. If h = 4.7 ft., then $d = 1.4\sqrt{4.7} \approx 3.0$ The horizon is approximately 3.0 miles.

14. Given $d = 1.4\sqrt{h}$. If d = 4, then $4 = 1.4\sqrt{h}$ and so

$$4 = 1.4\sqrt{h}$$
$$\frac{4}{1.4} = \sqrt{h}$$
$$\left(\frac{4}{1.4}\right)^2 = h$$
$$h \approx 8.2$$

The periscope must extend approximately 8.2 ft. above the water.

15. a. $(5 + 4i) + (7 - 12i) = 5 + 4i + 7 - 12i =$
$(5 + 7) + (4i - 12i) = 12 - 8i$.

b. $(-6 - 40i) - (-8 + 28i) = -6 - 40i + 8 - 28i$
$= (-6 + 8) - 40i - 28i = 2 - 68i$.

c. $(-32 + \sqrt{-144}) - (64 + \sqrt{-81}) =$
$(-32 + 12i) - (64 + 9i) = -32 + 12i - 64 - 9i$
$= -96 + 3i$

d. $(-8 + \sqrt{-8}) + (6 - \sqrt{-32}) = (-8 + 2\sqrt{2}\,i) +$
$(6 - 4\sqrt{2}\,i) = -8 + 2\sqrt{2}\,i + 6 - 4\sqrt{2}\,i =$
$-2 - 2\sqrt{2}\,i$

16. a. $(2 - 7i)(-3 + 4i) = -6 + 21i + 8i - 28i^2 =$
$-6 + 29i - 28(-1) = -6 + 29i + 28 = 22 + 29i$

b. $(-5 + 6i)(2 + i) = -10 + 12i - 5i + 6i^2 =$
$-10 + 7i + 6(-1) = -10 + 7i - 6 = -16 + 7i$

c. $(5 - \sqrt{-27})(-6 + \sqrt{-12}) =$
$(5 - 3\sqrt{3}\,i)(-6 + 2\sqrt{3}\,i) =$
$-30 + 18\sqrt{3}\,i + 10\sqrt{3}\,i - (3\sqrt{3})(2\sqrt{3})\,i^2 =$
$-30 + 28\sqrt{3}\,i - 6(3)(-1) = -12 + 28\sqrt{3}\,i$

d. $(2 + \sqrt{-128})(3 - \sqrt{-98}) =$
$(2 + 8\sqrt{2}\,i)(3 - 7\sqrt{2}\,i) =$
$6 + 24\sqrt{2}\,i - 14\sqrt{2}\,i - (8\sqrt{2})(7\sqrt{2})i^2 =$
$6 + 10\sqrt{2}\,i - 56(2)(-1) = 6 + 10\sqrt{2}\,i + 112 =$
$118 + 10\sqrt{2}\,i$

17. a. $\frac{3}{4i} = \frac{3}{4i} \cdot \frac{i}{i} = \frac{3i}{4i^2} = \frac{3i}{4(-1)} = -\frac{3}{4}i$

b. $\frac{-2}{5i^3} = \frac{-2}{5i^3} \cdot \frac{i}{i} = \frac{-2i}{5i^4} = \frac{-2i}{5 \cdot 1} = -\frac{2}{5}i$

c. $\frac{6}{2+i} = \frac{6}{2+i} \cdot \frac{2-i}{2-i} = \frac{6(2-i)}{2^2 - i^2} = \frac{6(2-i)}{4-(-1)} =$
$\frac{12-6i}{5} = \frac{12}{5} - \frac{6}{5}i$

d. $\frac{7}{3-i} = \frac{7}{3-i} \cdot \frac{3+i}{3+i} = \frac{7(3+i)}{3^2 - i^2} = \frac{21+7i}{9-(-1)} =$
$\frac{21+7i}{10} = \frac{21}{10} + \frac{7}{10}i$

e. $\dfrac{3}{5+\sqrt{-4}} = \dfrac{3}{5+2i}\cdot\dfrac{5-2i}{5-2i} = \dfrac{3(5-2i)}{5^2-(2i)^2} =$

$\dfrac{15-6i}{25-4i^2} = \dfrac{15-6i}{25-(-4)} = \dfrac{15-6i}{29} = \dfrac{15}{29} - \dfrac{6}{29}i$

f. $\dfrac{2}{3-\sqrt{-9}} = \dfrac{2}{3-3i}\cdot\dfrac{3+3i}{3+3i} = \dfrac{2(3+3i)}{3^2-(3i)^2} =$

$\dfrac{6+6i}{9-9i^2} = \dfrac{6+6i}{9-9(-1)} = \dfrac{6(1+i)}{18} = \dfrac{1+i}{3} =$

$\dfrac{1}{3} + \dfrac{1}{3}i$

Chapter 9 Test

1. $\sqrt{250x^3y^5} = \sqrt{(5xy^2)^2(10xy)} = 5xy^2\sqrt{10xy}$

2. $\dfrac{\sqrt[3]{24x^{15}y^4}}{\sqrt[3]{y}} = \sqrt[3]{\dfrac{24x^{15}y^4}{y}} = \sqrt[3]{24x^{15}y^3} =$

$\sqrt[3]{(2x^5y)^3\cdot 3} = 2x^5y\sqrt[3]{3}$

3. $\sqrt{\dfrac{3a^5}{48a^7}} = \sqrt{\dfrac{1}{16a^2}} = \sqrt{\left(\dfrac{1}{4a}\right)^2} = \dfrac{1}{4a}$

4. $\sqrt[3]{27x^3} = \sqrt[3]{(3x)^3} = 3x$

5. $\sqrt{18x^4y^9} = \sqrt{(3x^2y^4)^2(2y)} = 3x^2y^4\sqrt{2y}$

6. a. $16^{\frac{1}{4}} = \sqrt[4]{16} = 2$

b. $36^{-\frac{3}{2}} = \dfrac{1}{36^{\frac{3}{2}}} = \dfrac{1}{\left(\sqrt{36}\right)^3} = \dfrac{1}{6^3} = \dfrac{1}{216}$

c. $\left(\dfrac{-8}{27}\right)^{-\frac{2}{3}} = \left(\dfrac{-27}{8}\right)^{\frac{2}{3}} = \left(\sqrt[3]{\dfrac{-27}{8}}\right)^2 = \left(\dfrac{-3}{2}\right)^2 = \dfrac{9}{4}$

7. $\sqrt{12} - \sqrt{27} = 2\sqrt{3} - 3\sqrt{3} = (2-3)\sqrt{3} = -\sqrt{3}$

8. $2\sqrt[3]{40} - \sqrt[3]{5000} + 4\sqrt[3]{625} =$

$2\sqrt[3]{8\cdot 5} - \sqrt[3]{1000\cdot 5} + 4\sqrt[3]{125\cdot 5} =$

$2\cdot 3\sqrt[3]{5} - 10\sqrt[3]{5} + 4\cdot 5\sqrt[3]{5} = (6-10+20)\sqrt[3]{5} =$

$16\sqrt[3]{5}$

9. $2\sqrt{48y^5} - 3y\sqrt{12y^3} =$

$2\sqrt{(4y^2)^2(3y)} - 3y\sqrt{(2y)^2(3y)} =$

$2\cdot 4y^2\sqrt{3y} - 3y\cdot 2y\sqrt{3y} = (8y^2-6y^2)\sqrt{3y} =$

$2y^2\sqrt{3y}$

10. $2\sqrt{xy}(3\sqrt{x} + \sqrt{xy^3}) = 2\sqrt{xy}(3\sqrt{x}) +$

$2\sqrt{xy}(\sqrt{xy^3}) = 6\sqrt{x^2y} + 2\sqrt{x^2y^4} =$

$6x\sqrt{y} + 2xy^2$

11. a. $\dfrac{1}{\sqrt{5}}\cdot\dfrac{\sqrt{5}}{\sqrt{5}} = \dfrac{\sqrt{5}}{5}$

b. $\dfrac{6}{\sqrt[3]{9}}\cdot\dfrac{\sqrt[3]{3}}{\sqrt[3]{3}} = \dfrac{6\sqrt[3]{3}}{\sqrt[3]{27}} = \dfrac{6\sqrt[3]{3}}{3} = 2\sqrt[3]{3}$

c. $\dfrac{-4\sqrt{2}}{\sqrt{5}+3}\cdot\dfrac{\sqrt{5}-3}{\sqrt{5}-3} = \dfrac{-4\sqrt{2}(\sqrt{5}-3)}{5-9} =$

$\dfrac{-4\sqrt{10}+12\sqrt{2}}{-4} = \dfrac{-4(\sqrt{10}+3\sqrt{2})}{-4} =$

$\sqrt{10} + 3\sqrt{2}$

d. $\dfrac{3x-1}{\sqrt{3x}-1}\cdot\dfrac{\sqrt{3x}+1}{\sqrt{3x}+1} = \dfrac{(3x-1)(\sqrt{3x}+1)}{3x-1} =$

$\sqrt{3x} + 1$

12. $\sqrt[3]{6x+4} - 4 = 0$

$\sqrt[3]{6x+4} = 4$ Cube both sides.

$6x + 4 = 4^3 = 64$

$6x = 60$

$x = 10$

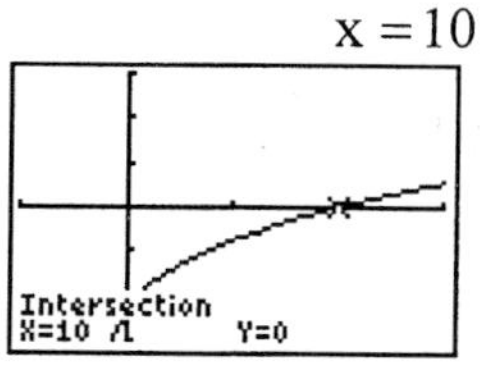

13. $1-\sqrt{x}=\sqrt{x-3}$ Square both sides.

$\left(1-\sqrt{x}\right)^2 = x-3$

$1-2\sqrt{x}+x = x-3$

$-2\sqrt{x} = -4$

$\sqrt{x} = 2$

x = 4, but 4 is extraneous. No solution.

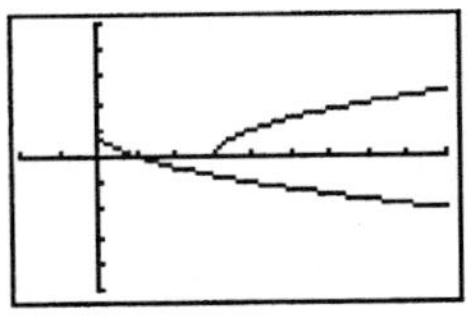

14. P(-2,5), Q(22,12)

$d(PQ) = \sqrt{(-2-22)^2+(5-12)^2} =$

$\sqrt{(-24)^2+(-7)^2} = \sqrt{576+49} = \sqrt{625} = 25$

15. The height h is the leg of a right triangle whose other leg is 45 in. and whose hypotenuse is 53 in. Thus, h will satisfy

$h^2+45^2 = 53^2$

$h^2+2025 = 2809$

$h^2 = 784$

$h = 28$

The height of the crate is 28 in.

16. $f(x) = 2\sqrt{x}+1$

a. $f(3) = 2\sqrt{3}+1 \approx 4.464$

b. $f(9) = 2\sqrt{9}+1 = 2\cdot 3+1 = 7$

c. $f(-20) = 2\sqrt{-20}+1 = 2\cdot 2\sqrt{5}\,i+1 =$

$4i\sqrt{5}+1$

d. $f(13) = 2\sqrt{13}+1 \approx 8.211$

17. a. $f(x) = \sqrt{x-6}+2$

domain = $[6,\infty)$ range = $[2,\infty)$

b. $f(x) = -\sqrt{x}$

domain = $[0,\infty)$ range = $(-\infty, 0]$

18. a. (2 + 4i) + (-3 + 7i) = 2 + (-3) + 4i + 7i =

2 + (-3) + (4 + 7)i = -1 + 11i

b. $(3-\sqrt{-9})-(-1+\sqrt{-16}) =$

$(3-3i)-(-1+4i) = 3-3i+1-4i = 4-7i$

19. a. $2i(3-4i) = 2i(3)-2i(4i) = 6i-8i^2 =$

$6i-(-8) = 8+6i$

b. $(3+2i)(-4-i) = -12-8i-3i-2i^2 =$

$-12-11i-2(-1) = -12-11i+2 = -10-11i$

20. a. $\frac{1}{i\sqrt{2}}\cdot\frac{i\sqrt{2}}{i\sqrt{2}} = \frac{i\sqrt{2}}{2i^2} = \frac{i\sqrt{2}}{2(-1)} = -\frac{\sqrt{2}}{2}i$

b. $\frac{2+i}{3-i} = \frac{2+i}{3-i}\cdot\frac{3+i}{3+i} = \frac{6+3i+2i+i^2}{3^2-i^2} =$

$\frac{6+5i-1}{9-(-1)} = \frac{5+5i}{10} = \frac{5(1+i)}{10} = \frac{1+i}{2} = \frac{1}{2}+\frac{1}{2}i$

Chapter 10 Quadratics

Section 10.1 Square Root Property and Completing the Square

Vocabulary

1. $ax^2 + bx + c = 0$; a

3. roots (or solutions)

5. STOre

Concepts

7. a. imaginary b. real
   c. real d. imaginary

9. The equations a. $x^2 = 49$, c. $x^2 = 6$, d. $x^2 = 8$ have real solutions since they involve square roots of positive numbers.

11. $x^2 = 25$, $x = \pm\sqrt{25} = \pm 5$

13. $x^2 = 100$, $x = \pm\sqrt{100} = \pm 10$

15. a. $x^2 + 10x + \mathbf{25} =$ b. $(x + 5)^2$
    (You add the square of 10/2).

17. a. $x^2 - 6x + \mathbf{9} =$ b. $(x - 3)^2$
    (You add the square of -6/2).

19. a. $x^2 - 5x + \mathbf{\frac{25}{4}} =$ b. $(x - \frac{5}{2})^2$
    (You add the square of -5/2).

Practice

21. $3x^2 - 16 = 0$

$$x^2 = \frac{16}{3}$$

$$x = \pm\sqrt{\frac{16}{3}} = \pm\frac{4}{\sqrt{3}} = \pm\frac{4\sqrt{3}}{3}$$

23. $(x+1)^2 = 27$

$$x + 1 = \pm\sqrt{27} = \pm 3\sqrt{3}$$

$$x = -1 \pm 3\sqrt{3}$$

25. $(x+3)^2 = -18$

$$x + 3 = \pm\sqrt{-18}$$

$$x + 3 = \pm 3i\sqrt{2}$$

$$x = -3 \pm 3i\sqrt{2}$$

27. $(2x+3)^2 = 40$

$$2x + 3 = \pm\sqrt{40} = \pm 2\sqrt{10}$$

$$2x = -3 \pm 2\sqrt{10}$$

$$x = \frac{-3}{2} \pm \sqrt{10}$$

29. $(3x+1)^2 = -13$

$$3x + 1 = \pm\sqrt{-13} = \pm i\sqrt{13}$$

$$3x = -1 \pm i\sqrt{13}$$

$$x = -\frac{1}{3} \pm \frac{\sqrt{13}}{3}i$$

31. $x^2 + 2x - 8 = 0$

$$x^2 + 2x = 8$$

$$x^2 + 2x + 1 = 8 + 1 = 9$$

$$(x+1)^2 = 9$$

$$x + 1 = \pm 3$$

$$x = -1 \pm 3 = 2, -4$$

33. $2x^2 - x = -1$

$$x^2 - \frac{1}{2}x = -\frac{1}{2}$$

$$x^2 - \frac{1}{2}x + \left(\frac{1}{4}\right)^2 = -\frac{1}{2} + \left(\frac{1}{4}\right)^2 = -\frac{8}{16} + \frac{1}{16}$$

$$\left(x - \frac{1}{4}\right)^2 = -\frac{7}{16}$$

$$x - \frac{1}{4} = \pm\sqrt{\frac{-7}{16}} = \pm\frac{\sqrt{7}}{4}i$$

$$x = \frac{1}{4} \pm \frac{\sqrt{7}}{4}i$$

35. $3x^2 - 11x = -10$

$$x^2 - \frac{11}{3}x = \frac{-10}{3}$$

$$x^2 - \frac{11}{3}x + \left(\frac{11}{6}\right)^2 = -\frac{10}{3} + \left(\frac{11}{6}\right)^2$$

$$\left(x - \frac{11}{6}\right)^2 = \frac{-120}{36} + \frac{121}{36} = \frac{1}{36}$$

$$x - \frac{11}{6} = \pm\sqrt{\frac{1}{36}} = \pm\frac{1}{6}$$

$$x = \frac{11}{6} \pm \frac{1}{6} = 2, \frac{5}{3}$$

37. $5x^2 = -3x - 2$

$$5x^2 + 3x = -2$$

$$x^2 + \frac{3}{5}x = -\frac{2}{5}$$

$$x^2 + \frac{3}{5}x + \left(\frac{3}{10}\right)^2 = -\frac{2}{5} + \left(\frac{3}{10}\right)^2$$

$$\left(x + \frac{3}{10}\right)^2 = -\frac{40}{100} + \frac{9}{100} = -\frac{31}{100}$$

$$x + \frac{3}{10} = \pm\sqrt{\frac{-31}{100}} = \pm\frac{i\sqrt{31}}{10}$$

$$x = -\frac{3}{10} \pm \frac{\sqrt{31}}{10}i$$

39. $5x^2 + 2x = 6$

$$x^2 + \frac{2}{5}x = \frac{6}{5}$$

$$x^2 + \frac{2}{5}x + \left(\frac{1}{5}\right)^2 = \frac{6}{5} + \left(\frac{1}{5}\right)^2$$

$$\left(x + \frac{1}{5}\right)^2 = \frac{30}{25} + \frac{1}{25} = \frac{31}{25}$$

$$x + \frac{1}{5} = \pm\sqrt{\frac{31}{25}} = \pm\frac{\sqrt{31}}{5}$$

$$x = -\frac{1}{5} \pm \frac{\sqrt{31}}{5}$$

41. a. and b. Written opinions, answers may vary

43. a. $x^2 - x - 12 = 0$

$$(x - 4)(x + 3) = 0$$

$$x - 4 = 0 \text{ or } x + 3 = 0$$

$$x = 4 \text{ or } x = -3$$

b. $x^2 - x - 12 = 0$

$$x^2 - x = 12$$

$$x^2 - x + \left(\frac{1}{2}\right)^2 = 12 + \left(\frac{1}{2}\right)^2$$

$$\left(x - \frac{1}{2}\right)^2 = 12 + \frac{1}{4} = \frac{49}{4}$$

$$x - \frac{1}{2} = \pm\sqrt{\frac{49}{4}} = \pm\frac{7}{2}$$

$$x = \frac{1}{2} \pm \frac{7}{2} = 4, -3$$

c.

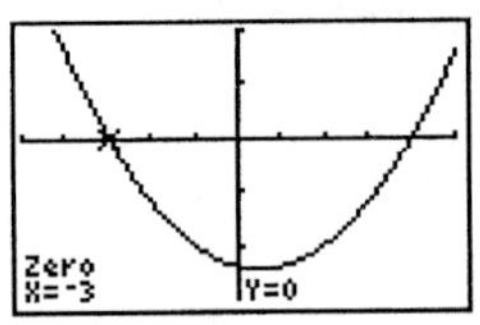

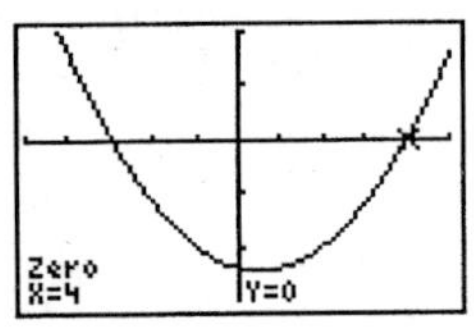

45. Roots: - 4, 6 Factors: x + 4, x – 6
Equation: $(x + 4)(x - 6) = x^2 - 2x - 24 = 0$

47. Roots: 3, -3, 4, - 4
Factors: $x - 3$, $x + 3$, $x - 4$ and $x + 4$
Equation: $(x - 3)(x + 3)(x - 4)(x + 4) = 0$ or
$x^4 - 25x^2 + 144 = 0$

Applications

49. Let x = width, so x + 4 = length. Then

$$x(x+4)=96$$
$$x^2+4x=96$$
$$x^2+4x-96=0$$
$$(x+12)(x-8)=0$$
$$x=-12 \text{ or } x=8$$

The shed is 8' by 8 + 4 = 12'.

51. a. The last row will have these entries.
fare: .25 + .05x, Number of passengers: 3000 – 80x, Daily revenue: (.25 + .05x)·(3000 – 80x).
b. For a daily revenue of $994,

$$(.25+.05x)(3000-80x)=994$$
$$750+150x-20x-4x^2=994$$
$$-4x^2+130x-244=0$$
$$2x^2-65x+122=0$$
$$(2x-61)(x-2)=0$$
$$x=2 \quad \text{(30.5 increases is not possible)}$$

c. Fare = .25 + 2(.05) = $.35

53. $$1000(1+r)^2+2000(1+r)=3368.10$$
$$(1+r)^2+2(1+r)=3.3681$$
$$1+2r+r^2+2+2r=3.3681$$
$$r^2+4r+3=3.3681$$
$$r^2+4r=.3681$$
$$r^2+4r+4=4.3681$$
$$(r+2)^2=4.3681$$
$$r+2=\pm\sqrt{4.3681}$$
$$r=-2\pm\sqrt{4.3681}=-2\pm 2.09=.09,-4.09$$

The rate r is .09 or 9%.

Review

55. a. x-intercept: (4,0) b. y-intercept: (0, -2)
c. $m=\frac{\Delta y}{\Delta x}=\frac{0-(-2)}{4-0}=\frac{1}{2}$ Using slope-intercept form, the equation of the line is $y=\frac{1}{2}x-2$.

57. linear function

Section 10.2 The Quadratic Formula and the Discriminant

Vocabulary

1. $\dfrac{-b\pm\sqrt{b^2-4ac}}{2a}$

3. approximate

5. rational

7. rational; double

Concepts

9. The x-coordinates are -3 and 4.

11. The solutions, or roots, of $x^2 - x - 12 = 0$ are the same as the x-coordinates of the x-intercepts of $y = x^2 - x - 12$.

Practice

13. $x^2 - 3x + 2 = 0$; a = 1, b = -3, c = 2

$$x=\frac{-(-3)\pm\sqrt{(-3)^2-4\cdot 1\cdot 2}}{2\cdot 1}$$
$$x=\frac{3\pm\sqrt{9-8}}{2}$$
$$x=\frac{3\pm 1}{2}=2 \text{ or } 1$$

15. $x^2 + 16 = 0$; $a = 1, b = 0, c = 16$

$$x = \frac{0 \pm \sqrt{0^2 - 4 \cdot 1 \cdot 16}}{2 \cdot 1}$$

$$x = \frac{\pm\sqrt{-64}}{2} = \frac{\pm 8i}{2} = \pm 4i$$

17. $y^2 - 18y = -81$ or $y^2 - 18y + 81 = 0$
$a = 1, b = -18, c = 81$

$$x = \frac{-(-18) \pm \sqrt{(-18)^2 - 4 \cdot 1 \cdot 81}}{2 \cdot 1}$$

$$x = \frac{18 \pm \sqrt{0}}{2} = 9, \text{ a double root}$$

19. $2x^2 = -25$ or $2x^2 + 25 = 0$; $a = 2, b = 0, c = 25$

$$x = \frac{0 \pm \sqrt{0^2 - 4 \cdot 2 \cdot 25}}{2 \cdot 2}$$

$$x = \frac{\pm\sqrt{-200}}{4} = \frac{\pm 10\sqrt{2}}{4} = \frac{\pm 5\sqrt{2}}{2}i$$

21. $x^2 + 3x + 3 = 0$; $a = 1, b = 3, c = 3$

$$x = \frac{-3 \pm \sqrt{3^2 - 4 \cdot 1 \cdot 3}}{2 \cdot 1}$$

$$x = \frac{-3 \pm \sqrt{-3}}{2} = \frac{-3 \pm \sqrt{3}\, i}{2} = -\frac{3}{2} \pm \frac{\sqrt{3}}{2}i$$

23. $4w^2 + 6w + 1 = 0$; $a = 4, b = 6, c = 1$

$$w = \frac{-6 \pm \sqrt{6^2 - 4 \cdot 4 \cdot 1}}{2 \cdot 4}$$

$$w = \frac{-6 \pm \sqrt{20}}{8} = \frac{-6 \pm 2\sqrt{5}}{8} = \frac{-3 \pm \sqrt{5}}{4}$$

25. $3x^2 + 2x + 1 = 0$; $a = 3, b = 2, c = 1$

$$x = \frac{-2 \pm \sqrt{2^2 - 4 \cdot 3 \cdot 1}}{2 \cdot 3}$$

$$x = \frac{-2 \pm \sqrt{-8}}{6} = \frac{-2 \pm 2i\sqrt{2}}{6} = \frac{2(-1 \pm i\sqrt{2})}{2 \cdot 3}$$

$$x = \frac{-1 \pm i\sqrt{2}}{3} = -\frac{1}{3} \pm \frac{\sqrt{2}}{3}i$$

27. $16x^2 + 16x + 3 = 0$; $a = 16, b = 16, c = 3$

$$x = \frac{-16 \pm \sqrt{16^2 - 4 \cdot 16 \cdot 3}}{2 \cdot 16}$$

$$x = \frac{-16 \pm \sqrt{64}}{32} = \frac{-16 \pm 8}{32} = -\frac{3}{4} \text{ or } -\frac{1}{4}$$

29. $2x^2 + 3x = -3$ or $2x^2 + 3x + 3 = 0$
$a = 2, b = 3, c = 3$

$$x = \frac{-3 \pm \sqrt{3^2 - 4 \cdot 2 \cdot 3}}{2 \cdot 2}$$

$$x = \frac{-3 \pm \sqrt{-15}}{4} = \frac{-3 \pm i\sqrt{15}}{4} = -\frac{3}{4} \pm \frac{\sqrt{15}}{4}i$$

31. $-3x = \frac{x^2}{2} + 2$ or $-6x = x^2 + 4$, $0 = x^2 + 6x + 4$
$a = 1, b = 6, c = 4$

$$x = \frac{-6 \pm \sqrt{6^2 - 4 \cdot 1 \cdot 4}}{2 \cdot 1}$$

$$x = \frac{-6 \pm \sqrt{20}}{2} = \frac{-6 \pm 2\sqrt{5}}{2} = \frac{2(-3 \pm \sqrt{5})}{2}$$

$$x = -3 \pm \sqrt{5}$$

33. $5x^2 = 2x - 1$ or $5x^2 - 2x + 1 = 0$
$a = 5, b = -2, c = 1$

$$x = \frac{-(-2) \pm \sqrt{(-2)^2 - 4 \cdot 5 \cdot 1}}{2 \cdot 5}$$

$$x = \frac{2 \pm \sqrt{-16}}{10} = \frac{2 \pm 4i}{10} = \frac{1}{5} \pm \frac{2}{5}i$$

35. $4t + 3 = 4t^2$ or $4t^2 - 4t - 3 = 0$
$a = 4, b = -4, c = -3$

$$t = \frac{-(-4) \pm \sqrt{(-4)^2 - 4 \cdot 4 \cdot -3}}{2 \cdot 4}$$

$$t = \frac{4 \pm \sqrt{64}}{8} = \frac{4 \pm 8}{8} = \frac{3}{2} \text{ or } -\frac{1}{2}$$

37. $5x^2 + x + 2 = 0$; $a = 5$, $b = 1$, $c = 2$
$b^2 - 4ac = 1^2 - 4 \cdot 5 \cdot 2 = -39 < 0$
Two imaginary(complex conjugate) solutions

39. $x(2x - 3) = 20$ or $2x^2 - 3x - 20 = 0$
$a = 2$, $b = -3$, $c = -20$
$b^2 - 4ac = (-3)^2 - 4(2)(-20) = 169 = 13^2$
Two distinct, rational solutions.

41. $3x^2 + 10x - 2 = 0$; $a = 3$, $b = 10$, $c = -2$
$b^2 - 4ac = 10^2 - 4(3)(-2) = 124 > 0$
Two distinct, irrational solutions.

43. $x(x - 3) = -10$ or $x^2 - 3x + 10 = 0$
$a = 1$, $b = -3$, $c = 10$
$b^2 - 4ac = (-3)^2 - 4(1)(10) = -31 < 0$
Two complex conjugate solutions

45. $9x^2 + 4 = -kx$ or $9x^2 + kx + 4 = 0$
$a = 9$, $b = k$, $c = 4$
$b^2 - 4ac = k^2 - 4 \cdot 9 \cdot 4 = k^2 - 144$ For equal roots the discriminant must be 0. $k^2 - 144 = 0$ gives $k^2 = 144$ or $k = \pm 12$

47. $(k + 4)x^2 + 2kx + 9 = 0$
$a = k + 4$, $b = 2k$, $c = 9$
$b^2 - 4ac = (2k)^2 - 4(k + 4)9 = 4k^2 - 36(k + 4)$ $= 4k^2 - 36k - 144$. For equal roots the discriminant must be 0. $4k^2 - 36k - 144 = 4(k^2 - 9k - 36) = 4(k - 12)(k + 3) = 0$ gives $k = 12$ or $k = -3$

49. $9x^2 - kx + 25 = 0$; $a = 9$, $b = -k$, $c = 25$
$b^2 - 4ac = (-k)^2 - 4 \cdot 9 \cdot 25 = k^2 - 900$ For equal roots the discriminant must be 0. $k^2 - 900 = 0$ gives $k^2 = 900$ or $k = \pm 30$.

51. $(k + 15)x^2 + (k - 30)x + 4 = 0$
$a = k + 15$, $b = k - 30$, $c = 4$
$b^2 - 4ac = (k - 30)^2 - 4(k + 15) \cdot 4 = (k - 30)^2 - 16(k + 15) = k^2 - 60k + 900 - 16k - 240 = k^2 - 76k + 660 = 0$ For equal roots the discriminant must be 0. $k^2 - 76k + 660 = (k - 66)(k - 10) = 0$ gives $k = 66$ or 10.

53. $1776x^2 - 1492x + 1984 = 0$
$a = 1776$, $b = -1492$, $c = 1984$
$b^2 - 4ac = (-1492)^2 - 4(1776)(1984) < 0$
Thus, the roots are not real.

55. $kx^2 - 4x = 7$ or $kx^2 - 4x - 7 = 0$
$a = k$, $b = -4$, $c = -7$
$b^2 - 4ac = (-4)^2 - 4k(-7) = 16 + 28k$. For imaginary roots the discriminant must be less than 0. $16 + 28k < 0$ gives $k < \frac{-16}{28} = \frac{-4}{7}$.

57. $3.4x^2 + 10.2x = 16.5$ $\{-4.17, 1.17\}$
Use [-10,10] by [-10,25] and intersect.

59. $\frac{3x}{x-1} - \frac{x}{x+2} = 5$ $\{-1.52, 2.19\}$
Use [-9.4, 9.4] by [-6.2 6.2] and intersect.

61. $(x + 1)(x + 3) = 2x^2 + 3x$ $\{-1.30, 2.30\}$
Use [-5, 5] by [-10, 30] and intersect.

63. $.4x + \frac{6}{x} = 6$ $\{1.08, 13.92\}$
Use [-5, 15] by [-10,10] and intersect.

65. $x^2 + 2ix - 5 = 0$; $a = 1$, $b = 2i$, $c = -5$

$$x = \frac{-2i \pm \sqrt{(2i)^2 - 4 \cdot 1 \cdot -5}}{2 \cdot 1}$$

$$x = \frac{-2i \pm \sqrt{4i^2 + 20}}{2} = \frac{-2i \pm \sqrt{-4 + 20}}{2}$$

$$x = \frac{-2i \pm \sqrt{16}}{2} = \frac{-2i \pm 4}{2} = -i \pm 2$$

67. $12x^2 - 5x - 2 = 0$ or $(3x - 2)(4x + 1) = 0$
$a = 12$, $b = -5$, $c = -2$
Solutions are $\frac{2}{3}$ and $-\frac{1}{4}$. Then the sum of the solutions is $\frac{2}{3} + \left(-\frac{1}{4}\right) = \frac{8}{12} - \frac{3}{12} = \frac{5}{12} = -\frac{-5}{12} = -\frac{b}{a}$. The product of the solutions is

$\frac{2}{3}\cdot\frac{-1}{4}=\frac{-2}{12}=\frac{c}{a}$.

69. $2x^2 + 5x + 1 = 0$; $a = 2$, $b = 5$, $c = 1$

$x=\frac{-5\pm\sqrt{5^2-4\cdot2\cdot1}}{2\cdot2}=\frac{-5\pm\sqrt{17}}{4}$. The sum of the solutions is $\frac{-5+\sqrt{17}}{4}+\frac{-5-\sqrt{17}}{4}=$ $\frac{-10}{4}=-\frac{5}{2}=-\frac{b}{a}$. The product of the solutions is $\frac{-5+\sqrt{17}}{4}\cdot\frac{-5-\sqrt{17}}{4}=\frac{(-5)^2-17}{16}$ $=\frac{8}{16}=\frac{1}{2}=\frac{c}{a}$.

71. a. $x^2+2\sqrt{3}\,x+2=0$

$x^2+2\sqrt{3}\,x=-2$

$x^2+2\sqrt{3}\,x+(\sqrt{3})^2=-2+3=1$

$(x+\sqrt{3})^2=1$

$x+\sqrt{3}=\pm1,\ x=-\sqrt{3}\pm1$

b. $a = 1$, $b = 2\sqrt{3}$, $c = 2$

$b^2 - 4ac = \left(2\sqrt{3}\right)^2 - 4(1)(2) = 12 - 8 = 4 = 2^2$

c. You might expect rational roots.
d. There is no contradiction since the coefficients must all be rational for a perfect square discriminant to guarantee rational roots.

Review

73. $f(x) = x^2 + 5x - 6$

a.

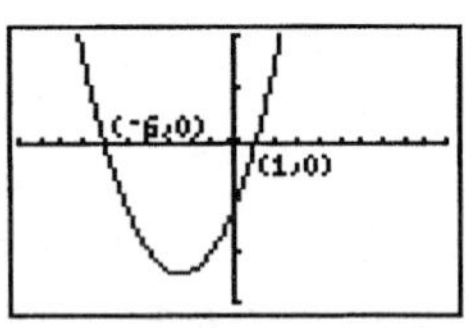

b. The x-intercepts are (-6,0) and (1,0).

c. $x^2+5x-6=0$

$(x+6)(x-1)=0$

$x=-6$ or $x=1$

d. The solutions in c. are the x-coordinates of the x-intercepts.

75. $y = |3x + 6| - 2$

a.

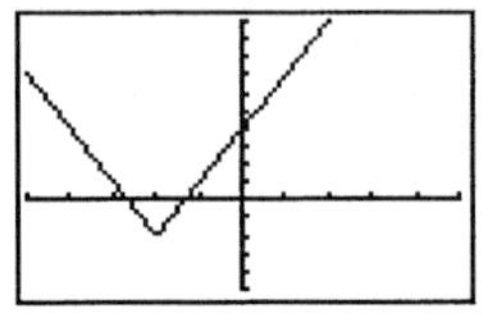

b. The x-intercepts are $\left(\frac{-4}{3},0\right)$ and $\left(\frac{-8}{3},0\right)$.

c. $|3x+6|-2=0$

$|3x+6|=2$

$3x+6=2$ or $3x+6=-2$

$3x=-4$ or $3x=-8$

$x=\frac{-4}{3}$ or $x=\frac{-8}{3}$

d. The solutions in c. are the x-coordinates of the x-intercepts.

Section 10.3 Applications of Quadratic Functions and Solution of Equations Using Quadratic Methods

Vocabulary

1. x-intercepts

3. vertex; highest; lowest

Concepts

From the graph of $h = -16t^2 + 1454$:

5. (4, 1198)

7. (4, 1198)

9. After 6 seconds, the ball is **878** feet from the ground.

Practice

11. $f(x) = -x^2 + 8x - 16$
   a. (4, 0)
   b. (0, -16)
   c. (4, 0)
   d.

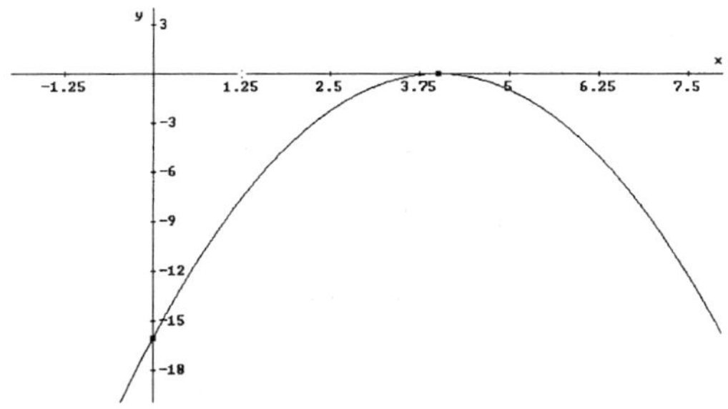

   e. $f(2) = -(2^2) + 8(2) - 16 = -4$
   f. If $f(x) = -1$, $-x^2 + 8x - 16 = -1$ or $x^2 - 8x + 15 = 0$. Thus, $(x - 3)(x - 5) = 0$ so that $x = 3$ or $x = 5$.

13. From the table, if $A \approx 78.5$, then $r = 5$. As an ordered pair this is (5, 78.5) and, using function notation, $f(5) = 78.5$.

15. The area and radius of a circle are both positive quantities.

For Exercises 17 – 19, $y = -16x^2 + 750x$

17. The rocket will attain a maximum height of approximately 8789 ft. after approximately 23 seconds.

19. After 20 sec., the rocket is at a height of 8600 ft. You might use 'value', STOre, or TABLE.

Applications

21. $y = 48x - 16x^2$
   a. The graph in [0,5] by [0,50] is

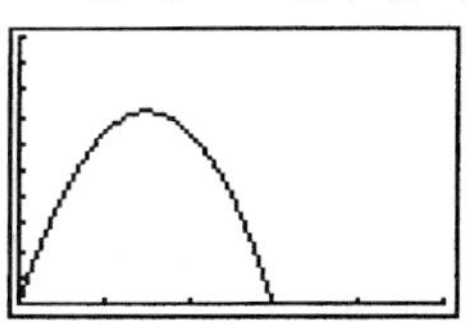

   b. Using 'maximum', the maximum height of the ball is 36 ft.

   c. It reaches this maximum height in 1.5 sec.

   d. The ball is on the ground when $y = 0$. $48x - 16x^2 = 16x(3 - x) = 0$ gives $x = 0, 3$. The ball reaches the ground in 3 sec.

23. Let x = length of each side perpendicular to the river. Then $1000 - 2x$ = length of the side parallel to the river. Area = $x(1000 - 2x) = 1000x - 2x^2$. Graphing $y = 1000x - 2x^2$ in [0,600] by [0, 150,000]and using 'maximum' gives a maximum area of 125,000 ft.^{2} when the field is 250' by 500'.

Practice

25. $x^4 - 10x^2 + 9 = 0$ Let $u = x^2$.

$$u^2 - 10u + 9 = 0$$
$$(u - 9)(u - 1) = 0$$
$$u = 9 \text{ or } u = 1$$
$$x^2 = 9 \text{ or } x^2 = 1 \quad \text{so} \quad x = \pm 3 \text{ or } x = \pm 1$$

27. $x^4 - 29x^2 = -100$ or $x^4 - 29x^2 + 100 = 0$ Let $u = x^2$.

$$u^2 - 29u + 100 = 0$$
$$(u - 25)(u - 4) = 0$$
$$u = 25 \text{ or } u = 4$$
$$x^2 = 25 \text{ or } x^2 = 4 \text{ so } x = \pm 5 \text{ or } x = \pm 2$$

29. $x^4 = 8x^2 - 7$ or $x^4 - 8x^2 + 7 = 0$ Let $u = x^2$.

$$u^2 - 8u + 7 = 0$$
$$(u - 7)(u - 1) = 0$$
$$u = 7 \text{ or } u = 1$$
$$x^2 = 7 \text{ or } x^2 = 1 \text{ so } x = \pm\sqrt{7} \text{ or } x = \pm 1$$

31. $2x^4 + 24 = 26x^2$ or $x^4 - 13x^2 + 12 = 0$ Let $u = x^2$.

$$u^2 - 13u + 12 = 0$$
$$(u - 12)(u - 1) = 0$$
$$u = 12 \text{ or } u = 1$$
$$x^2 = 12 \text{ or } x^2 = 1 \text{ so } x = \pm 2\sqrt{3} \text{ or } x = \pm 1$$

33. $2x - x^{\frac{1}{2}} - 1 = 0$ Let $u = x^{\frac{1}{2}}$.

$2u^2 - u - 1 = 0 \quad (2u + 1)(u - 1) = 0$

$u = -\frac{1}{2}$ or $u = 1$

$x^{\frac{1}{2}} = -\frac{1}{2}$ (no solution) or $x^{\frac{1}{2}} = 1$, so $x = 1$.

35. $3x - 4x^{\frac{1}{2}} + 1 = 0$ Let $u = x^{\frac{1}{2}}$.

$3u^2 - 4u + 1 = 0 \quad (3u - 1)(u - 1) = 0$

$u = \frac{1}{3}$ or $u = 1$

$x^{\frac{1}{2}} = \frac{1}{3}$ or $x^{\frac{1}{2}} = 1$ so $x = \left(\frac{1}{3}\right)^2 = \frac{1}{9}$ or $x = 1$.

37. $x^{\frac{2}{3}} - 7x^{\frac{1}{3}} + 12 = 0$ Let $u = x^{\frac{1}{3}}$.

$u^2 - 7u + 12 = 0 \quad (u - 4)(u - 3) = 0$

$u = 4$ or $u = 3$

$x^{\frac{1}{3}} = 4$ or $x^{\frac{1}{3}} = 3$, $x = 4^3 = 64$ or $x = 3^3 = 27$.

39. $x^{\frac{2}{3}} + 4x^{\frac{1}{3}} - 5 = 0$ Let $u = x^{\frac{1}{3}}$.

$u^2 + 4u - 5 = 0 \quad (u + 5)(u - 1) = 0$

$u = -5$ or $u = 1$

$x^{\frac{1}{3}} = -5$ or $x^{\frac{1}{3}} = 1$

$x = (-5)^3 = -125$ or $x = 1^3 = 1$

41. $x - 4 + \frac{3}{x} = 0$ Multiply both sides by x.

$x^2 - 4x + 3 = 0 \quad (x - 3)(x - 1) = 0$

$x = 3$ or $x = 1$

43. $x + \frac{15}{x} = 8$ Multiply both sides by x.

$x^2 + 15 = 8x$ or $x^2 - 8x + 15 = 0$

$(x - 5)(x - 3) = 0$ so $x = 5$ or $x = 3$.

45. $\frac{6}{x - 2} - \frac{12}{x - 1} = -1$ LCD = $(x - 2)(x - 1)$

Multiply both sides by the LCD.

$6(x - 1) - 12(x - 2) = -(x - 1)(x - 2)$

$6x - 6 - 12x + 24 = -(x^2 - 3x + 2)$

$-6x + 18 = -x^2 + 3x - 2$

$x^2 - 9x + 20 = 0 \quad (x - 4)(x - 5) = 0$

$x = 4$ or $x = 5$

47. $\frac{3}{x} + \frac{4}{x + 1} = 2$ LCD = $x(x + 1)$

Multiply both sides by the LCD.

$3(x + 1) + 4x = 2x(x + 1)$

$3x + 3 + 4x = 2x^2 + 2x$

$0 = 2x^2 - 5x - 3 \quad 0 = (2x + 1)(x - 3)$

$x = -\frac{1}{2}$ or $x = 3$

49. $4x^{-4} + 1 = 5x^{-2}$ Multiply both sides by x^4.

$4 + x^4 = 5x^2$ or $x^4 - 5x^2 + 4$. Let $u = x^2$.

$u^2 - 5u + 4 = 0 \quad (u - 4)(u - 1) = 0$

$u = 4$ or $u = 1$

$x^2 = 4$ or $x^2 = 1$ so $x = \pm 2$ or $x = \pm 1$

51. $x + \frac{x + 5}{x - 3} = 0$ Multiply both sides by $x - 3$.

$x(x - 3) + x + 5 = 0$

$x^2 - 3x + x = -5$ or $x^2 - 2x = -5$

$x^2 - 2x + 1 = -5 + 1 = -4$

$(x - 1)^2 = -4$, $x - 1 = \pm\sqrt{-4} = \pm 2i$, $x = 1 \pm 2i$

Review

53. $y = ½\, x - 3$

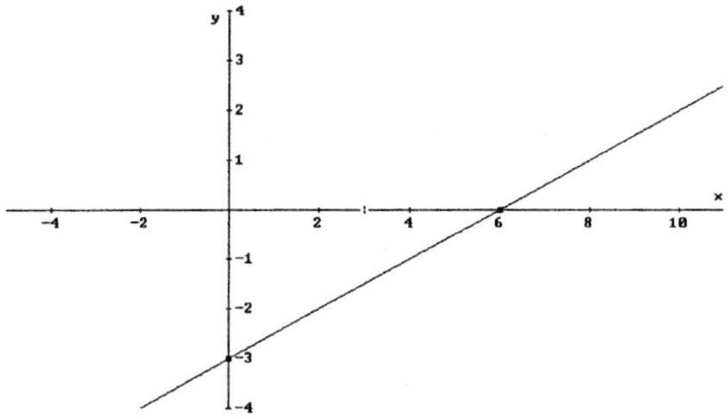

55. $y = (x - 3)^2 - 3$

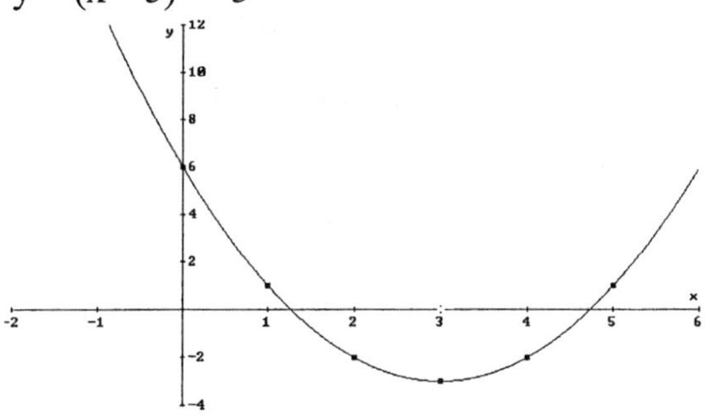

57. $y = \sqrt{x} - 3$

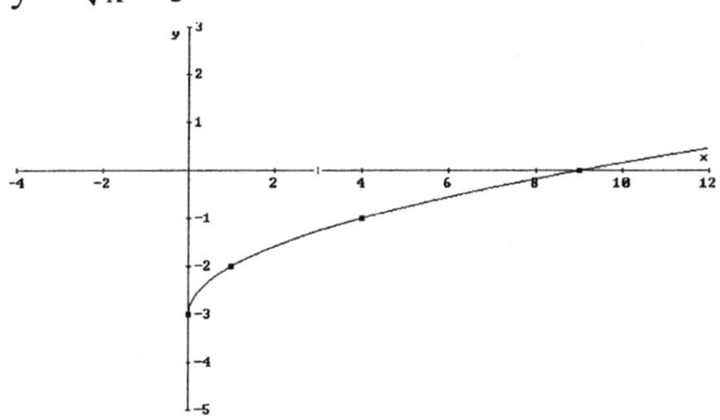

59. $y = |x| - 3$

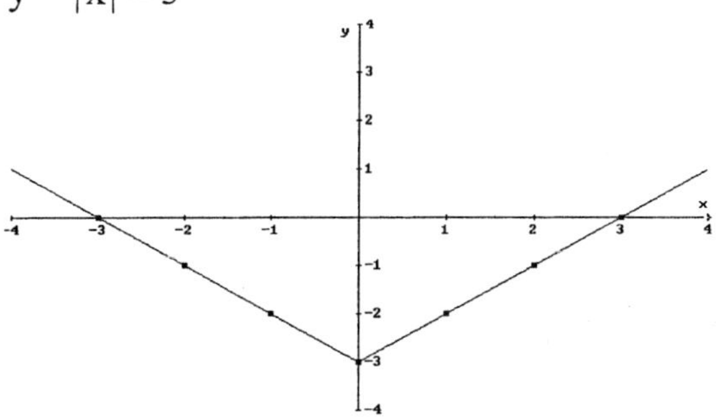

Section 10.4 Graphs of Quadratic Functions

Vocabulary

1. $f(x) = ax^2 + bx + c$; a

3. $x = h$

5. down

Concepts

7. $f(x) = (x - h)^2$ e.

9. $f(x) = -x^2$ b.

11. $f(x) = x^2 - k$ c.

Practice

13. $y = -x^2$

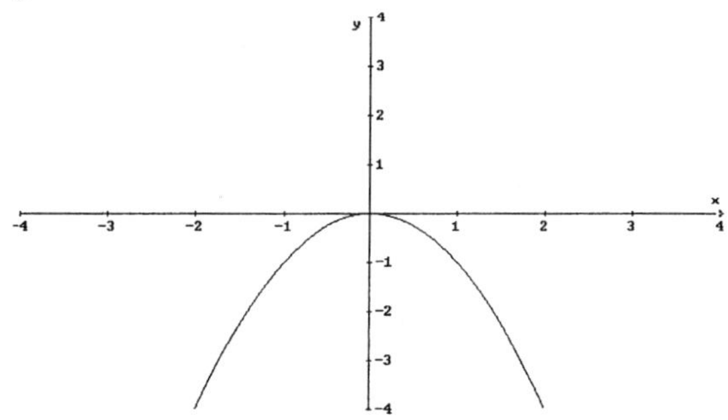

15. $y = x^2 - 3$

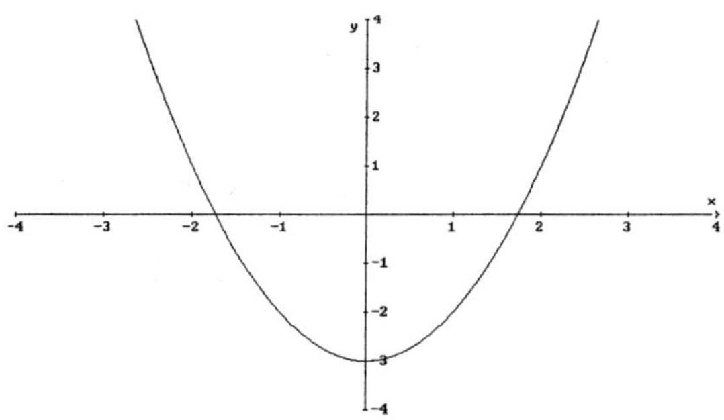

17. $y = (x + 2)^2$

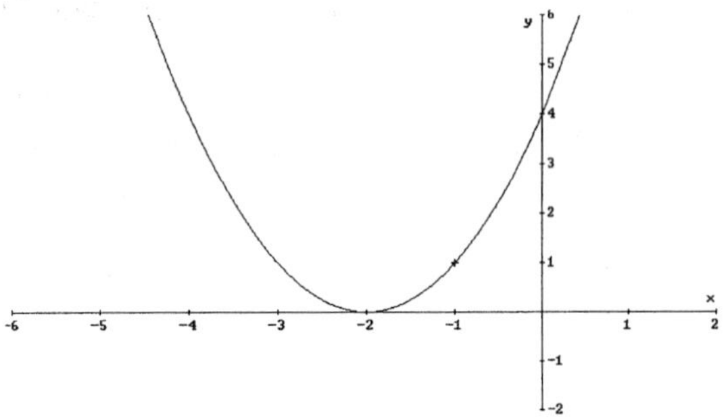

19. $y = 5x + x^2$

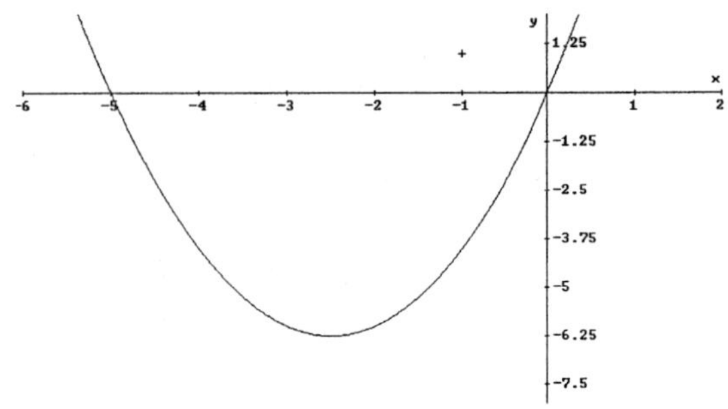

21. $y = x^2 - x - 6$

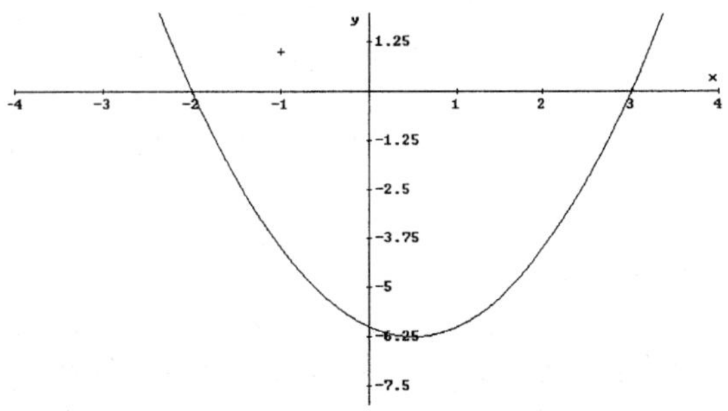

23. $y = -2x^2 + 4x + 3$

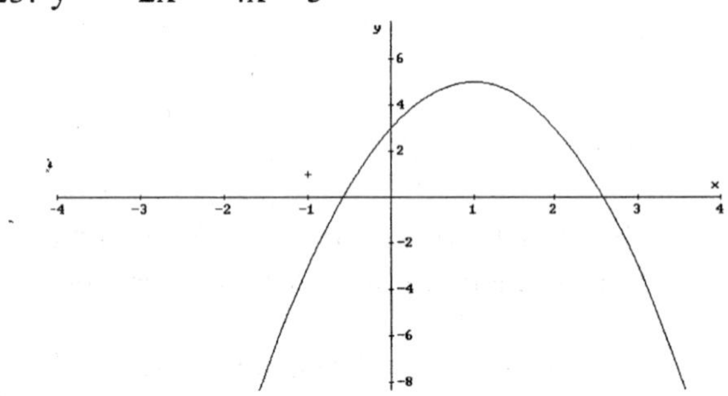

25. $y = 2x^2 - 8x + 7 = 2(x^2 - 4x) + 7$
$y = 2(x^2 - 4x + 4) + (7 - 8) = 2(x - 2)^2 - 1$
opens up, vertex: (2,-1), axis: x = 2

27. $y = -3x^2 - 6x = -3(x^2 + 2x)$
$y = -3(x^2 + 2x + 1) + 3 = -3(x + 1)^2 + 3$
opens down, vertex: (-1,3), axis: x = -1

29. $y = 3x^2 - 3 = 3(x - 0)^2 - 3$
opens up, vertex:(0,-3), axis: x = 0

31. $y = 3x^2 + 6x = 3(x^2 + 2x) = 3(x^2 + 2x + 1) - 3$
$y = 3(x + 1)^2 - 3$
opens up, vertex: (-1,-3), axis: x = -1

33. $y = 5x^2 + 20x + 25 = 5(x^2 + 4x) + 25$
$y = 5(x^2 + 4x + 4) + (25 - 20) = 5(x + 2)^2 + 5$
opens up, vertex: (-2,5), axis: x = -2

35. $2x^2 - 5x - 3 = 0$ Solutions: {-0.5, 3}

37. $2x^2 - 0.5x - 2 = 0$ Solutions: {-0.88, 1.13}

39. If a < 0, changes in k affect the range, but not the domain. Changes in h have no effect on either one. In this case, k will be the largest y value in the range. Range = (- ∞, k]

41. y = ax + b compared to y = x.
a. Differing values of *a* affect how the line slants, both in steepness and direction.

b. Differing values of *b* translate the graph vertically, up if b > 0, down if b < 0.

43. $y = a\sqrt{x - h} + k$ compared to $y = \sqrt{x}$

a. Differing values of *a* determine if the graph opens up (a > 0) or down (a < 0), and the amount of vertical stretch (|a|>1) or shrink (|a|<1).

b. Differing values of *k* translate the graph vertically, up if k > 0, down if k < 0.

c. Differing values of *h* translate the graph horizontally, right if h > 0, left if h < 0.

Applications

45. $R = -\frac{x^2}{5} + 80x - 1000$

a. The graph of R in [0,500] by [0, 8000] is:

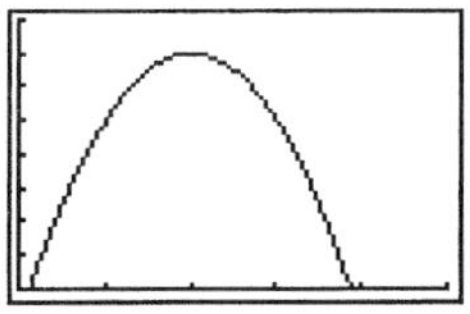

b. The maximum point is (200, 7000).
c. To obtain the maximum revenue, 200 stereos must be sold.
d. The maximum revenue is \$7,000.

Review

47. $3(x-2) > 2(x+7)$

$3x - 6 > 2x + 14, \quad x > 20$

Interval: $(20, \infty)$ ◂- - - - -o———▸ 20

49. $3x + 2 < 8$ or $2x - 3 > 11$

$3x < 6$ or $2x > 14$

$x < 2$ or $x > 7$

Intervals: $(-\infty, 2) \cup (7, \infty)$ ◂——o - - - -o——▸ 2 7

Section 10.5 Quadratic and Other Inequalities

Vocabulary

1. critical

3. open; closed

Concepts

5. $\frac{x-3}{x} \geq 0$ Set the denominator equal to 0 and solve for x. The excluded value is 0.

7. $\frac{x^2 + x - 12}{(x-2)(x+3)} < 0$ Set the denominator equal to 0 and solve for x. $(x-2)(x+3) = 0$ gives $x = 2, -3$. The excluded values are 2 and -3.

9. $\mathbb{R}$

11. ϕ

13. c.

15. a.

17. $x^2 - 5x + 4 < 0$
$x^2 - 5x + 4 = 0, \ (x-4)(x-1) = 0, x = 1,4$
The critical points are 1 and 4. Testing a point in each region gives these results.

\+ – +
◂——|——|——▸ Interval: (1,4)
1 4 ◂- - - - -o———o - - - - -▸ 1 4

19. $x^2 - 8x + 15 > 0$
$x^2 - 8x + 15 = 0, (x-3)(x-5) = 0, x = 3,5.$
The critical points are 3 and 5. Testing a point in each region gives these results.

\+ – +
◂——|——|——▸ Intervals: $(-\infty, 3) \cup (5, \infty)$
3 5 ◂———o - - - - -o———▸ 3 5

21. $x^2 + x - 12 \leq 0$
$x^2 + x - 12 = 0, (x+4)(x-3) = 0, x = 3, -4.$
The critical points are 3 and -4. Testing a point in each region gives these results.

\+ – +
◂——|——|——▸ Interval: [- 4, 3]
- 4 3 ◂- - - - -●———● - - - - -▸ - 4 3

23. $x^2 + 2x \geq 15$ or $x^2 + 2x - 15 \geq 0$
$x^2 + 2x - 15 = 0, (x+5)(x-3) = 0, x = 3, -5$
The critical points are 3 and -5. Testing a point in each region gives these results.

\+ – +
◂——|——|——▸ Intervals: $(-\infty, -5] \cup [3, \infty)$
-5 3 ◂———● - - - - -●———▸ -5 3

25. $-x^2 + 8x > 18$ or $-x^2 + 8x - 18 > 0$
Rewrite as $x^2 - 8x + 18 < 0$.
$x^2 - 8x + 18 = 0$, no real solutions. Testing $x = 0$ gives a positive number, so there are no real solutions to the inequality: ϕ

27. $x^2 \geq 9$ or $x^2 - 9 \geq 0$
$x^2 - 9 = 0$, $(x - 3)(x + 3) = 0$, $x = 3, -3$
The critical points are 3 and -3. Testing a point in each region gives these results.

+ − +

-3 3

Intervals: $(-\infty, -3] \cup [3, \infty)$

-3 3

29. $2x^2 - 50 < 0$
$2x^2 - 50 = 0$, $2(x - 5)(x + 5) = 0$, $x = 5, -5$
The critical points are 5 and -5. Testing a point in each region gives these results.

+ − +

-5 5

Interval: $(-5, 5)$

-5 5

31. $\frac{1}{x} < 2$ or $\frac{1}{x} - 2 < 0$ or $\frac{1-2x}{x} < 0$
Set numerator and denominator equal to 0.
$1 - 2x = 0$, $x = \frac{1}{2}$ and $x = 0$. The critical points are ½ and 0. Testing a point in each region gives these results.

− + −

0 ½

Intervals: $(-\infty, 0) \cup (\frac{1}{2}, \infty)$

0 ½

33. $\frac{4}{x} \geq 2$ or $\frac{4}{x} - 2 \geq 0$ or $\frac{4-2x}{x} \geq 0$
Set numerator and denominator equal to 0.
$4 - 2x = 0$, $x = 2$ and $x = 0$. The critical points are 0 and 2. Testing a point in each region gives these results.

− + −

0 2

Interval: $(0, 2]$

0 2

35. $\frac{x^2 - x - 12}{x - 1} < 0$ or $\frac{(x-4)(x+3)}{x-1} < 0$
Set numerator and denominator equal to 0.
$(x - 4)(x + 3) = 0$, $x = 4, -3$; $x - 1 = 0$, $x = 1$
The critical points are -3, 1 and 4. Testing a point in each region gives these results.

− + − +

-3 1 4

Intervals: $(-\infty, -3) \cup (1, 4)$

-3 1 4

37. $\frac{x^2 + x - 20}{x + 2} \geq 0$ or $\frac{(x+5)(x-4)}{x+2} \geq 0$
Set numerator and denominator equal to 0.
$(x + 5)(x - 4) = 0$, $x = -5, 4$; $x + 2 = 0$, $x = -2$
The critical points are -5, -2 and 4. Testing a point in each region gives these results.

− + − +

-5 -2 4

Intervals: $[-5, -2) \cup [4, \infty)$

-5 -2 4

39. $\frac{6x^2 - 5x + 1}{2x + 1} > 0$ or $\frac{(3x-1)(2x-1)}{2x+1} > 0$
Set numerator and denominator equal to 0.
$(3x - 1)(2x - 1) = 0$, $x = \frac{1}{3}, \frac{1}{2}$; $2x + 1 = 0$, $x = -\frac{1}{2}$. The critical points are $-\frac{1}{2}, \frac{1}{3}, \frac{1}{2}$. Testing a point in each region gives these results.

− + − +

-½ ⅓ ½

Intervals: $(-\frac{1}{2}, \frac{1}{3}) \cup (\frac{1}{2}, \infty)$

-½ ⅓ ½

41. $\frac{3}{x - 2} < \frac{4}{x}$ or $\frac{3}{x-2} - \frac{4}{x} < 0$
$\frac{3x - 4(x-2)}{x(x-2)} < 0$ or $\frac{8-x}{x(x-2)} < 0$
Set numerator and denominator equal to 0.
$x(x - 2) = 0$, $x = 2, 0$; $8 - x = 0$, $x = 8$ The critical points are 0, 2 and 8. Testing a point in each region gives these results.

+ − + −

0 2 8

Intervals: $(0, 2) \cup (8, \infty)$

0 2 8

43. $\frac{-5}{x+2} \geq \frac{4}{2-x}$ or $\frac{-5}{x+2} - \frac{4}{2-x} \geq 0$

$\frac{-5(2-x)-4(x+2)}{(x+2)(2-x)} \geq 0$

$\frac{-10+5x-4x-8}{(x+2)(2-x)} \geq 0$ or $\frac{x-18}{(x+2)(2-x)} \geq 0$

Set numerator and denominator equal to 0, giving 18, 2 and -2 as critical numbers. Testing a point in each region gives these results.

+ – + –

-2 2 18

Intervals: (- ∞,-2) ∪ (2,18]

-2 2 18

45. $\frac{x}{x+4} \leq \frac{1}{x+1}$ or $\frac{x}{x+4} - \frac{1}{x+1} \leq 0$

$\frac{x(x+1)-(x+4)}{(x+1)(x+4)} \leq 0$ or $\frac{x^2-4}{(x+1)(x+4)} \leq 0$

$\frac{(x-2)(x+2)}{(x+1)(x+4)} \leq 0$

Setting numerator and denominator equal to 0 gives the critical points - 4,-2,-1 and 2. Testing a point in each region gives these results.

+ – + – +

-4 -2 -1 2

Intervals: (-4,-2] ∪ (-1,2]

-4 -2 -1 2

47. $\frac{x}{x+16} > \frac{1}{x+1}$ or $\frac{x}{x+16} - \frac{1}{x+1} > 0$

$\frac{x(x+1)-(x+16)}{(x+16)(x+1)} > 0$ or $\frac{x^2-16}{(x+16)(x+1)} > 0$

$\frac{(x-4)(x+4)}{(x+16)(x+1)} > 0$

Setting numerator and denominator equal to 0 gives the critical points -16, -4, -1 and 4. Testing a point in each region gives these results.

+ – + – +

-16 -4 -1 4

Intervals: (- ∞, -16) ∪ (- 4, -1) ∪ (4, ∞)

-16 -4 -1 4

49. $(x+2)^2 > 0$

$(x+2)^2 = 0$, $x = -2$ The critical number is -2. Testing a point in each region gives these results.

+ +

-2

Intervals: (- ∞, -2) ∪ (-2, ∞)

-2

51. $2x - y > 2$ Graph $2x - y = 2$. Test (0,0). Since $2(0) - 0 < 2$, the solutions are on the opposite side of $2x - y = 2$ from (0,0). The solution set does not include the line.

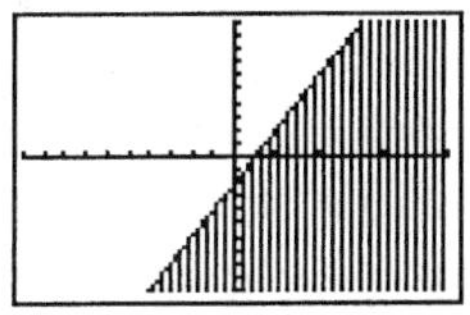

53. $4x + 3y \leq 12$ Graph $4x + 3y = 12$. Test (0,0). Since $4(0) + 3(0) \leq 12$, the solutions are on the same side of $4x + 3y = 12$ as (0,0). The solution set includes the line.

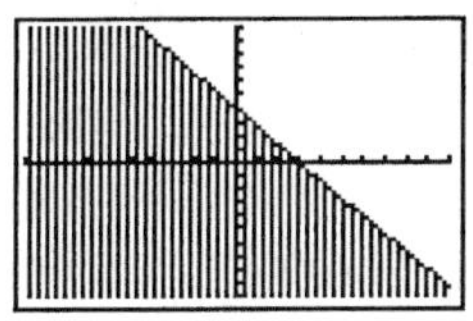

55. $3(x + y) + x < 6$ Graph $3(x + y) + x = 6$. Test (0,0). Since $3(0 + 0) + 0 < 6$, the solutions are on the same side of the line as (0,0). The solution set does not include the line.

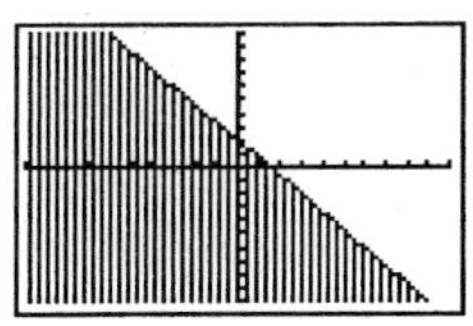

57. $4x - 4(x + 2y) \geq -6y - 8$ Graph the line and test (0,0). Since $4(0) - 4(0 + 2\cdot0) > -6(0) - 8$, the solutions lie on the same side of the line as (0,0). The solution set includes the line.

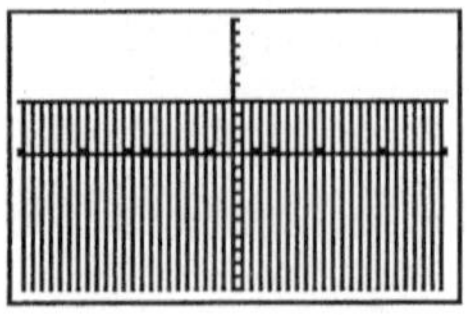

59. $y < x^2 + 1$ Graph $y = x^2 + 1$. Test (0,0). Since $0 < 0^2 + 1$, the solutions are on the same side of the curve as (0,0). The solution set does not include the curve.

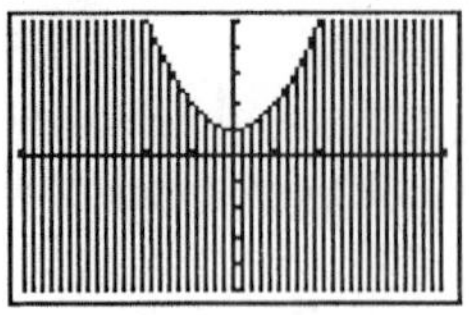

61. $y < x^2 + 5x + 6$ Graph $y = x^2 + 5x + 6$. Test (0,0). Since $0 < 0^2 + 5(0) + 6$, the solutions lie on the same side of the curve as (0,0). The solution set does not include the curve.

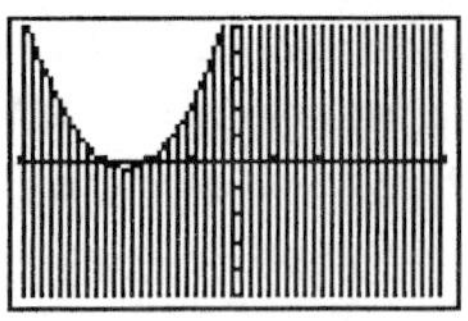

63. $x \geq y^2 - 3$ Graph $x = y^2 - 3$. Test (0,0). Since $0 \geq 0^2 - 3$, the solutions lie on the same side of the curve as (0,0). The solution set <u>includes</u> the curve.

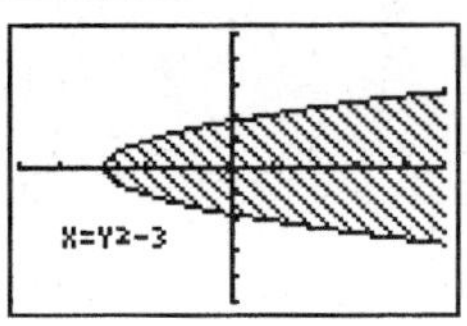

65. $-x^2 - y + 6 > -x$ Graph $-x^2 - y + 6 = -x$. Test (0,0). Since $-0^2 - 0 + 6 > -0$, the solutions lie on the same side of the curve as (0,0). The solution set does not include the curve.

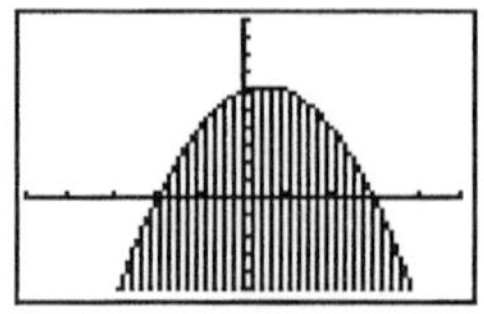

67. $y < |x + 4|$ Graph $y = |x + 4|$. Test (0,0). Since $0 < |0 + 4|$, the solutions lie on the same side of the curve as (0,0). The solution set does not include the curve.

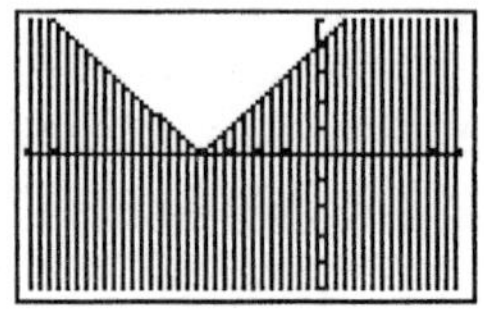

69. $y \leq -|x| + 2$ Graph $y = -|x| + 2$. Test (0,0). Since $0 < -|0| + 2 = 2$, the solutions lie on the same side of the curve as (0,0). The solution set <u>includes</u> the curve.

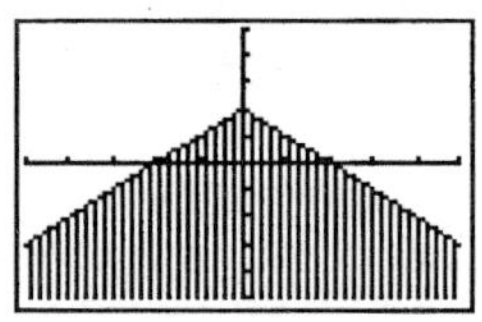

Review

71. $A = \pi r^2$ If $r = 8$, $A = \pi(8^2) = 64\pi$ in.^{2}

73. If x = age of the car, $V = \frac{k}{x}$. If v = 9000 when x = 3, $9000 = \frac{k}{3}$ or $k = 27000$. Thus, $V = \frac{27000}{x}$. If x = 9, $V = \frac{27000}{9} = \$3000$.

Chapter 10 Summary

1. a. $12x^2 + x - 6 = 0$

$(4x+3)(3x-2) = 0$

$x = -\frac{3}{4}$ or $x = \frac{2}{3}$

b. $15x^2 + 2x - 8 = 0$

$(3x-2)(5x+4) = 0$

$3x - 2 = 0$ or $5x + 4 = 0$

$x = \frac{2}{3}$ or $x = -\frac{4}{5}$

c. $6x^2 + 17x + 5 = 0$

$(3x+1)(2x+5) = 0$

$3x + 1 = 0$ or $2x + 5 = 0$

$x = -\frac{1}{3}$ or $x = -\frac{5}{2}$

d. $(x+2)^2 = 36$

$x + 2 = \pm\sqrt{36} = \pm 6$

$x = -2 \pm 6 = -8, 4$

2. a. $x^2 + 6x + 8 = 0$

$x^2 + 6x = -8$

$x^2 + 6x + 3^2 = -8 + 9 = 1$

$(x+3)^2 = 1, \quad x + 3 = \pm 1, \; x = -3 \pm 1 = -4, -2$

b. $2x^2 - 9x + 7 = 0$

$2x^2 - 9x = -7$

$x^2 - \frac{9}{2}x = -\frac{7}{2}, \quad x^2 - \frac{9}{2}x + \left(\frac{9}{4}\right)^2 = -\frac{7}{2} + \frac{81}{16}$

$\left(x - \frac{9}{4}\right)^2 = -\frac{56}{16} + \frac{81}{16} = \frac{25}{16}$

$x - \frac{9}{4} = \pm\sqrt{\frac{25}{16}} = \pm\frac{5}{4}, \; x = \frac{9}{4} \pm \frac{5}{4} = \frac{7}{2}, 1$

3. a. $x^2 - 8x - 9 = 0 \quad a = 1, b = -8, c = -9$

$$x = \frac{-(-8) \pm \sqrt{(-8)^2 - 4\cdot 1\cdot -9}}{2\cdot 1}$$

$$x = \frac{8 \pm \sqrt{100}}{2} = \frac{8 \pm 10}{2} = 9, -1$$

b. $x^2 - 10x = 0 \quad a = 1, b = -10, c = 0$

$$x = \frac{-(-10) \pm \sqrt{(-10)^2 - 4\cdot 1\cdot 0}}{2\cdot 1}$$

$$x = \frac{10 \pm \sqrt{100}}{2} = \frac{10 \pm 10}{2} = 10, 0$$

c. $2x^2 + 13x - 7 = 0 \quad a = 2, b = 13, c = -7$

$$x = \frac{-13 \pm \sqrt{13^2 - 4\cdot 2\cdot -7}}{2\cdot 2}$$

$$x = \frac{-13 \pm \sqrt{225}}{4} = \frac{-13 \pm 15}{4} = \frac{-28}{4}, \frac{2}{4} = -7, \frac{1}{2}$$

d. $3x^2 - 7 = 20x$ or $3x^2 - 20x - 7 = 0$
$a = 3, b = -20, c = -7$

$$x = \frac{-(-20) \pm \sqrt{(-20)^2 - 4\cdot 3\cdot -7}}{2\cdot 3}$$

$$x = \frac{20 \pm \sqrt{484}}{6} = \frac{20 \pm 22}{6} = \frac{42}{6}, \frac{-2}{6} = 7, -\frac{1}{3}$$

4. a. $3x^2 + 4x - 3 = 0 \quad a = 3, b = 4, c = -3$
The discriminant $b^2 - 4ac = 4^2 - 4(3)(-3) = 52 > 0$. Two distinct irrational solutions.

b. $4x^2 - 5x + 7 = 0 \quad a = 4, b = -5, c = 7$
The discriminant $b^2 - 4ac = (-5)^2 - 4(4)(7) = -87 < 0$. Two complex conjugate solutions.

5. a. $(k - 8)x^2 + (k + 16)x = -49$ Rewrite as $(k - 8)x^2 + (k + 16)x + 49 = 0$. Then $a = k - 8$, $b = k + 16$, $c = 49$. The discriminant $b^2 - 4ac = (k + 16)^2 - 4(k - 8)(49) = k^2 + 32k + 256 - 196(k - 8) = k^2 - 164k + 1824$. For equal roots the discriminant must be 0.

$k^2 - 164k + 1824 = 0$

$(k-12)(k-152) = 0$

$k = 12$ or $k = 152$

6. a. $3x^2 + 4x = k + 1$
Rewrite as $3x^2 + 4x - (k + 1) = 0$. Then a = 3, b = 4, c = -(k + 1). The discriminant is then $b^2 - 4ac = 4^2 - 4(3)(-k - 1) = 16 - 12(-k - 1)$ = 28 + 12k. For real solutions, the discriminant must be non-negative.

$$28 + 12k \geq 0$$

$$12k \geq -28$$

$$k \geq -\frac{28}{12} = -\frac{7}{3}$$

7. Let x = width of the rectangle. Then x + 2 = length and the area is x(x + 2). If length and width are doubled, they become 2x, 2(x + 2) and the area is $2x \cdot 2(x + 2) = 4x^2 + 8x$. Thus,

$$4x^2 + 8x = x(x + 2) + 72$$

$$4x^2 + 8x = x^2 + 2x + 72 \text{ or } 3x^2 + 6x - 72 = 0$$

$$x^2 + 2x - 24 = 0 \text{ or } (x + 6)(x - 4) = 0$$

$$x = -6 \text{ or } x = 4$$

The original dimensions are 4 cm. by 6 cm.

8. $h = 112t - 16t^2$
a. The graph in [0,10] by [0,200] is

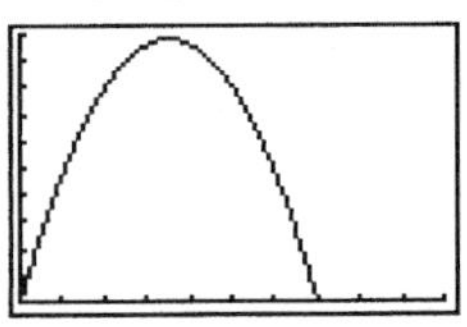

b. It touches the ground when h = 0.

$$0 = 112t - 16t^2$$

$$16t^2 - 112t = 0$$

$$16t(t - 7) = 0,\ t = 0,7$$

It touches the ground after 7 seconds.
c. Height at t = 2 secs. is $112(2) - 16(2^2)$ = 160 ft.
d. If h = 160, $160 = 112t - 16t^2$. Thus,

$$16t^2 - 112t + 160 = 0$$

$$16(t^2 - 7t + 10) = 0 \text{ or } 16(t - 2)(t - 5) = 0$$

$$t = 2,5$$

It is at a height of 160 ft. after 5 secs. also.
e. The maximum height is 196 ft.

9. a. $x - 13x^{\frac{1}{2}} + 12 = 0$ Let $u = x^{\frac{1}{2}}$.

$$u^2 - 13u + 12 = 0 \text{ or } (u - 12)(u - 1) = 0$$

$$u = 12 \text{ or } u = 1$$

$$x^{\frac{1}{2}} = 12,\ x = 12^2 = 144 \text{ or } x^{\frac{1}{2}} = 1, x = 1$$

b. $a^{\frac{2}{3}} + a^{\frac{1}{3}} - 6 = 0$ Let $u = a^{\frac{1}{3}}$.

$$u^2 + u - 6 = 0 \text{ or } (u + 3)(u - 2) = 0$$

$$u = -3 \text{ or } u = 2$$

$$a^{\frac{1}{3}} = -3,\ a = (-3)^3 = -27 \text{ or } a^{\frac{1}{3}} = 2,\ a = 2^3 = 8$$

c. $\frac{1}{x+1} - \frac{1}{x} = -\frac{1}{x+1}$ LCD = x(x + 1)
Multiply both sides by the LCD.

$$x - (x + 1) = -x$$

$$-1 = -x \text{ or } 1 = x$$

d. $\frac{6}{x+2} - \frac{6}{x+1} = 5$ LCD = (x + 2)(x + 1)
Multiply both sides by the LCD.

$$6(x + 1) - 6(x + 2) = 5(x + 1)(x + 2)$$

$$6x + 6 - 6x - 12 = 5(x^2 + 3x + 2)$$

$$-6 = 5x^2 + 15x + 10 \text{ or } 0 = 5x^2 + 15x + 16$$

$$x = \frac{-15 \neq \sqrt{15^2 - 4 \cdot 5 \cdot 16}}{2 \cdot 5} = \frac{-15 \pm \sqrt{-95}}{10} = \frac{-15 \pm \sqrt{95}\, i}{10}$$

10. a. $y = 2x^2 - 3 = 2(x - 0)^2 - 3$
vertex: (0,-3), axis of symmetry: x = 0

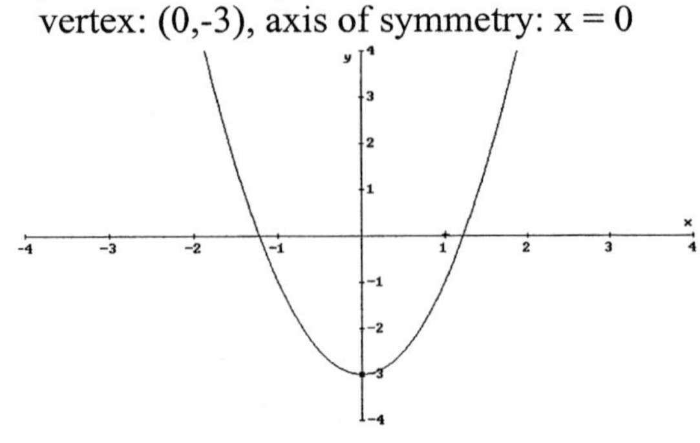

b. $y = -2x^2 - 1 = -2(x - 0)^2 - 1$
vertex: (0, -1), axis of symmetry: x = 0

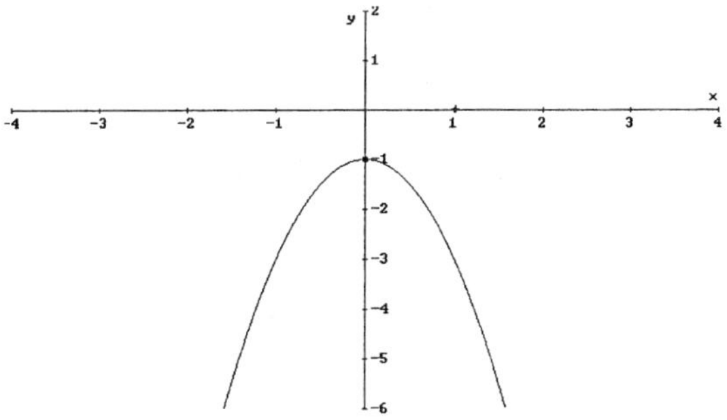

c. $y = -4x^2 + 16x - 15 = -4(x^2 - 4x) - 15$
$y = -4(x^2 - 4x + 4) - 15 + 16 =$
$-4(x - 2)^2 + 1$. vertex: (2, 1), axis: x = 2

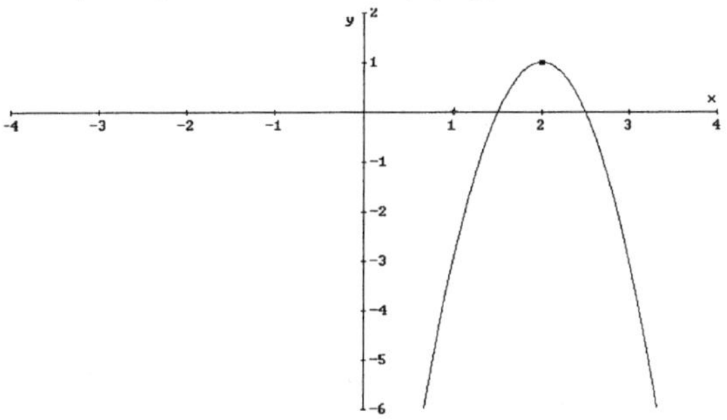

d. $y = 5x^2 + 10x - 1 = 5(x^2 + 2x) - 1$
$5(x^2 + 2x + 1) - 1 - 5 = 5(x + 1)^2 - 6$
vertex: (-1,-6), axis of symmetry: x = -1

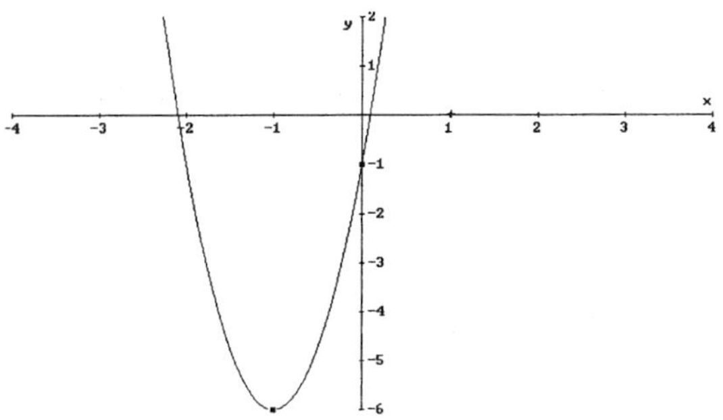

11. a. Let x = width of the rectangle. Then $50 - x$ = length. Let y = area = $x(50 - x) = 50x - x^2$. The maximum occurs at the vertex since the parabola opens down. $y = -x^2 + 50x = -(x^2 - 50x) = -(x^2 - 50x + 25^2) + 625 = -(x - 25)^2 + 625$. The vertex is at (25, 625). The rectangle should be 25' x 25'
b. The maximum area is 625 $ft.^2$

12. a. $x^2 + 2x - 35 > 0$ or $(x + 7)(x - 5) > 0$
$(x + 7)(x - 5) = 0$, x = -7, 5. The critical points are -7 and 5. Testing a point in each region gives these results.

\+ – +
◄——|——|——► Intervals: $(-\infty,-7) \cup (5, \infty)$
-7 5
◄——o - - - - -o——►
-7 5

b. $x^2 + 7x - 18 < 0$ or $(x + 9)(x - 2) < 0$
$(x + 9)(x - 2) = 0$, x = -9,2. The critical points are -9 and 2. Testing a point in each region gives these results.

\+ – +
◄——|——|——► Interval: (-9, 2)
-9 2
◄ - - - - -o——o - - - - -►
-9 2

13. a. $(x - 2)(x + 1)(x + 4) < 0$
Solving $(x - 2)(x + 1)(x + 4) = 0$ gives the critical numbers 2, -1 and - 4. Testing a point in each region gives this result.

– + – + Intervals:
◄——|——|——|——► $(-\infty,-4) \cup (-1, 2)$
- 4 -1 2
◄——o - - -o——o - - -►
- 4 -1 2

b. $\dfrac{3}{x} \le 5$ or $\dfrac{3}{x} - 5 \le 0$ or $\dfrac{3 - 5x}{x} \le 0$

Set the numerator and denominator equal to 0

$3 - 5x = 0$, $x = 3/5$ and $x = 0$. The critical points are 0 and 3/5. Testing a point in each region gives these results.

– + –
◄——|——|——► Intervals: $(-\infty, 0) \cup [3/5, \infty)$
0 3/5
◄——o - - - - -●——►
0 3/5

c. $\frac{2x^2 - x - 28}{x - 1} > 0$ or $\frac{(2x + 7)(x - 4)}{x - 1} > 0$
Setting the numerator and denominator equal to 0 gives the critical points -3.5, 1 and 4. Testing a point in each region gives these results.

$$\begin{array}{cccc} - & + & - & + \end{array}$$
◄——|——|——|——►
-3.5 1 4

Intervals: (-3.5, 1) ∪ (4, ∞)

◄ - - - -O———O - - - - -O———►
-3.5 1 4

14. a. $3x + y \geq 3$ Graph $3x + y = 3$. Test (0,0). Since $3(0) + 0 < 3$, the solutions are on the opposite side of the line from (0,0). The solution set includes the line.

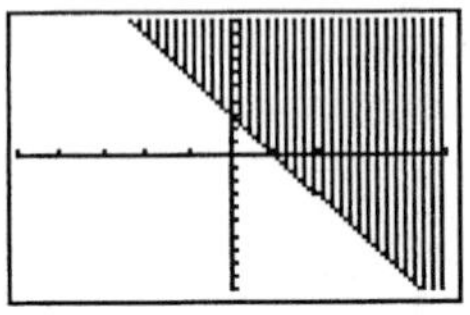

b. $y < 4$ Graph the line $y = 4$. Test (0,0). Since $0 < 4$, the solutions are on the same side of the line as (0,0). The solution set does not include the line.

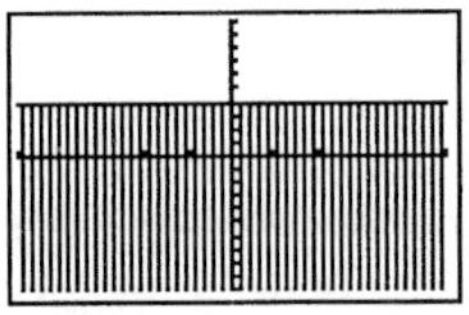

c. $y < \frac{1}{2}x^2 - 1$ Graph $y = \frac{1}{2}x^2 - 1$. Test (0,0) Since $0 > \frac{1}{2}(0^2) - 1$, the solutions are on the opposite of the curve from (0,0). The solution set does not include the curve.

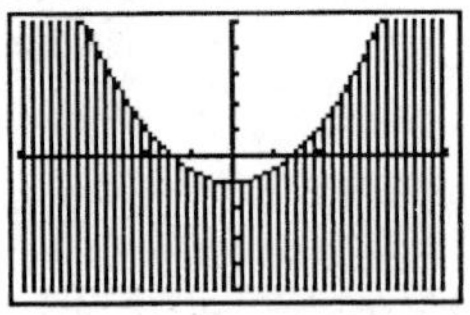

d. $y \geq -|x|$ Graph $y = -|x|$. Test (0,1). Since $1 \geq -|0|$, the solutions are on the same side of the curve as (0,1). The solution set includes the curve.

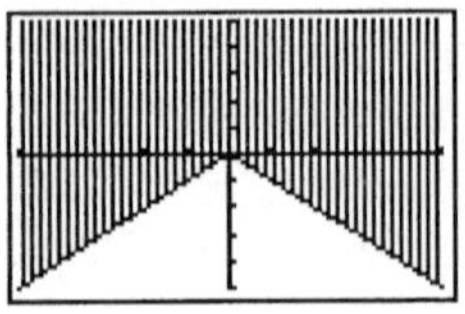

Chapter 10 Test

1. a. $x^2 + 3x + 2 = 0$
$(x + 2)(x + 1) = 0$
$x + 2 = 0$ or $x + 1 = 0$, so $x = -2$ or $x = -1$

b. $12x^2 + 5x = 2$ or $12x^2 + 5x - 2 = 0$
$(4x - 1)(3x + 2) = 0$
$4x - 1 = 0$ or $3x + 2 = 0$
$x = \frac{1}{4}$ or $x = -\frac{2}{3}$

2. a. $(x - 3)^2 = -2$
$x - 3 = \pm\sqrt{-2} = \pm\sqrt{2}\,i$, $x = 3 \pm \sqrt{2}\,i$

b. $3x^2 = 27$, $x^2 = 9$, $x = \pm 3$

3. a. $x^2 + 3x - 18 = 0$ $a = 1, b = 3, c = -18$
$$x = \frac{-3 \pm \sqrt{3^2 - 4 \cdot 1 \cdot -18}}{2 \cdot 1} = \frac{-3 \pm \sqrt{81}}{2}$$
$$x = \frac{-3 \pm 9}{2} = \frac{-12}{2}, \frac{6}{2} = -6, 3$$

b. $x(6x + 19) = -15$ or $6x^2 + 19x + 15 = 0$
$a = 6, b = 19, c = 15$
$$x = \frac{-19 \pm \sqrt{19^2 - 4 \cdot 6 \cdot 15}}{2 \cdot 6} = \frac{-19 \pm \sqrt{1}}{12}$$
$$x = \frac{-19 \pm 1}{12} = \frac{-20}{12}, \frac{-18}{12} = -\frac{5}{3}, -\frac{3}{2}$$

c. $x^2 + 4x + 1 = 0$
$x^2 + 4x = -1$
$x^2 + 4x + 2^2 = -1 + 4 = 3$
$(x + 2)^2 = 3$, $x + 2 = \pm\sqrt{3}$, $x = -2 \pm \sqrt{3}$

d. $x^2 - 5x - 3 = 0$ $a = 1, b = -5, c = -3$

$$x = \frac{-(-5) \pm \sqrt{(-5)^2 - 4 \cdot 1 \cdot -3}}{2 \cdot 1} = \frac{5 \pm \sqrt{37}}{2}$$

4. $3x^2 + 5x + 17 = 0$ $a = 3, b = 5, c = 17$
Discriminant: $b^2 - 4ac = 5^2 - 4(3)(17) < 0$
The solutions are imaginary.

5. $4x^2 - 2kx + k - 1 = 0$
$a = 4, b = -2k, c = k - 1$ The discriminant is $b^2 - 4ac = (-2k)^2 - 4(4)(k - 1) = 4k^2 - 16(k - 1) = 4k^2 - 16k + 16$. For equal roots the discriminant must be 0.

$4k^2 - 16k + 16 = 0$

$k^2 - 4k + 4 = 0$

$(k - 2)^2 = 0,\ k = 2$

6. Let x = length of the shortest leg. Then x +14 = length of the other leg. By the Pythagorean Theorem,

$x^2 + (x + 14)^2 = 26^2$

$x^2 + x^2 + 28x + 196 = 676$

$2x^2 + 28x - 480 = 0$

$x^2 + 14x - 240 = 0$

$(x - 10)(x + 24) = 0$

$x = 10$ or -24

The shortest leg is 10 inches.

7. $2x - 3x^{\frac{1}{2}} + 1 = 0$ Let $u = x^{\frac{1}{2}}$.
$2u^2 - 3u + 1 = 0$ or $(2u - 1)(u - 1) = 0$

$u = \frac{1}{2}$ or $u = 1$

$x^{\frac{1}{2}} = \frac{1}{2},\ x = \left(\frac{1}{2}\right)^2 = \frac{1}{4}$ or $x^{\frac{1}{2}} = 1,\ x = 1$

8. $y = \frac{1}{2}x^2 + 5$
a. $y = \frac{1}{2}(x - 0)^2 + 5$
b. vertex: (0,5)
c. Axis of symmetry: $x = 0$

d.

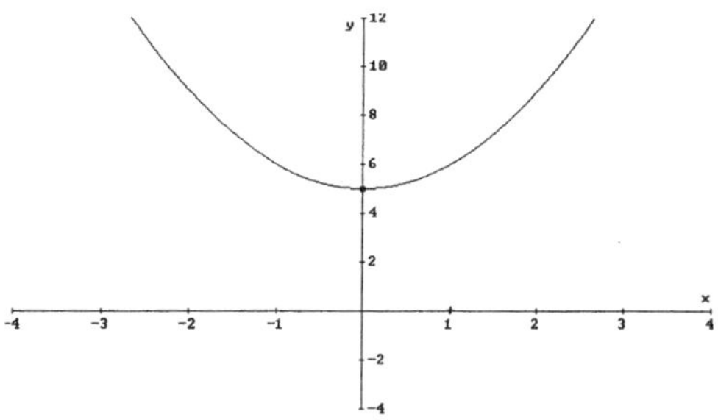

9. $y \le -x^2 + 3$ Graph $y = -x^2 + 3$. Test (0,0). Since $0 \le -0^2 + 3$, the solutions are on the same side of the curve as (0,0). The solution set <u>includes</u> the curve.

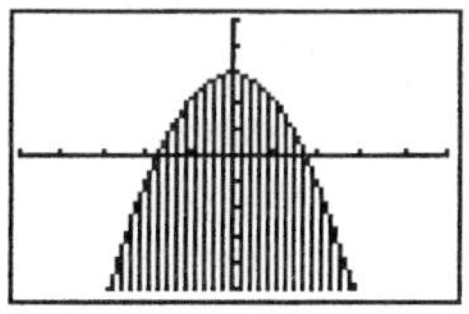

10. $x^2 - 2x - 8 > 0$ or $(x - 4)(x + 2) > 0$
$(x - 4)(x + 2) = 0$ gives $x = -2, 4$ as critical points. Testing a point in each region gives these results.

+   −   +
◄——|——|——►
-2   4

Intervals:
$(-\infty, -2) \cup (4, \infty)$
◄——o - - - - -o——►
-2   4

11. $\frac{x - 2}{x + 3} \le 0$ Setting the numerator and denominator equal to 0 gives 2 and -3 as the critical points. Testing a point in each region gives these results.

+   −   +
◄——|——|——►
-3   2

Interval: $(-3, 2]$,
◄- - - - -o——● - - - - -►
-3   2

Cumulative Review: Chapters 9 and 10

Vocabulary/Concepts

1. radicand
2. real number (exact); irrational
3. legs; hypotenuse

4. $i, -1, -i$, and 1
5. solutions (or roots)
6. x
7. parabola; vertex

8. a. $\sqrt{64} = 8$ is rational.
   b. $\sqrt{32} = 4\sqrt{2}$ is irrational.
   c. $\sqrt[5]{32} = 2$ is rational.
   d. $\sqrt[4]{18}$ is irrational.

9. a. $3\sqrt{ab} = 3(ab)^{\frac{1}{2}}$
   b. $\sqrt{3ab} = (3ab)^{\frac{1}{2}}$
   c. $\sqrt[3]{x^5} = x^{\frac{5}{3}}$
   d. $\sqrt[6]{\frac{1}{3}p^5q} = \left(\frac{1}{3}p^5q\right)^{\frac{1}{6}}$

10. a. F, $\sqrt{3} + \sqrt{3} = 2\sqrt{3} \neq 3$
    b. T
    c. F, $\sqrt{25-9} = \sqrt{16} = 4$, $\sqrt{25} - \sqrt{9} = 5 - 3 = 2$

11. a. $y = \sqrt{x+5}$, graph IV
    b. $y = \sqrt{x} + 5$, graph I
    c. $y = -\sqrt{x+5}$, graph III
    d. $y = \sqrt{x-5}$, graph II

12. a. $f(1) = 2$ b. $f(3) \approx 2.732$
    c. The number -1 is not a part of the domain.

13. a. $\sqrt{-4} = \sqrt{4}\sqrt{-1} = 2i$
    b. $i^{15} = (i^2)^7\, i = (-1)^7\, i = -i$
    c. $6 + \sqrt{-25} = 6 + \sqrt{25}\sqrt{-1} = 6 + 5i$
    d. $(8 + 3i) - 2(5 - 3i) = 8 + 3i - 10 + 6i =$
    $-2 + 9i$

14. a. Roots are real. b. Roots are imaginary.
    b. Roots are real. d. Roots are imaginary.

15. a. $y = 3(x-2)^2 + 4$, vertex: (2, 4)
    b. $y = -2(x+4)^2 - 5$, vertex: (- 4, -5)

16. Given $y = x^2$. Shift right 3 units, up 4 units.
    New equation: $y = (x-3)^2 + 4$

17. Given $y = x^2$. Shift left 5 units, down 6 units.
    New equation: $y = (x+5)^2 - 6$

18. Given $y = x^2$. Vertical stretch by a factor 1/6.
    New equation: $y = \frac{1}{6}x^2$

19. $y < (x-1)^2 + 3$ is graph b.

20. The solution of $\frac{x+2}{x-1} \leq 0$ is graph a.

Practice

21. $\sqrt{175a^4b^3c^5} = \sqrt{(5a^2bc^2)^2(7bc)} = 5a^2bc^2\sqrt{7bc}$

22. $\sqrt[4]{\frac{5x}{16z^5}} = \frac{\sqrt[4]{5x}}{\sqrt[4]{16z^5}} = \frac{\sqrt[4]{5x}}{2z\sqrt[4]{z}} \cdot \frac{\sqrt[4]{z^3}}{\sqrt[4]{z^3}} = \frac{\sqrt[4]{5xz^3}}{2z\sqrt[4]{z^4}} =$
    $\frac{\sqrt[4]{5xz^3}}{2z^2}$

23. $\sqrt{18} + \sqrt{400} - \sqrt{243} = \sqrt{9 \cdot 2} + \sqrt{20^2} - \sqrt{9^2 \cdot 3}$
    $= 3\sqrt{2} + 20 - 9\sqrt{3}$

24. $\sqrt[3]{6x^2y^4}\sqrt[3]{9xy^2} = \sqrt[3]{54x^3y^6} = \sqrt[3]{(3xy^2)^3(2)} =$
    $3xy^2\sqrt[3]{2}$

25. $\frac{\sqrt[3]{2}}{\sqrt[3]{3}}\frac{\sqrt[3]{3^2}}{\sqrt[3]{3^2}} = \frac{\sqrt[3]{18}}{\sqrt[3]{3^3}} = \frac{\sqrt[3]{18}}{3}$

26. $\frac{\sqrt{3}+1}{\sqrt{3}-1} \cdot \frac{\sqrt{3}+1}{\sqrt{3}+1} = \frac{3 + 2\sqrt{3} + 1}{\left(\sqrt{3}\right)^2 - 1^2} = \frac{4 + 2\sqrt{3}}{3-1} =$
    $\frac{2(2+\sqrt{3})}{2} = 2 + \sqrt{3}$

27. $\dfrac{-4i}{2-3i} = \dfrac{-4i}{2-3i} \cdot \dfrac{2+3i}{2+3i} = \dfrac{-8i-12i^2}{2^2-(3i)^2} =$

$\dfrac{-8i-12(-1)}{4-9i^2} = \dfrac{12-8i}{4-9(-1)} = \dfrac{12-8i}{13} = \dfrac{12}{13} - \dfrac{8}{13}i$

28. $f(x) = \sqrt{2x+2}$, $g(x) = \sqrt{3x+3} - 1$

a. $f(1) = \sqrt{2 \cdot 1 + 2} = \sqrt{4} = 2$

b. $g(2) = \sqrt{3 \cdot 2 + 3} - 1 = \sqrt{9} - 1 = 3 - 1 = 2$

c. $f(5) = \sqrt{2 \cdot 5 + 2} = \sqrt{12} = 2\sqrt{3}$

29. $\sqrt{x} + 6 = -x + 8$ or $\sqrt{x} = -x + 2$ Square both sides. $x = (-x+2)^2 = x^2 - 4x + 4$

$0 = x^2 - 5x + 4$

$0 = (x-4)(x-1)$

$x = 1$ or $x = 4$, but 4 is extraneous.

30. $y = \sqrt{x} + 3$ Domain: $[0, \infty)$ Range: $[3, \infty)$

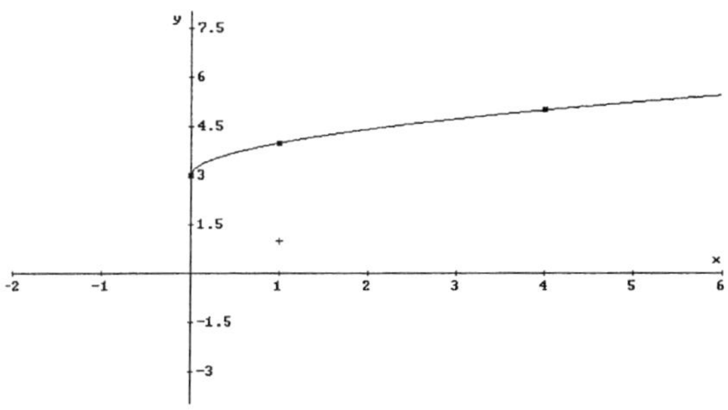

31. $2x^2 - 5x - 3 = 0$

a. $(2x + 1)(x - 3) = 0$ so $2x + 1 = 0$ or $x - 3 = 0$. Thus, $x = -0.5$ or $x = 3$.

b. $2x^2 - 5x = 3$

$x^2 - \dfrac{5}{2}x = \dfrac{3}{2}$

$x^2 - \dfrac{5}{2}x + \left(\dfrac{5}{4}\right)^2 = \dfrac{3}{2} + \dfrac{25}{16}$

$\left(x - \dfrac{5}{4}\right)^2 = \dfrac{49}{16}, \; x - \dfrac{5}{4} = \pm\sqrt{\dfrac{49}{16}} = \pm\dfrac{7}{4}$

$x = \dfrac{5}{4} \pm \dfrac{7}{4} = \dfrac{12}{4}, -\dfrac{2}{4}$ or $x = 3, -\dfrac{1}{2}$

c. $a = 2$, $b = -5$, $c = -3$

$x = \dfrac{-(-5) \pm \sqrt{(-5)^2 - 4 \cdot 2 \cdot -3}}{2 \cdot 2} = \dfrac{5 \pm \sqrt{49}}{4} =$

$\dfrac{5 \pm 7}{4} = 3, -\dfrac{1}{2}$

32. Roots: $-3, \dfrac{2}{3}$ Factors: $(x + 3), (x - \dfrac{2}{3})$

Equation: $(x + 3)(x - \dfrac{2}{3}) = 0$ Multiply by 3.

$(x + 3)(3x - 2) = 0$ or $3x^2 + 7x - 6 = 0$

33. a. x-intercepts: (2,0), (3,0)

b. y-intercept: (0,6)

c. $f(5) = 6$

d. If $f(x) = 6$, then $x = 0$ or $x = 5$.

34. $x^4 = 6x^2 - 5$ or $x^4 - 6x^2 + 5 = 0$ Let $u = x^2$.

$u^2 - 6u + 5 = 0$ or $(u-5)(u-1) = 0$

$u = 5$ or $u = 1$

$x^2 = 5, x = \pm\sqrt{5}$ or $x^2 = 1, x = \pm 1$

35. $x^{\frac{2}{3}} - 2x^{\frac{1}{3}} - 3 = 0$ Let $u = x^{\frac{1}{3}}$. Then

$u^2 - 2u - 3 = 0$ or $(u-3)(u+1) = 0$

$u = 3$ or $u = -1$

$x^{\frac{1}{3}} = 3, x = 3^3 = 27$ or $x^{\frac{1}{3}} = -1, x = (-1)^3 = -1$

36. $\dfrac{1}{x+2} + \dfrac{24}{x+3} = 13$ LCD = $(x + 2)(x + 3)$

Multiply both sides by the LCD.

$x + 3 + 24(x+2) = 13(x+3)(x+2)$

$x + 3 + 24x + 48 = 13(x^2 + 5x + 6)$

$25x + 51 = 13x^2 + 65x + 78$

$0 = 13x^2 + 40x + 27$

$0 = (13x + 27)(x + 1)$

$x = -\dfrac{27}{13}$ or $x = -1$

37. $3(x-4)^2+2=5$

a. $3(x-4)^2=3$

$(x-4)^2=1,\ x-4=\pm1,\ x=4\pm1=5$ or 3

b.

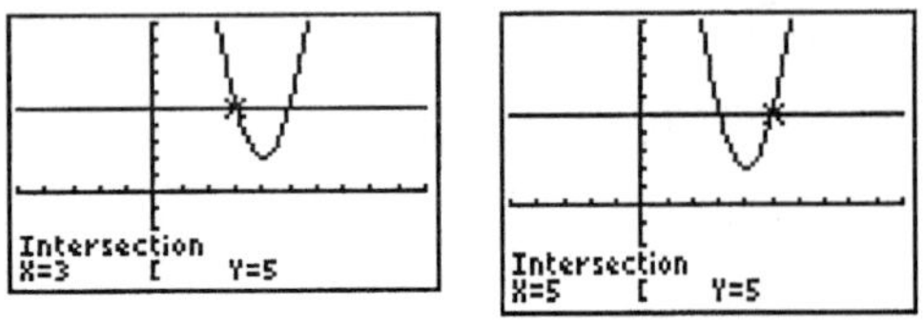

38. $y=(x+3)^2-4$

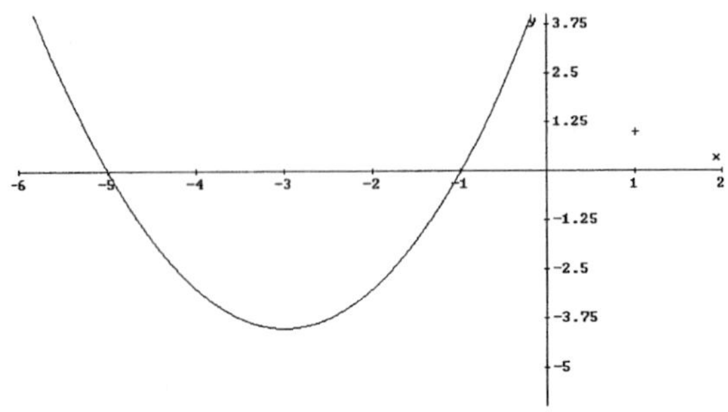

39. $y=-2x^2-1$

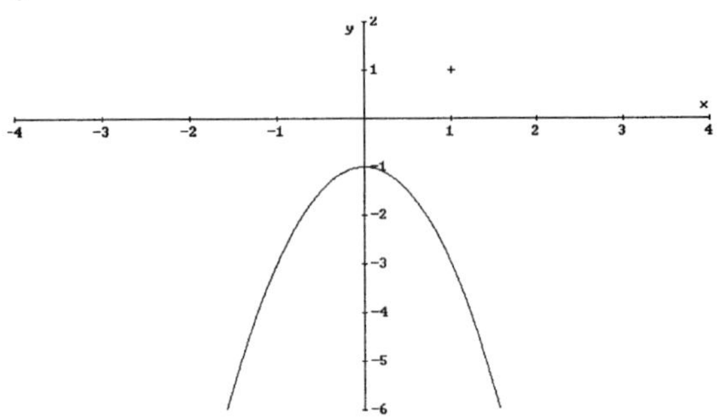

40. Let $y = 2x^2 - 8x + 13$

a. Graph opens up since $a = 2 > 0$

b. Vertex: (2,5) (See part d.)

c. Axis of symmetry: $x = 2$

d. $y=2(x^2-4x)+13$

$y=2(x^2-4x+4)+(13-8)$

$y=2(x-2)^2+5$

41. $(x-8)^2+3=8$

a. Using 'intersect' and the window [-5,15] by [-5,15], the solutions are 5.76, 10.24.

b. $(x-8)^2=5,\ x-8=\pm\sqrt{5},\ x=8\pm\sqrt{5}$

42. $\sqrt{x}-\sqrt{5x+4}+6=\sqrt{18x-48}+\sqrt{16x}$

Graphing in [-5,15] by [-5,25] shows that there is no solution.

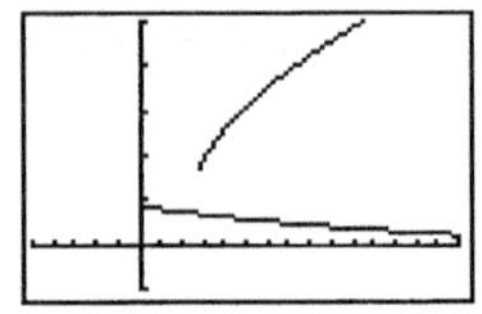

43. $\sqrt[4]{x}=\sqrt{\frac{x}{4}}$ Solution set: {0,16}, using the window [-2,30] by [-2,4] and 'intersect'

44. $\sqrt{1.3x}+4=1.2x+3.2$ Solution set: {2.02}, using ZStandard and 'intersect'

45. The values of the two sides might be slightly different since 2.02 is an approximate solution.

46. $2.3x^2-5.2x-7=0$

Solution set: {-0.95,3.21}, using 'zero' in ZStandard.

47. $y=50x-16x^2$

a. Graphing in [0,5] by [0,50] gives:

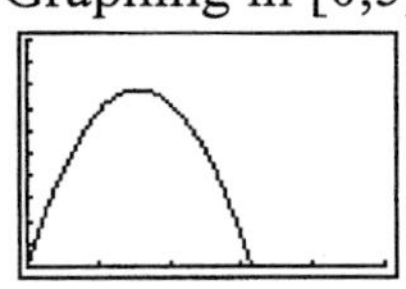

b. Maximum height = 39 feet.

c. 1.6 seconds to reach maximum height

d. The ball is on the ground when $y = 0$:

$-16x^2-50x=-2x(8x-25)=0$. Thus, $8x-25=0$, so $x=25/8\approx3.1$ seconds.

Applications

48. Let x = distance from the base of the pole to the anchor stake. Then x is one leg of a right triangle whose other leg is 17 and whose hypotenuse is 18. Thus, $x^2+17^2=18^2$ so that $x^2=324-289=35$ and $x=\sqrt{35}\approx5.9$ ft.

49. $v^2=64d$ If $d = 4$, $v^2=64(4)=256$ so that $v = 16$ ft./sec.

50. $2000(1 + r)^2 + 3000(1 + r) = 5572.80$
Divide by 2000:

$$(1+r)^2 + \frac{3}{2}(1+r) = 2.7864$$

$$1 + 2r + r^2 + \frac{3}{2} + \frac{3}{2}r = 2.7864$$

$$r^2 + \frac{7}{2}r = .2864$$

$$r^2 + \frac{7}{2}r + \left(\frac{7}{4}\right)^2 = .2864 + \frac{49}{16} = 3.3489$$

$$\left(r + \frac{7}{4}\right)^2 = 3.3489$$

$$r + \frac{7}{4} = \pm\sqrt{3.3489} = \pm 1.83$$

$$r = -1.75 \pm 1.83 = -3.58 \text{ or } .08$$

The rate r = .08 or 8%.

51. $R = -\frac{x^2}{1000} + 10x$

The maximum revenue occurs at the vertex of the parabola.

$$R = -\frac{1}{1000}(x^2 - 10000x)$$

$$R = -\frac{1}{1000}(x^2 - 10000x + 5000^2) + 25{,}000$$

$$R = -\frac{1}{1000}(x - 5000)^2 + 25{,}000$$

The vertex is (5000, 25000)

a. For maximum revenue, 5000 stereos must be sold.

b. The maximum revenue is \$25,000.

Chapter 11 Functions and Their Inverses

Section 11.1 Algebra of Functions

Vocabulary

1. excluded

3. identity

Concepts

5. $f = \{(-2,1),(5,3),(7,-1),(-5,-2),(4,-3)\}$
$g = \{(-5,6),(4,1),(-2,8),(7,6),(5,3)\}$
$f+g = \{(-2,1+8),(5,3+3),(7,-1+6),(-5,-2+6),(4,-3+1)\}$
$= \{(-2,9),(5,6),(7,5),(-5,4),(4,-2)\}$
$f-g = \{(-2,-7),(5,0),(7,-7),(-5,-8),(4,-4)\}$
$f \cdot g = \{(-2,8),(5,9),(7,-6),(-5,-12),(4,-3)\}$
$f \div g =$
$$\left\{\left(-2,\frac{1}{8}\right),(5,1),\left(7,-\frac{1}{6}\right),\left(-5,-\frac{1}{3}\right),(4,-3)\right\}$$

In Exercises 7 - 9, $f(x) = 4x^2 - 8x$, $g(x) = 4x$.

7. a. $f(x) + g(x)$ Domain: $\mathbb{R}$ Range: $[-1, \infty)$

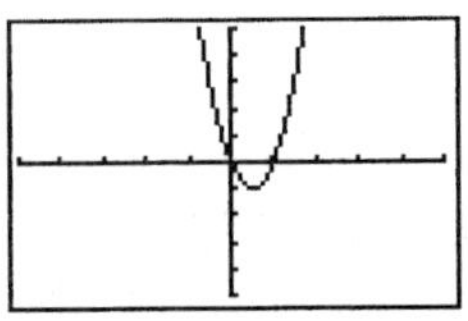

b. $f(x) - g(x)$ Domain: $\mathbb{R}$ Range: $[-9, \infty)$

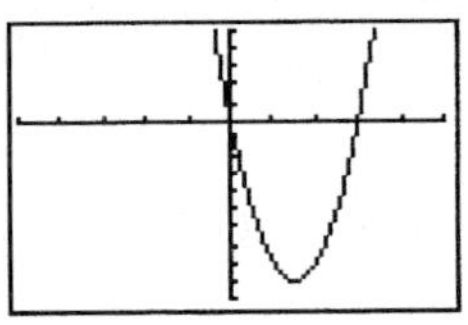

c. $f(x) \cdot g(x)$ Domain: $\mathbb{R}$ Range: $\mathbb{R}$

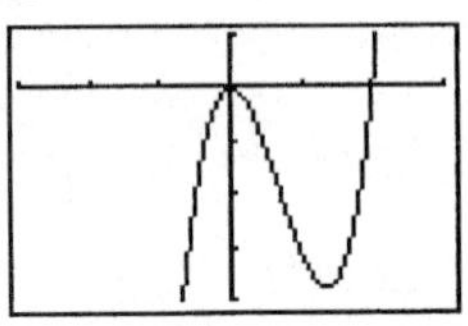

9. The graph does not exist when $x = 0$, since this makes the denominator equal to 0.

11. The point of discontinuity is (0.-2).

Practice

In Exercises 13 - 15, $f(x) = 3x - 2$, $g(x) = 2x^2 + 1$

13. $(f-g)(x) = (3x-2) - (2x^2+1) = 3x - 2 - 2x^2 - 1 = 2x^2 + 3x - 3$ Domain: $\mathbb{R}$

15. $(f \div g)(x) = \dfrac{3x-2}{2x^2+1}$ Domain: $\mathbb{R}$

In Exercises 17 - 19, $f(x) = x^2 - 1$, $g(x) = x^2 - 4$

17. $(f-g)(x) = (x^2-1) - (x^2-4) = -1 + 4 = 3$
Domain: $\mathbb{R}$

19. $(f \div g)(x) = \dfrac{x^2-1}{x^2-4} = \dfrac{x^2-1}{(x-2)(x+2)}$
Domain: $\{x: x \neq 2 \text{ or } -2\}$

21. $f(x) = 9x^6 - 12x^4$, $g(x) = 3x^2$
$$\frac{f(x)}{g(x)} = \frac{9x^6 - 12x^4}{3x^2} = \frac{9x^6}{3x^2} - \frac{12x^4}{3x^2} = 3x^4 - 4x^2$$
Domain: $\{x: x \neq 0\}$

23. $f(x) = 2x^2 + 25x + 63$, $g(x) = 2x + 7$
$$\frac{f(x)}{g(x)} = \frac{2x^2+25x+63}{2x+7} = \frac{(2x+7)(x+9)}{2x+7} =$$
$x + 9$
Domain: $\left\{x : x \neq -\frac{7}{2}\right\}$

In Exercises 25 - 35, $f(x) = 2x + 1$, $g(x) = x^2 - 1$.

25. $(f \circ g)(2) = f(g(2))$ and $g(2) = 2^2 - 1 = 3$
Thus, $f(g(2)) = f(3) = 2(3) + 1 = 7$

27. $(g \circ f)(-3) = g(f(-3))$ and $f(-3) = 2(-3) + 1 = -5$ Thus, $g(f(-3)) = g(-5) = (-5)^2 - 1 = 24$.

29. $(f \circ g)(0) = f(g(0))$ and $g(0) = 0^2 - 1 = -1$. Thus, $f(g(0)) = f(-1) = 2(-1) + 1 = -1$.

31. $(g \circ f)(½) = g(f(½))$ and $f(½) = 2(½) + 1 = 2$. Thus, $g(f(½) = g(2) = 2^2 - 1 = 3$.

33. $(f \circ g)(x) = f(g(x)) = f(x^2 - 1) = 2(x^2 - 1) + 1 = 2x^2 - 2 + 1 = 2x^2 - 1$.

35. $(g \circ f)(2x) = g(f(2x))$ and $f(2x) = 2(2x) + 1 = 4x + 1$. Thus, $g(f(2x)) = g(4x + 1) = (4x + 1)^2 - 1 = 16x^2 + 8x + 1 - 1 = 16x^2 + 8x$.

37. The operation of composition is associative. If you choose $f(x) = x + 2$, $g(x) = x^2$ and $h(x) = 3x + 1$, both compositions $= (3x + 1)^2 + 2$.

39. Let $f(x) = x^2 + 2x - 3$.
$f(a) = a^2 + 2a - 3$ and $f(h) = h^2 + 2h - 3$.
Thus, $f(a) + f(h) = a^2 + 2a + h^2 + 2h - 6$.
$f(a + h) = (a + h)^2 + 2(a + h) - 3 = a^2 + 2ah + 1 + 2a + 2h - 3 \neq f(a) + f(h)$.

41. Let $f(x) = x^3 - 1$ and $f(x + h) = (x + h)^3 - 1$.
$$\frac{f(x+h)-f(x)}{h} = \frac{(x+h)^3 - 1 - (x^3 - 1)}{h} =$$
$$\frac{x^3 + 3x^2h + 3xh^2 + h^3 - 1 - x^3 + 1}{h} =$$
$$\frac{h(3x^2 + 3xh + h^2)}{h} = 3x^2 + 3xh + h^2$$

Applications

43. If $V = ⅓Bh$ and B = area of the circular base $= \pi r^2$, then $V = ⅓\pi r^2 h$. Since the height is given as twice the base radius, $h = 2r$ and so $V = ⅓ \pi r^2 \cdot 2r = ⅔ \pi r^3$.

45. The radius as a function of time is given by $r(t) = 2t$, and $A(r) = \pi r^2$. Thus, $A(t) = A(r(t)) = A(2t) = \pi(2t)^2 = 4\pi t^2$. After 7 minutes, $A(7) = 4\pi\ (7^2) = 196\pi$ in.$^2 \approx 615.75$ in.^{2}, rounded off to the nearest hundredth.

Review

47. Points (3,5) and (-2,1)
$$m = \frac{\Delta y}{\Delta x} = \frac{5-1}{3-(-2)} = \frac{4}{5}$$

49. a. $f(x) = 4x + 2$ Domain: $\mathbb{R}$ Range: $\mathbb{R}$
b. $f(x) = \sqrt{x-5} + 2$
Domain: $[5, \infty)$ Range: $[2, \infty)$
c. $f(x) = (x - 5)^2 + 2$
Domain: $\mathbb{R}$ Range: $[2, \infty)$

Section 11.2 Inverse Functions

Vocabulary

1. inverse relation

3. horizontal line

Concepts

5. T To find the inverse of a relation, switch the coordinates of the ordered pairs. There are no restrictions.

7. The inverse relation: {(5,4),(8,2),(7,-3), (-9,6)}

9. The inverse relation: {(2,- 4),(2,-3),(2,0), (2,1)}

11. The inverse relation is given by

lb	2.2	4.4	6.6	7.7	8.8
kg	1	2	3	3.5	4

13. Not one-to-one, fails the horizontal line test.

15. One-to-one by the horizontal line test.

17. Not one-to-one, fails the horizontal line test.

19. One-to-one by the horizontal line test.

21. Not one-to-one, $y = -1$ results from $x = 1$ and $x = 5$.

23. One-to-one, distinct x-values produce distinct y-values.

25. $y = 3x + 1$
Interchange x and y: $x = 3y + 1$
Solve for y: $y = \dfrac{x-1}{3} = f^{-1}(x)$

27. $x + 4 = 5y$
Interchange x and y: $y + 4 = 5x$
Solve for y: $y = 5x - 4 = f^{-1}(x)$

29. $y = \dfrac{x-4}{5}$
Interchange x and y: $x = \dfrac{y-4}{5}$
Solve for y: $y = 5x + 4 = f^{-1}(x)$

31. $4x - 5y = 20$
Interchange x and y: $4y - 5x = 20$
Solve for y: $y = \dfrac{5x+20}{4} = f^{-1}(x)$

33. $y = 4x + 3$
Interchange x and y: $x = 4y + 3$
Solve for y: $y = \dfrac{x-3}{4} = f^{-1}(x)$

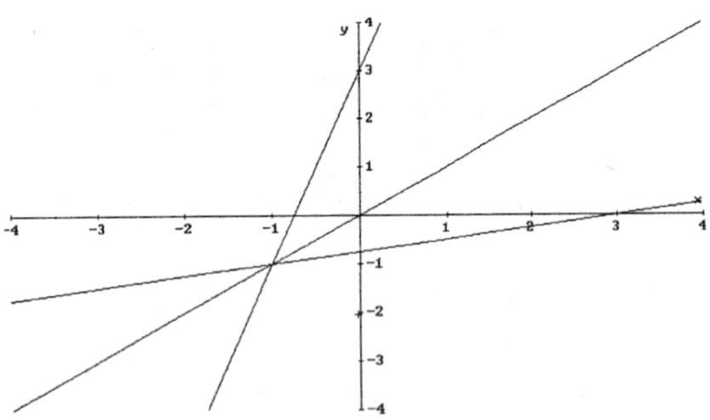

35. $x = \dfrac{y-2}{3}$
Interchange x and y: $y = \dfrac{x-2}{3} = f^{-1}(x)$

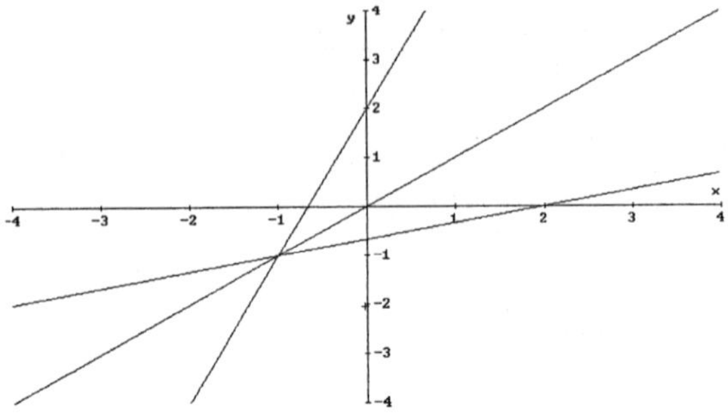

37. $3x - y = 5$
Interchange x and y: $3y - x = 5$
Solve for y: $y = \dfrac{x+5}{3} = f^{-1}(x)$

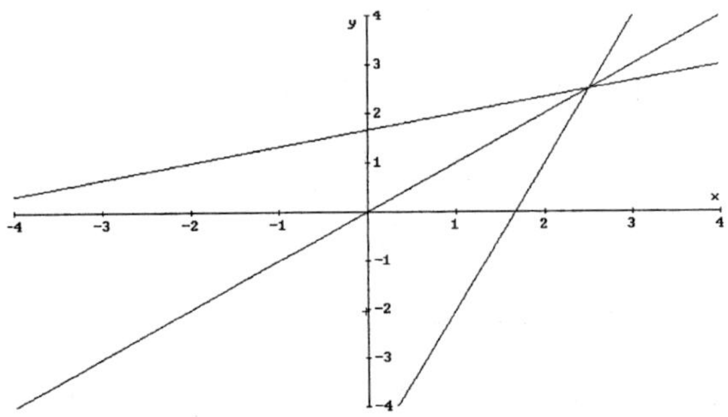

39. $3(x + y) = 2x + 4$
Interchange x and y: $3(y + x) = 2y + 4$
Solve for y: $y = -3x + 4 = f^{-1}(x)$

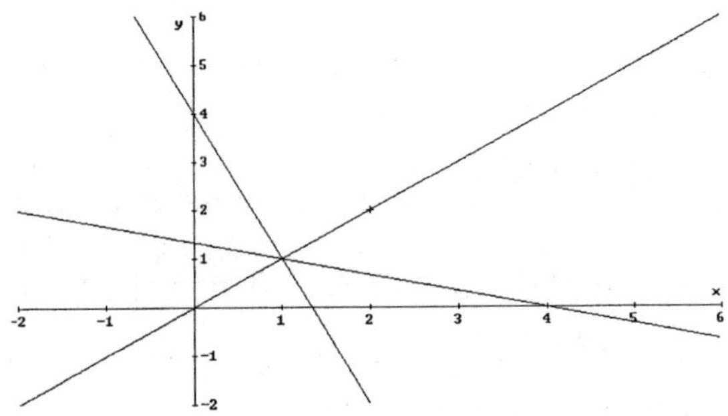

41. $y = 4x + 3,\ y = \dfrac{x-3}{4}$

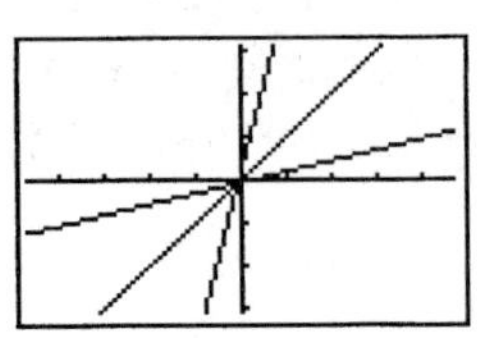

43. $x = \dfrac{y-2}{3}$, $y = \dfrac{x-2}{3}$

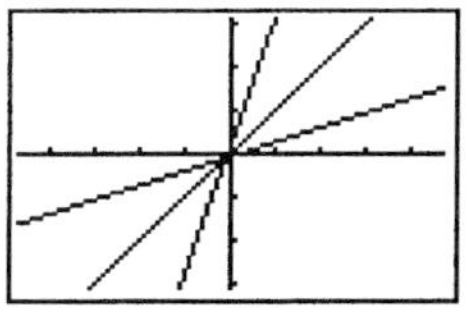

45. $3x - y = 5$, $y = \dfrac{x+5}{3}$

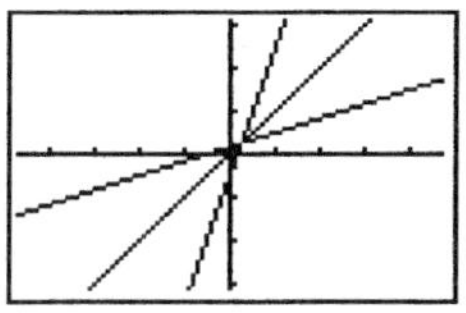

47. $3(x + y) = 2x + 4$, $y = -3x + 4$

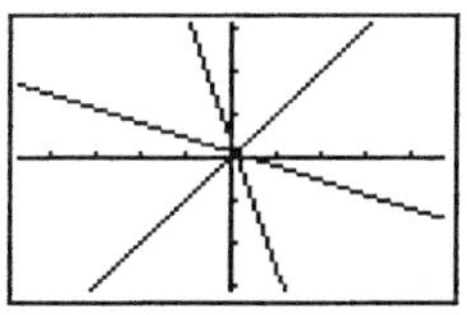

49. Let $y = 2x + 6 = f(x)$ and $y = \frac{1}{2}x - 3 = g(x)$
$f(g(x)) = f(\frac{1}{2}x - 3) = 2(\frac{1}{2}x - 3) + 6 = (x - 6) + 6 = x$ and $g(f(x)) = g(2x + 6) = \frac{1}{2}(2x + 6) - 3 = (x + 3) - 3 = x$. Since $f \circ g = g \circ f = I$, then f and g are inverse functions.

51. Let $y = \dfrac{x+5}{3} = f(x)$ and $y = 3x - 5 = g(x)$.

$f(g(x)) = f(3x - 5) = \dfrac{(3x-5)+5}{3} = \dfrac{3x}{3} = x$

and $g(f(x)) = g(\dfrac{x+5}{3}) = 3\left(\dfrac{x+5}{3}\right) - 5 =$
$(x + 5) - 5 = x$. Since $f \circ g = g \circ f = I$, then f and g are inverse functions.

53. $y = x^2 + 4$
Interchange x and y: $x = y^2 + 4$
Solve for y: $y^2 = x - 4$ or $y = \pm\sqrt{x-4}$.
The inverse relation is <u>not</u> a function.

55. $x = y^2 - 4$
Interchange x and y: $y = x^2 - 4$ is the inverse relation and it is a function.

57. $y = x^3$
Interchange x and y: $x = y^3$
Solve for y: $y = \sqrt[3]{x}$ is the inverse relation and it <u>is</u> a function.

59. $y = \pm\sqrt{x}$
Interchange x and y: $x = \pm\sqrt{y}$
Solve for y: $y = x^2$ is the inverse relation and it <u>is</u> a function.

61. $y = \sqrt{x}$, $x \ge 0$, $y \ge 0$
Interchange x and y: $x = \sqrt{y}$, $y \ge 0$, $x \ge 0$
Solve for y: $y = x^2$, $x \ge 0$ is the inverse relation and it <u>is</u> a function.

Review

63. a. $4^3 = 4 \cdot 4 \cdot 4 = 64$
b. $-4^4 = -4 \cdot 4 \cdot 4 \cdot 4 = -256$
c. $4^{-4} = \dfrac{1}{4^4} = \dfrac{1}{256}$
d. $4^{\frac{1}{2}} = \sqrt{4} = 2$
e. $-4^{\frac{1}{2}} = -\sqrt{4} = -2$
f. $(-4)^{-\frac{1}{2}} = \dfrac{1}{(-4)^{\frac{1}{2}}} = \dfrac{1}{\sqrt{-4}} = \dfrac{1}{2i} = \dfrac{i}{2i^2} = -\dfrac{1}{2}i$

65. a. $f(x) = x^2$ Domain: $\mathbb{R}$ Range: $[0, \infty)$

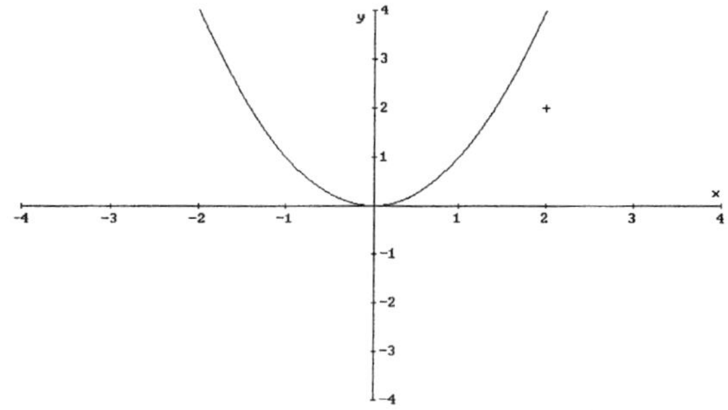

b. $f(x) = x^3$ Domain: $\mathbb{R}$ Range: $\mathbb{R}$

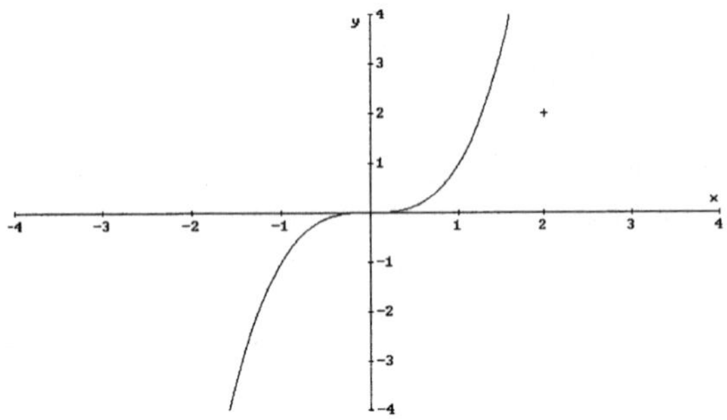

c. $f(x) = \sqrt{x}$ Domain: $[0, \infty)$ Range: $[0, \infty)$

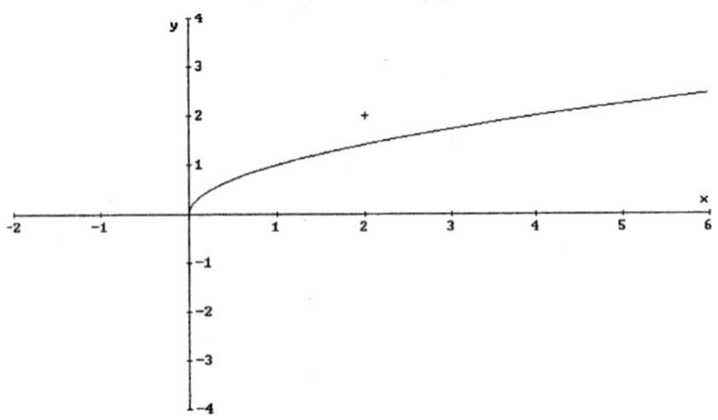

Section 11.3 Exponential Functions

Vocabulary

1. exponential

3. decay

Concepts

5. a. $25^{\frac{1}{2}} = \sqrt{25} = 5$

   b. $16^{\frac{3}{4}} = \left(\sqrt[4]{16}\right)^3 = 2^3 = 8$

   c. $25^{-\sqrt{2}} \approx 0.0105$

   d. $\left(\frac{1}{3}\right)^{\sqrt{3}} \approx 0.1491$

7. a. $f(x) = x^4$, not an exponential function
   b. $f(x) = 4^x$, an exponential function
   c. $f(x) = \left(\frac{1}{2}\right)^x$, an exponential function.
   d. $f(x) = \frac{1}{2} x^2$, not an exponential function.

9. The exponential function $f(x) = b^x$ represents exponential decay if $0 < b < 1$.

11. Since any non-zero base to the 0 power is equal to 1, (0,1) will be the y-intercept for every exponential function $f(x) = b^x$.

13. a. $2^x = 16 = 2^4$, so $x = 4$
    b. $3^x = 243 = 3^5$, so $x = 5$

Practice

15. $f(x) = 3^x$

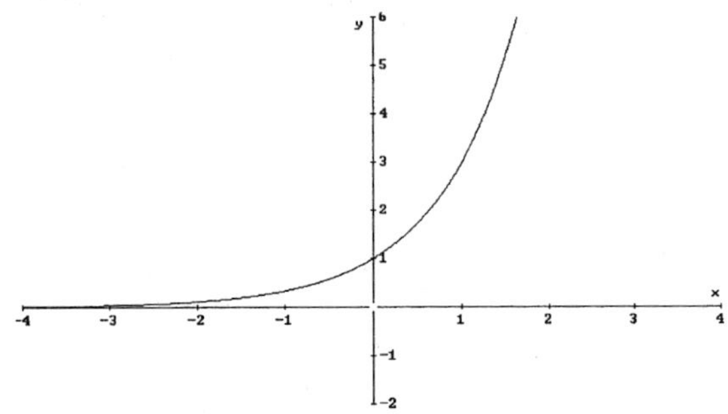

17. $f(x) = 5^x$

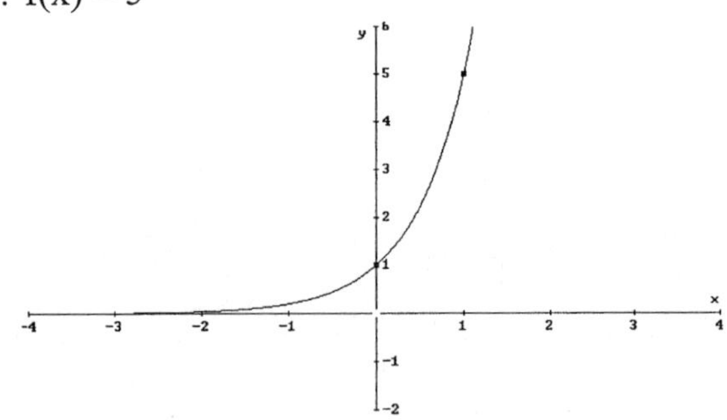

19. $f(x) = \left(\frac{1}{3}\right)^x$

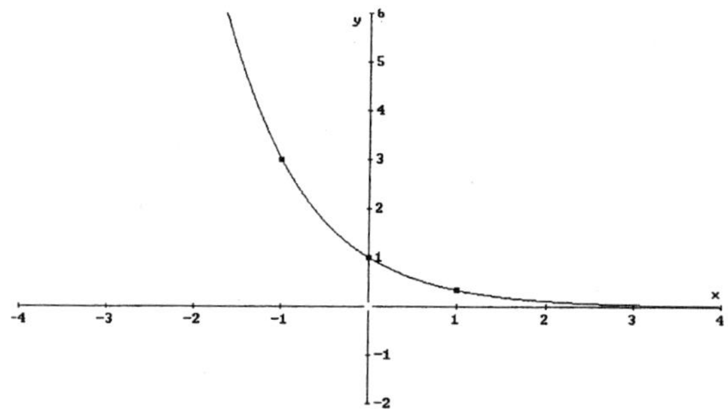

21. $f(x) = \left(\frac{1}{5}\right)^x$

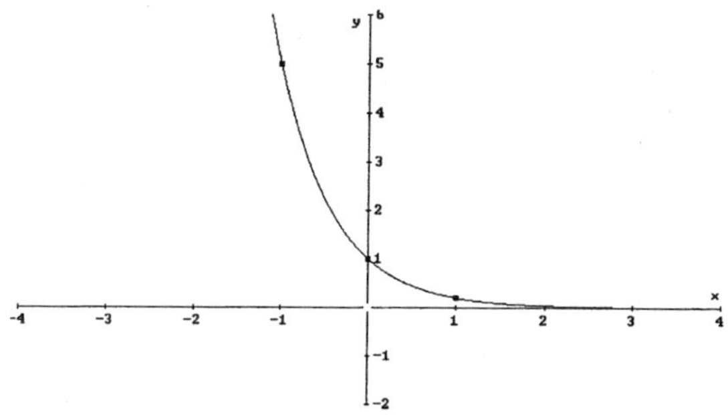

23. a. $f(x) = 2^x + 2$ Graphing in [-5,5] by [-2,10]

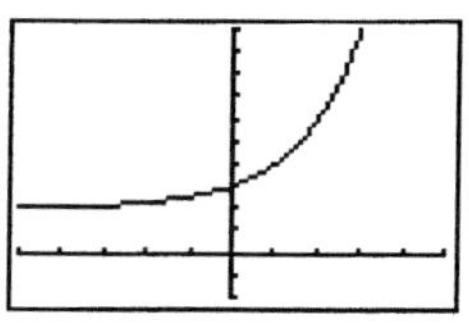

b. $f(x) = 3^x - 2$ Graphing in [-5,5] by [-5,10]

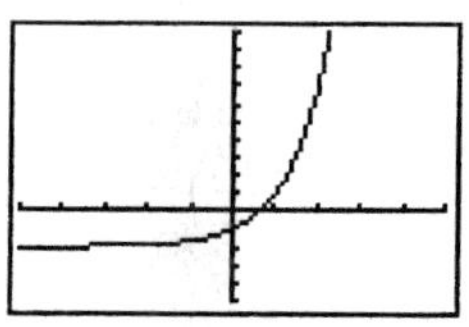

25. a. $f(x) = 3(2^x)$ Graphing in [-5,5] by [-2,10]

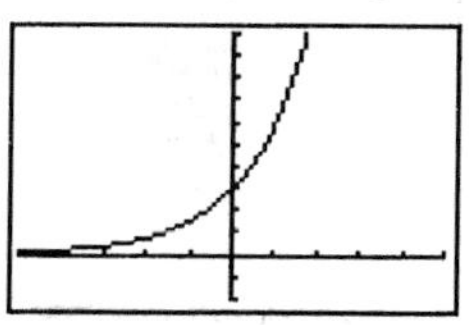

b. $f(x) = 2(3^x)$ Graphing in [-5,5] by [-2,10]

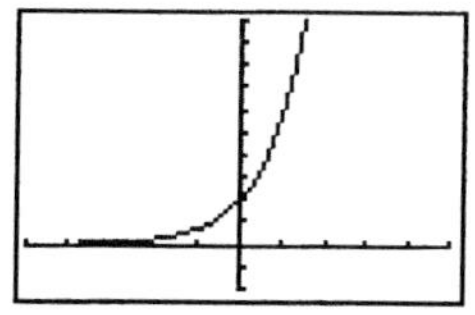

27. a. $f(x) = 2^{-x}$ Graphing in [-5,5] by [-2,10]

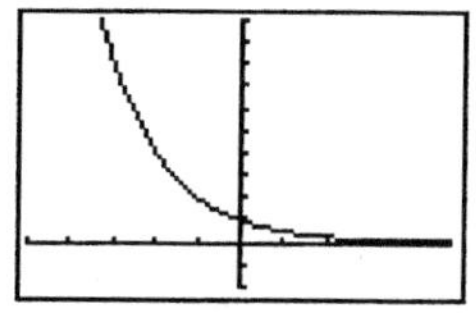

b. $f(x) = 3^{-x}$ Graphing in [-5,5] by [-2,10]

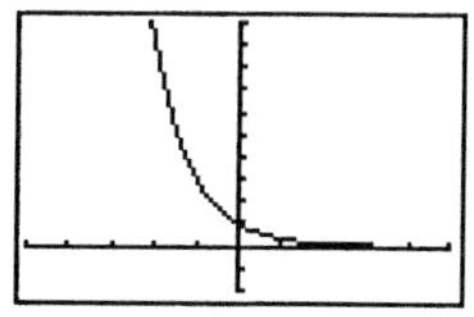

29. a. $f(x) = 2^{x+5}$ Graphing in [-10,5] by [-2,10]

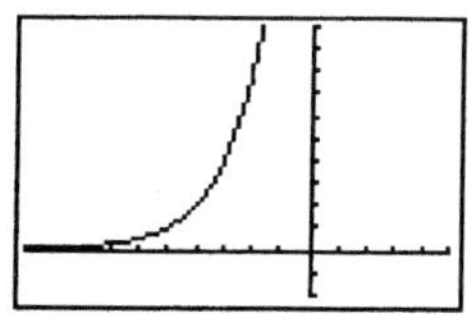

b. $f(x) = 3^{x-4}$ Graphing in [-5,10] by [-2,10]

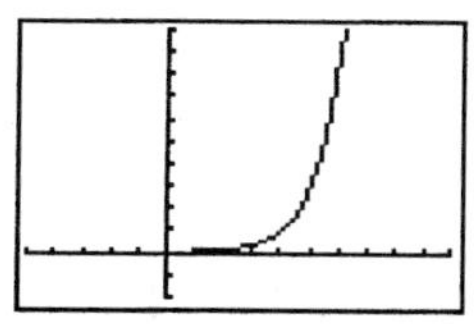

31. $2^3 = 2^{5x+7}$ Thus,

$3 = 5x + 7$

$-4 = 5x$ or $x = \frac{-4}{5}$

33. $9 = 3^{\frac{1}{2}x+5}$ or $3^2 = 3^{\frac{1}{2}x+5}$ Thus,

$2 = \frac{1}{2}x + 5$ or $-3 = \frac{1}{2}x$ so $x = -6$

35. $32 = 4^{6x-5}$ or $2^5 = (2^2)^{6x-5}$ Thus, $5 = 2(6x - 5) = 12x - 10$, so $15 = 12x$ and $x = \frac{15}{12} = \frac{5}{4}$.

37. $4^{5x-1} = 8^{3x+2}$ or $(2^2)^{5x-1} = (2^3)^{3x+2}$ Thus,
$2(5x - 1) = 3(3x + 2)$
$10x - 2 = 9x + 6$, so that $x = 8$.

39. $125^{6-2x} = 625^{3x+1}$ or $(5^3)^{6-2x} = (5^4)^{3x+1}$ Thus,
$3(6 - 2x) = 4(3x + 1)$
$18 - 6x = 12x + 4$
$14 = 18x$ and so $x = \frac{14}{18} = \frac{7}{9}$

Applications

41. $A = P\left(1 + \frac{r}{k}\right)^{kt}$

$P = 10{,}000$, $r = .08$, $k = 4$, $t = 10$

$A = 10000\left(1 + \frac{.08}{4}\right)^{4(10)} = \$22{,}080.40$

43. For $P = 1000$, $k = 4$, $t = 5$ and $r = .055$,

$A = 1000\left(1 + \frac{.055}{4}\right)^{4(5)} = \$1314.07$

For $P = 1000$, $k = 4$, $t = 5$ and $r = .05$,

$A = 1000\left(1 + \frac{.05}{4}\right)^{4(5)} = \$1282.04$

The \$1000 principal will earn \$32.03 more at 5.50% than at 5%.

45. $A = P\left(1 + \frac{r}{k}\right)^{kt}$

$P = \$1$, $k = 1$, $t = 300$, $r = .05$

$A = 1\left(1 + \frac{.05}{1}\right)^{1(300)} = (1.05)^{300} = \$2{,}273{,}996.13$

The dollar will be worth \$2,273,996.13

47. $A = Pe^{rt}$, $P = 1000$, $r = .03$ and $t = 10$
Thus, $A = 1000e^{(.03)(10)} = 3644.62$. There will be 3645 fish in 10 years.

49. $A = Pe^{rt}$, $P = 5 \times 10^9$, $r = .018$ and $t = 30$
Thus, $A = (5 \times 10^9)e^{(.018)(30)} \approx 8{,}580{,}034{,}311$, the approximate world population in 30 yrs.

51. $P = (6 \times 10^6)(2.3)^t$ If $t = 4$, $P = (6 \times 10^6) \cdot (2.3)^4 = 167{,}904{,}600$ bacteria.

53. $A = A_0\left(\frac{2}{3}\right)^t$ If $t = 5$, $A = A_0\left(\frac{2}{3}\right)^5 = \frac{32}{243}A_0$

The amount present in 5 years is $\frac{32}{243}$ of the original amount.

55. $A = Pe^{rt}$, $P = 4570$, $r = -0.06$(since it is losing value) and $t = 6.5$ Then $A = 4570e^{(-.06)(6.5)} = 3094.15$. The trailer will be worth approximately \$3094.15.

Review

57. a. $7^0 = 1$ b. $25^2 = 25 \cdot 25 = 625$

c. $25^{-2} = \frac{1}{25^2} = \frac{1}{625}$ d. $25^{\frac{1}{2}} = \sqrt{25} = 5$

e. $25^{-\frac{1}{2}} = \frac{1}{25^{\frac{1}{2}}} = \frac{1}{\sqrt{25}} = \frac{1}{5}$

59. $\sqrt{3x+1} + 1 = x$

$\sqrt{3x+1} = x - 1$

$3x + 1 = (x-1)^2 = x^2 - 2x + 1$

$0 = x^2 - 5x = x(x-5)$

$x = 0$ or 5, but 0 is extraneous.

61. $h = -16t^2 + 32t + 48$
The maximum occurs at the vertex of the parabola. The ball reaches a maximum height of 64 feet after $t = 1$ second.

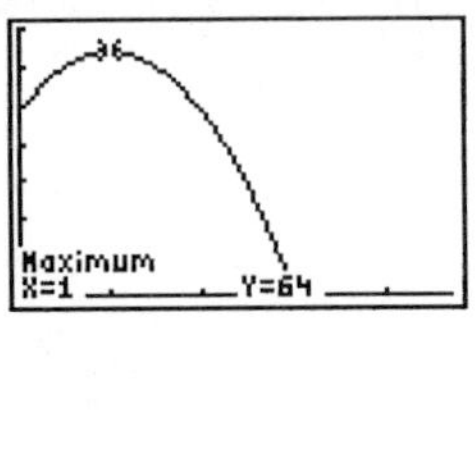

The vertex can also be found by completing the square in t.

Section 11.4 Logarithmic Functions

Vocabulary

1. logarithm

3. 10

5. change of base

Concepts

7. a. $8^2 = 64$ so $\log_8 64 = 2$

b. $4^{-2} = \frac{1}{16}$ so $\log_4 \left(\frac{1}{16}\right) = -2$

c. $\left(\frac{1}{2}\right)^{-5} = 32$ so $\log_{\frac{1}{2}} 32 = -5$

d. $10^{-4} = \frac{1}{10000}$ so $\log_{10}\left(\frac{1}{10000}\right) = -4$

Practice

9. $\log_2 8 = x$ so $2^x = 8 = 2^3$ and $x = 3$

11. $\log_4 64 = x$ so $4^x = 64 = 4^3$ and $x = 3$

13. $\log_{\frac{1}{2}}\left(\frac{1}{8}\right) = x$ so $\left(\frac{1}{2}\right)^x = \frac{1}{8} = \left(\frac{1}{2}\right)^3$ and $x = 3$

15. $\log_9 3 = x$ so $9^x = 3$ Then $(3^2)^x = 3^{2x} = 3$ and so $2x = 1$ or $x = ½$.

17. $\log_{\frac{1}{2}} 8 = x$ so $\left(\frac{1}{2}\right)^x = 8$ Then $(2^{-1})^x = 2^3$ so that $-x = 3$ or $x = -3$.

19. $\log_8 x = 2$ so that $8^2 = x$ or $x = 64$.

21. $\log_7 x = 1$ so that $7^1 = x$ or $x = 7$

23. $\log_{25} x = ½$ so that $x = 25^{\frac{1}{2}} = \sqrt{25} = 5$

25. $\log_5 x = -2$ so that $5^{-2} = x$ or $x = \frac{1}{5^2} = \frac{1}{25}$

27. $\log_{36} x = -½$ so that $x = 36^{-\frac{1}{2}} = \frac{1}{\sqrt{36}} = \frac{1}{6}$

29. $\log_{100}\left(\frac{1}{1000}\right) = x$ so $100^x = \frac{1}{1000}$ Thus,

$(10^2)^x = 10^{-3}$ so that $2x = -3$ or $x = -\frac{3}{2}$

31. $\log_{27} 9 = x$ so that $27^x = 9$. Thus, $(3^3)^x = 3^2$ so that $3x = 2$ or $x = ⅔$.

33. $\log_x(5^3) = 3$ so that $x^3 = 5^3$ and so $x = 5$.

35. $\log_x\left(\frac{9}{4}\right) = 2$ so that $x^2 = \frac{9}{4} = \left(\frac{3}{2}\right)^2$ and $x = \frac{3}{2}$

37. $\log_x\left(\frac{1}{64}\right) = -3$ so $x^{-3} = \frac{1}{64} = 4^{-3}$ and $x = 4$

39. $\log_{2\sqrt{2}} x = 2$ or $(2\sqrt{2})^2 = x$ so $x = 4 \cdot 2 = 8$

41. $2^{\log_2 5} = x$ Since $\log_2 5$ is the exponent on the base 2 that will produce 5, $x = 5$.

43. $x^{\log_4 6} = 6$ $\text{Log}_4 6$ is the exponent on the base 4 which will produce 6. Since 6 is the result here, x (the base) = 4.

45. $\log 10^3 = x$ so $10^x = 10^3$ or $x = 3$.

47. $10^{\log x} = 100 = 10^2$ so that $\log x = 2$ and $x = 10^2 = 100$.

49. $\log 3.25 = 0.5119$

51. $\log 0.00467 = -2.3307$

53. $\ln 0.93 = -0.0726$

55. $\ln 37.896 = 3.6348$

57. $\log(\ln 1.7)) = -0.2752$

59. ln(log 0.1) is not a real number since log(0.1) = -1 and ln(-1) is not real.

61. $\log y = 1.4023$ so $y = 10^{1.4023} = 25.2522$

63. $\ln y = 4.24$ so $y = e^{4.24} = 69.4079$

65. $\log y = -3.71$ so $y = 10^{-3.71} = 0.0002$

67. $\log y = \ln 8$ so $y = 10^{\ln 8} = 120.0719$

69. $y = \log_3 x$

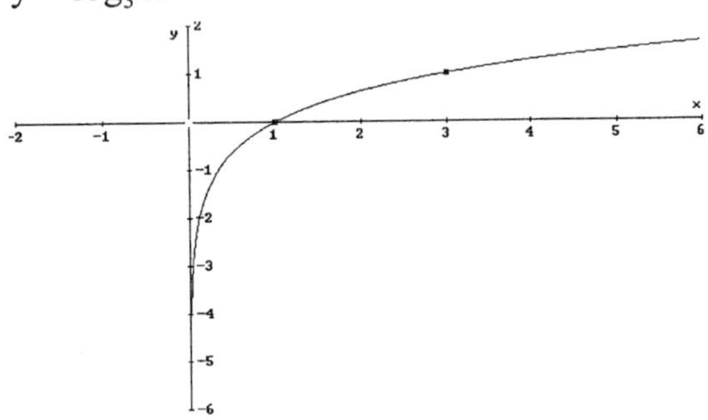

71. $y = \log_{\frac{1}{2}} x$

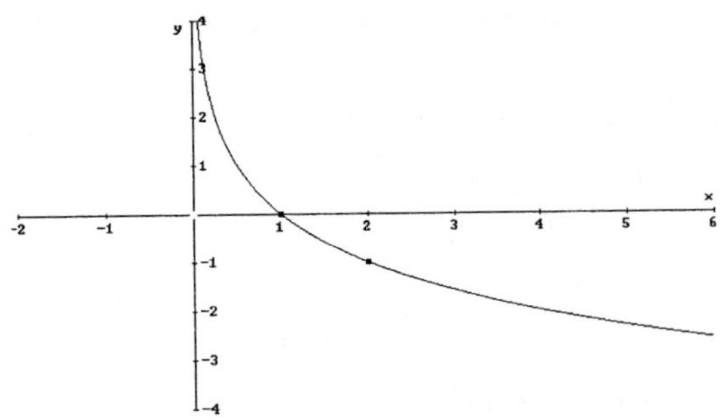

73. $y = 2^x$ and $y = \log_2 x$

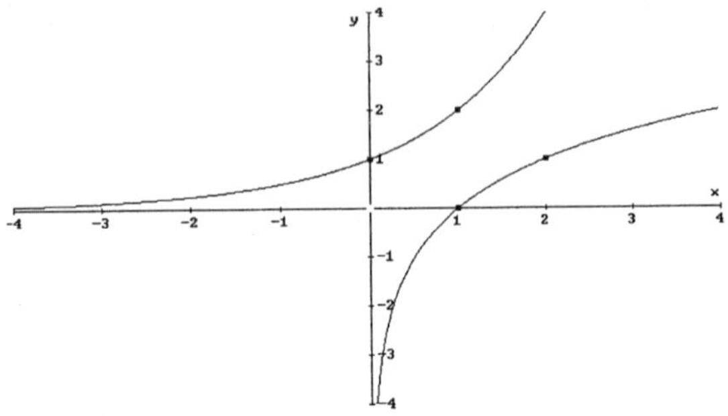

75. $y = \left(\frac{1}{4}\right)^x$ and $y = \log_{\frac{1}{4}} x$

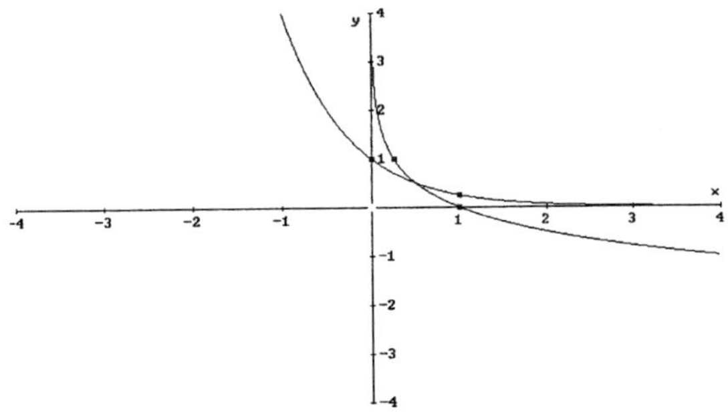

77. The graph of $y = k + \log x$ is a vertical translation of the graph of $y = \log x$.

79. The graph of $y = k \log x$ steepens ($|k|>1$) or flattens ($|k|<1$) the graph of $y = \log x$ and reflects it in the x-axis if $k < 0$.

81. $\log_3 7 = \dfrac{\log 7}{\log 3} = 1.7712$

83. $\log_{\frac{1}{3}} 3 = \dfrac{\log 3}{\log \frac{1}{3}} = -1$

85. $\log_3 8 = \dfrac{\log 8}{\log 3} = 1.8928$

87. $\log_{\sqrt{2}} \sqrt{5} = \dfrac{\log \sqrt{5}}{\log \sqrt{2}} = 2.3219$

Applications

89. The pH value = $-\log(1.7 \times 10^{-5}) \approx 4.77$.

91. Let x = hydrogen ion concentration.
For pH = 6.8, $-\log x = 6.8$ so $\log x = -6.8$ and $x = 10^{-6.8} \approx 1.58 \times 10^{-7}$.
For pH = 7.6, $-\log x = 7.6$, so $\log x = -7.6$ and $x = 10^{-7.6} \approx 2.51 \times 10^{-8}$.
The safe range is between concentrations of 2.51×10^{-8} and 1.58×10^{-7} gram-ions.

93. Using the $A = Pe^{rt}$ growth model and a growth rate r = .12, $2P = A = Pe^{0.12t}$, so that $2 = e^{0.12t}$. Thus, $\ln 2 = .12t$ and $t = \dfrac{\ln 2}{.12}$ and so $t \approx 5.8$. It will take about 5.8 years for the population to double.

95. If r = .12, the time t for the population to triple is $t = \dfrac{\ln 3}{.12} \approx 9.2$ years.

Review

97. $y = \sqrt{2x+5}$
a. The function is increasing.
b. The domain is where $2x + 5 \geq 0$, or $x \geq \dfrac{-5}{2}$
Thus, the domain is $[-\dfrac{5}{2}, \infty)$.

99. The reason for the extraneous solution x = 2 to the squared equation is that squaring both sides of an equation doesn't necessarily produce an equivalent equation. It is the solution to $\sqrt{2x+5} = -(x-5) = 5 - x$.

Section 11.5 Properties of Logarithms

Vocabulary

1. product

3. $p \log_b M$

Concepts

5. T

7. T

9. F; $7^{\log_7 7} = 7$

11. F: $\log_b\left(\dfrac{A}{B}\right) = \log_b A - \log_b B$

13. T

15. F; $\log_b 2 = \dfrac{1}{\log_2 b}$

17. T

19. T

21. T

23. F; $\log_{10} 10^3 = 3\left(10^{\log_{10} 1}\right)$

25. ln 1 = 0 so $\log_b$ (ln 1) is undefined.

27. The Quotient Property was used.

29. The Change of Base Property was used.

31. The Change of Base Property was used.

33. $\log_b xz = \log_b x + \log_b z$

35. $\log_b\left(\dfrac{x}{yz}\right) = \log_b x - \log_b yz =$
$\log_b x - (\log_b y + \log_b z) =$
$\log_b x - \log_b y - \log_b z$

37. $\log_b (x y^2 z^3) = \log_b x + \log_b y^2 + \log_b z^3 =$
$\log_b x + 2\log_b y + 3\log_b z$

39. $\log_b x^3 y^{\frac{1}{2}} = \log_b x^3 + \log_b y^{\frac{1}{2}} =$

$3\log_b x + \frac{1}{2}\log_b y$

41. $\log_b \sqrt{xy} = \log_b (xy)^{\frac{1}{2}} = \frac{1}{2}\log_b(xy) =$

$\frac{1}{2}(\log_b x + \log_b y) = \frac{1}{2}\log_b x + \frac{1}{2}\log_b y$

43.

$\log_b \sqrt[4]{\frac{x^3y^2}{z^4}} = \log_b\left(\frac{x^3y^2}{z^4}\right)^{\frac{1}{4}} = \frac{1}{4}\log_b\left(\frac{x^3y^2}{z^4}\right) =$

$\frac{1}{4}\left[\log_b(x^3y^2) - \log_b z^4\right] =$

$\frac{1}{4}\left[\log_b x^3 + \log_b y^2 - \log_b z^4\right] =$

$\frac{1}{4}\left[3\log_b x + 2\log_b y - 4\log_b z\right] =$

$\frac{3}{4}\log_b x + \frac{1}{2}\log_b y - \log_b z$

45. $\log_b x + \log_b (x + 2) - \log_b 8 = \log_b x(x + 2) -$
$\log_b 8 = \log_b \frac{x^2 + 2x}{8}$

47. $-2 \log_b x - 3 \log_b y + \log_b z = -\log_b x^2 -$
$\log_b y^3 + \log_b z = -(\log_b x^2 + \log_b y^3) + \log_b z$
$= -\log_b (x^2 y^3) + \log_b z = \log_b \frac{z}{x^2y^3}$

49. $3 \log_b (x + 1) - 2 \log_b (x + 2) + \log_b x =$
$\log_b (x + 1)^3 - \log_b (x + 2)^2 + \log_b x =$
$\log_b x(x + 1)^3 - \log_b (x + 2) = \log_b \frac{x(x+1)^3}{(x+2)^2}$

51. $\log_b (xy + y^2) - \log_b (xz + yz) + \log_b z =$
$\log_b z(xy + y^2) - \log_b (xz + yz) =$
$\log_b\left(\frac{zy(x+y)}{z(x+y)}\right) = \log_b y$.

Applications

53. db gain $= 20 \log \left(\frac{E_0}{E_I}\right) = 20 \log \left(\frac{30}{.1}\right) =$
$20 \log (300) \approx 49.5$

55. $R = \log \left(\frac{A}{P}\right)$, $A = 5000$ and $P = 0.2$. Thus,
$R = \log \left(\frac{5000}{.2}\right) = \log 25000 \approx 4.4$

57. $R = \log \left(\frac{A}{P}\right)$, $R = 4$, $P = 1/4 = .25$ Thus,
$4 = \log\left(\frac{A}{.25}\right)$ or $\frac{A}{.25} = 10^4$ and $A = .25\left(10^4\right) =$
2500 micrometers

59. $L = k \ln I$ If intensity is doubled, $L = k \ln(2I)$ $= k \ln 2 + k \ln I$. The loudness increases by $k \ln 2$.

61. Let $L_0 = k \ln I_0$. Tripling the loudness means that $3L_0 = 3k \ln I_0 = k \ln (I_0^3)$. The intensity must be cubed.

63. $t = -3.7 \ln (1 - C)$ For an 80% charge, $C = .80$, so $t = -3.7 \ln (.20) \approx 6$ hours.

65. $n = \frac{\log V - \log C}{\log\left(1 - \frac{2}{N}\right)}$, $C = \$37{,}500$, $V = \$8000$,
$N = 5$. Thus, $n = \frac{\log 8000 - \log 37000}{\log\left(1 - \frac{2}{5}\right)} \approx 3$

The computer is approximately 3 years old.

67. An earthquake measuring 7 on the Richter scale is 10 times as intense as one measuring 6 on the Richter scale. (See Exercise 58.)

Review

69. a. $\frac{3}{4}(x+5)-3 \leq 0$ Solution: $(-\infty, -1]$

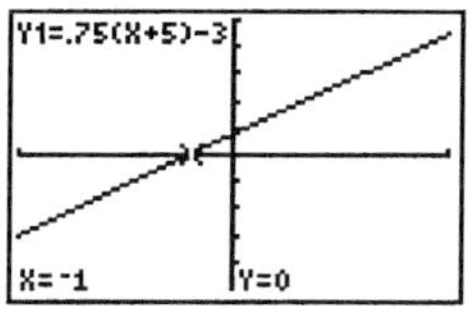

Window: [-5,5]by[-5,5]

b. $\frac{3}{4}|x+5|-3 \leq 0$ Solution: [-9 -1]

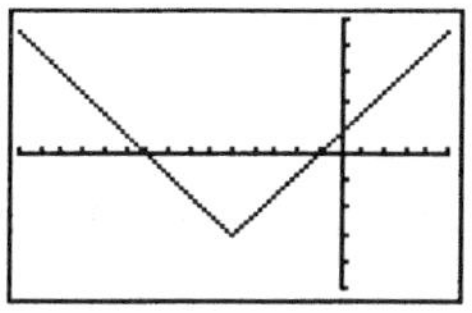

Window: [-15,5]by[-5,5]

c. $\frac{3}{4}(x+5)^2-3 \leq 0$ Solution: [-7, -3]

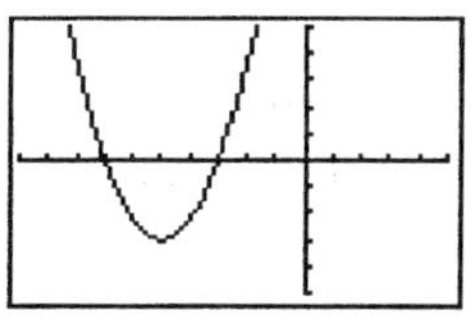

Window: [-10,5]by[-5,5]

71. Graphically solving an inequality in one variable gives a solution set that is a set of real numbers (intervals on the number line). Graphing an inequality in two variables gives a set of ordered pairs, a subset of the coordinate plane.

Section 11.6 Exponential and Logarithmic Equations

Concepts

1. For x = -4, log (2x + 1) = log [2(- 4) + 1] = log (-7), which is not real. Thus, - 4 cannot be a solution of the equation.

3. The solution is 2.398.

5. The solution is 23.208

Practice

7. $4^x = 5$ so that x log 4 = log 5 and $x = \frac{\log 5}{\log 4} =$ 1.1610.

9. $13^{x-1} = 2$ so that (x – 1)log 13 = log 2 and $x - 1 = \frac{\log 2}{\log 13}$, or $x = \frac{\log 2}{\log 13} + 1 = 1.2702$.

11. $2^{x+1} = 3^x$ so that (x + 1)log 2 = x log 3 or
x log 2 + log 2 = x log 3
log 2 = x log 3 – x log 2 = x (log 3 – log 2)
$x = \frac{\log 2}{\log 3 - \log 2} = 1.7095$.

13. $2^x = 3^x$ so that x log 2 = x log 3 or x log 2 – x log 3 = x(log 2 – log 3) = 0. Thus, x = 0.

15. $7^{x^2} = 10$ so that x^2 log 7 = log 10 = 1. Thus, $x^2 = \frac{1}{\log 7}$ so $x = \pm\sqrt{\frac{1}{\log 7}} = \pm 1.0878$

17. $8^{x^2} = 9^x$ so that x^2 log 8 = x log 9 or x^2 log 8 – x log 9 = x(x log 8 – log 9) = 0. Thus, x = 0 or $x = \frac{\log 9}{\log 8} = 1.0566$.

19. $2^{x^2-2x} = 8 = 2^3$ so that $x^2 - 2x = 3$ or $x^2 - 2x - 3 = (x-3)(x+1) = 0$. So x = 3 or -1

21. $3^{x^2-4x} = \frac{1}{81} = 3^{-4}$ so that $x^2 - 4x = -4$ or $x^2 - 4x + 4 = (x-2)^2 = 0$. Thus, x = 2.

23. log 2x = log 4 so that 2x = 4 or x = 2.

25. log (3x + 1) = log (x + 7) so that 3x + 1 = x + 7, 2x = 6 or x = 3.

27. log (2x – 3) – log (x + 4) = 0 or log (2x – 3) = log (x + 4) so that 2x – 3 = x + 4, or x = 7.

29. $\log\frac{4x+1}{2x+9} = 0$ so that $\frac{4x+1}{2x+9} = 10^0 = 1$. Thus, $4x + 1 = 2x + 9$, $2x = 8$ and $x = 4$.

31. $\log x + \log(x - 48) = 2$
$\log x(x - 48) = 2$, so that $x(x - 48) = 10^2 = 100$ or $x^2 - 48x - 100 = 0$
$(x - 50)(x + 2) = 0$, and $x = 50$ or -2, but -2 is extraneous, so 50 is the only solution.

33. $\log(x + 90) = 3 - \log x$ or $\log(x + 90) + \log x = 3$ so that $\log x(x + 90) = 3$ and $x^2 + 90x = 10^3 = 1000$. Then $x^2 + 90x - 1000 = 0$ or $(x + 100)(x - 10) = 0$ and $x = 10$ or $x = -100$, but -100 is extraneous, so 10 is the only solution.

35. $\log x^2 = (\log x)^2$ so that $2\log x = (\log x)^2$ and $(\log x)^2 - 2\log x = \log x(\log x - 2) = 0$. Thus, $\log x = 0$ and $x = 1$ or $\log x = 2$ and $x = 10^2 = 100$.

37. $\frac{\log(3x-4)}{\log x} = 2$ or $\log(3x-4) = 2\log x = \log x^2$
Thus, $x^2 = 3x - 4$ or $x^2 - 3x + 4 = 0$. Since the discriminant $b^2 - 4ac = -7$, there are no real solutions.

39. $\frac{\log(5x+6)}{2} = \log x$ or
$\log(5x+6) = 2\log x = \log x^2$
Thus, $x^2 = 5x + 6$ or $x^2 - 5x - 6 = 0$ and so $(x - 6)(x + 1) = 0$ and $x = -1$ or $x = 6$, but -1 is extraneous, so 6 is the only solution.

Applications

41. $A = A_0 2^{\frac{-t}{h}}$, $A = .75A_0$ and $h = 12.4$.

$$.75A_0 = A_0 2^{\frac{-t}{12.4}} \text{ or } .75 = 2^{\frac{-t}{12.4}}$$

$$\log(.75) = \left(\frac{-t}{12.4}\right)\log 2$$

$$t = -12.4\left(\frac{\log(.75)}{\log 2}\right) \approx 5.15 \text{ years}$$

43. $A = A_0 2^{\frac{-t}{h}}$, $A = .20A_0$ and $h = 18.4$.

$$.20A_0 = A_0 2^{\frac{-t}{18.4}} \text{ or } .20 = 2^{\frac{-t}{18.4}}$$

$$\log(.20) = -\frac{t}{18.4}\log 2$$

$$t = 18.4\frac{\log(.2)}{\log 2} \approx 42.72 \text{ days}$$

45. $A = A_0 2^{\frac{-t}{h}}$, $A = .60A_0$ and $h = 5700$

$$.60A_0 = A_0 2^{\frac{-t}{5700}} \text{ or } .60 = 2^{\frac{-t}{5700}}$$

$$\log(.60) = -\frac{t}{5700}\log 2$$

$$t = -5700\frac{\log(.60)}{\log 2} \approx 4200 \text{ years old.}$$

47. $A = P\left(1 + \frac{r}{k}\right)^{kt}$, $A = 800$, $P = 500$, $k = 2$ and $r = .085$. Then $800 = 500\left(1 + \frac{.085}{2}\right)^{2t}$ or

$$1.6 = \left(1 + \frac{.085}{2}\right)^{2t} \text{ so } \log(1.6) = 2t(1 + .0425)$$

$$t = \frac{\log(1.6)}{2\log(1.0425)} \approx 5.65 \text{ years.}$$

49. $A = P\left(1 + \frac{r}{k}\right)^{kt}$, $A = 2100$, $P = 1300$, $k = 4$, $r = .09$. Then, $2100 = 1300\left(1 + \frac{.09}{4}\right)^{4t}$ or

$\frac{21}{13} = \left(1+\frac{.09}{4}\right)^{4t}$ so $\log\left(\frac{21}{13}\right) = 4t\log(1.0225)$

$t = \frac{\log\left(\frac{21}{13}\right)}{4\log(1.0225)} \approx 5.39$ years.

It will take 4(5.39) = 21.56 or 22 quarters to get at least \$2100.

51. Use a principal of \$100 and a time of 1 year in all three cases.
At 4.9%, A = 100(1.049) = \$104.90
At 4.8% compounded quarterly, A =
$100\left(1+\frac{.048}{4}\right)^4 \approx \$104.89$.
At 4.7% compounded monthly, A =
$100\left(1+\frac{.047}{12}\right)^{12} \approx \$104.80$. The 4.9% investment is best.

53. $P = P_0 e^{kt}$, $P_0 = 30{,}000$. The population will double in 5 years so $60000 = 30000e^{k\cdot 5}$ or $2 = e^{5k}$. Thus, ln 2 = 5k, so $k = \frac{\ln 2}{5} \approx .13863$ and $P = 30000e^{(0.13863)t}$. For P = 1,000,000,
$1{,}000{,}000 = 30{,}000e^{(0.13863)t}$ or $\frac{100}{3} = e^{(0.13863)t}$

Then $\ln\left(\frac{100}{3}\right) = .13863t$ so $t = \frac{\ln\left(\frac{100}{3}\right)}{.13863} \approx$ 25.3 years.

55. $P = P_0e^{kt}$ Since the culture doubles in 24 hours, $2P_0 = P_0e^{k\cdot 24}$ or $2 = e^{24k}$. Thus, ln 2 = 24k and $k = \frac{\ln 2}{24} \approx .02888$ and $P = P_0e^{(0.02888)t}$.
When t = 36, $P = P_0e^{(0.02888)36} \approx (2.83)P_0$. The total population will be 283% of the original or an increase of 183%.

57. $T = 60 + 40e^{kt}$ After 3 minutes, $90 = 60 + 40e^{3k}$ so that $30 = 40e^{3k}$ and $e^{3k} = .75$.
Then 3k = ln (.75) and $k = \frac{\ln(.75)}{3} \approx -.0959$.

59. log x + log (x – 15) = 2 Solution: {20}
Use 'intersect' in [-5,25] by [-5,5].

61. $2^{x+1} = 7$ Solution: {1.81}
Use 'intersect' in [-5,5] by [-5,10].

63. To solve $2^x = 7$, take base 10 logs on both sides to get $\log(2^x) = \log 7$. Use properties of logs to write this as x log 2 = log 7. Now divide by log 2 to find x.

Review

65. $h = -16t^2 + 80t + 3$

a. Graphing in [-10,10] by [-10,110] gives

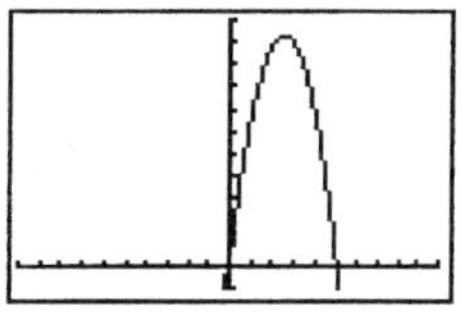

b. Using 'value', the ball is 99' high.

c. It is at the same height at t = 2 secs.
Use 'intersect' to solve $-16t^2 + 80t + 3 = 99$.

d. The ball is on the ground when h = 0, at $t \approx 5.04$ secs. Use 'root' or 'zero' to solve h = 0.

67. The equation of the line through (0,3) and (2.5, 103), using m = 40 from Exercise 66, is y = 40x + 3, in slope-intercept form.

Chapter 11 Summary

1. $f(x) = 2x,\ g(x) = x^2 - 1$
   a. $(f + g)(x) = 2x + (x^2 - 1) = x^2 + 2x - 1$
   b. $(f - g)(x) = 2x - (x^2 - 1) = 2x - x^2 + 1$
   c. $(f \cdot g)(x) = 2x(x^2 - 1) = 2x^3 - 2x$
   d. $(f \div g)(x) = \dfrac{2x}{x^2 - 1},\ x \neq \pm 1$

2. $f(x) = 2x,\ g(x) = x^2 + 1$
   a. $(f \circ g)(x) = f(g(x)) = f(x^2 + 1) = 2(x^2 + 1) = 2x^2 + 2$
   b. $(g \circ f)(x) = g(f(x)) = g(2x) = (2x)^2 + 1 = 4x^2 + 1$
   c. $(f \circ g)(2) = f(g(2)) = f(5) = 10$
   d. $(g \circ f)(-1) = g(f(-1)) = g(-2) = 5$

3. $A(r) = \pi r^2,\ r = f(t) = 5t$. Thus, $A(t) = A(f(t)) = \pi(5t)^2 = 25\pi t^2$. Then $A(40) = 25\pi(40)^2 = 125{,}664\ m^2$, the area 40 minutes after the leak began.

4.

mi	.25	.50	.75
ft	1320	2640	3960
mi	1	1.25	
ft	5280	6600	

5. a. $y = 2(x - 3)$

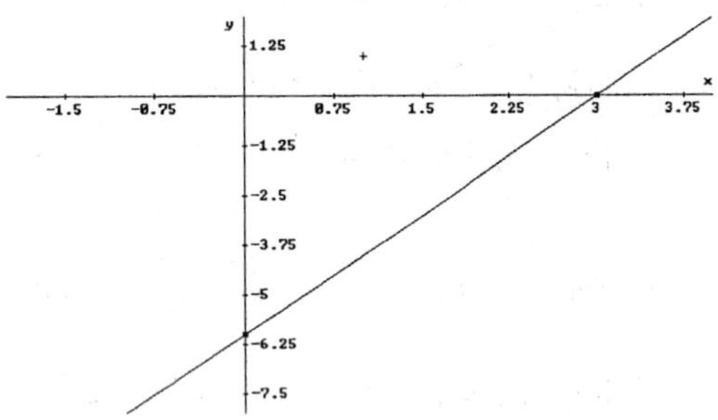

The function is one-to-one.

b. $y = x^2 - 4$

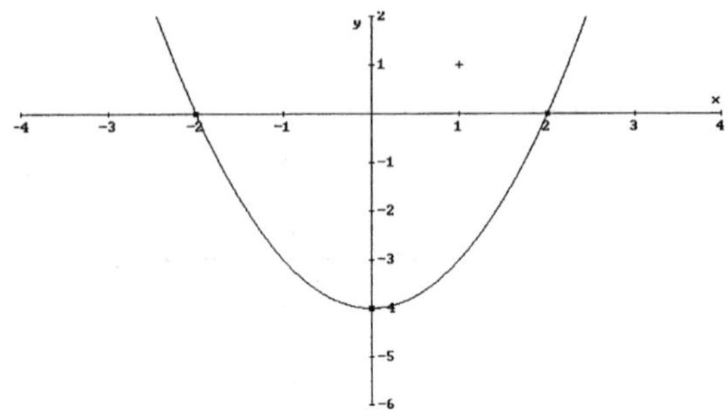

The function is not one-to-one

6. a. {(4,1), (2,-3),(3,2),(1,4)} is a one-to-one function since distinct values of x result in distinct values of y.
   b. {(2,3),(3,2),(4,2),(1,5)} is not a one-to-one function since y = 2 for both x = 3 and x = 4.

7. a. $y = 6x - 3$
   Interchange x and y: $x = 6y - 3$
   Solve for y: $y = \dfrac{x + 3}{6}$

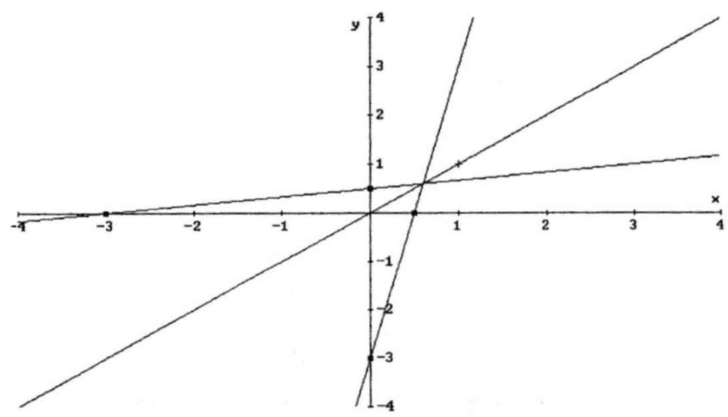

b. $y = \frac{1}{2}x + 3$
   Interchange x and y: $x = \frac{1}{2}y + 3$
   Solve for y: $y = 2(x - 3)$

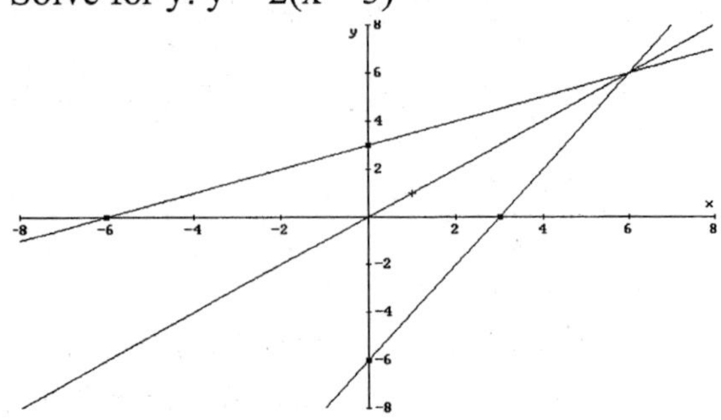

8. a. $f(x) = \left(\frac{6}{5}\right)^x$

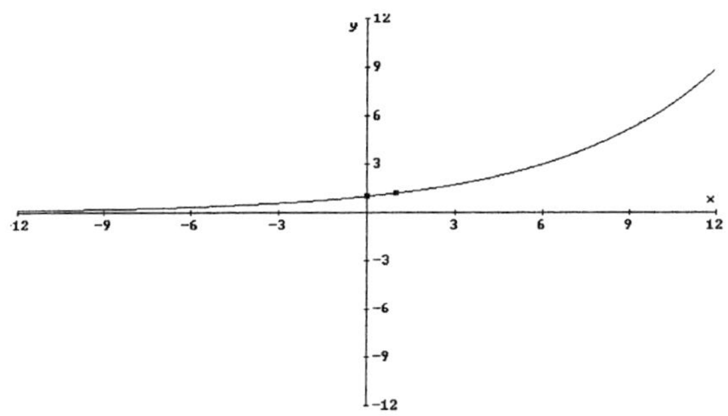

b. $f(x) = \left(\frac{1}{5}\right)^x$

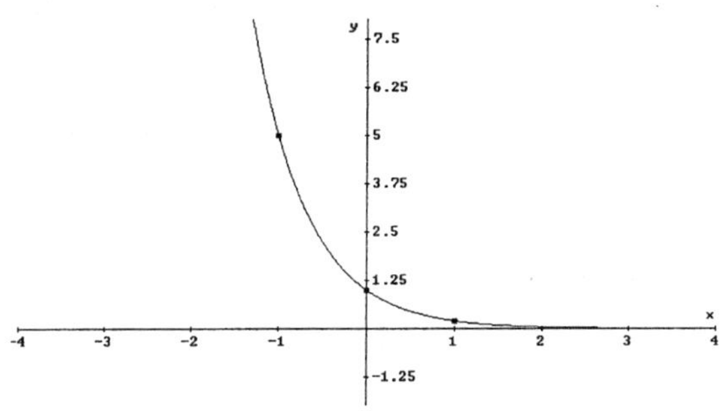

c. $f(x) = e^x$

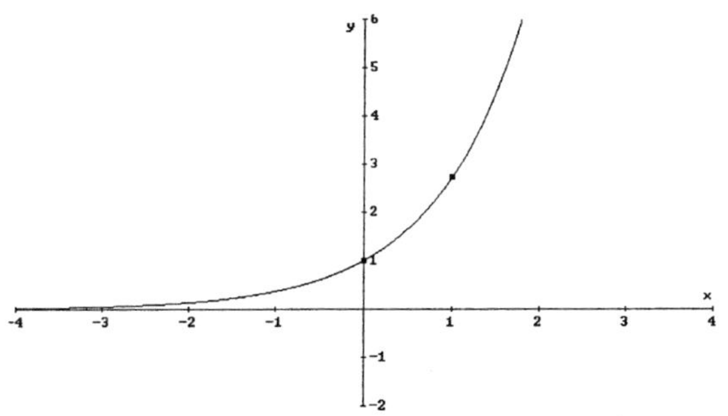

9. a. $2^{x+5} = 2^{x^2-4x}$ so that $x + 5 = x^2 - 4x$ or $x^2 - 3x - 5 = 0$ and $x = \dfrac{5 \pm \sqrt{5^2 - 4 \cdot 1 \cdot -5}}{2 \cdot 1}$ or $x = \dfrac{5 \pm \sqrt{65}}{2}$

b. $3^{x^2} = 9^{2x}$ or $3^{x^2} = (3^2)^{2x} = 3^{4x}$ Thus, $x^2 = 4x$ or $x^2 - 4x = x(x - 4) = 0$, so $x = 0$ or $x = 4$.

c. $2^{x^2+4x} = \frac{1}{8} = 2^{-3}$ so that $x^2 + 4x = -3$ or $x^2 + 4x + 3 = (x + 1)(x + 3) = 0$. Thus, $x = -1$ or $x = -3$.

d. $\left(\frac{1}{5}\right)^x = 5^{x-3}$ or $(5^{-1})^x = 5^{x-3}$ so that $-x = x - 3$ or $2x = 3$ and $x = 1.5$.

10. $3^x = -x^x + 5$ Solution: {1.203}
Use 'intersect' in [-5,5] by [-10,10].

11. $A = P\left(1 + \frac{r}{k}\right)^{kt}$, P = 1000, r = .06, k = 2, t = 1

$A = 1000\left(1 + \frac{.06}{2}\right)^{2 \cdot 1} = 1000(1.03)^2 =$ \$1060.90.

12. $A = Pe^{rt}$, P = 1000, r = .06, t = 1
$A = 1000e^{(.06)\,1} = \$1061.84$

13. a. $\log_3 9 = 2$ since $3^2 = 9$

b. $\log_9 \frac{1}{3} = -\frac{1}{2}$ since $9^{-\frac{1}{2}} = \frac{1}{3}$

c. $\log_5 (.04) = \log_5 \frac{1}{25} = -2$ since $5^{-2} = \frac{1}{25}$

d. $\ln e^9 = 9$

14. a. $\log_2 x = 3$ so $x = 2^3 = 8$

b. $\log_3 x = -2$ so $x = 3^{-2} = \frac{1}{9}$

c. $\log_x 9 = 2$ so $x^2 = 9$ or $x = \pm 3$

d. $\log_x .125 = -3$ or $x^3 = .125 = \frac{1}{8}$, so $x = .5$

e. $\log_3 \sqrt{3} = x$ so $3^x = \sqrt{3} = 3^{\frac{1}{2}}$ and $x = ½$.

f. $\log_2 \sqrt{2} = x$ or $2^2 = \sqrt{2} = 2^{\frac{1}{2}}$ and $x = ½$.

15. a. $\log_3 9 = \dfrac{\log 9}{\log 3} = 2$

b. $\log_5 7 = \dfrac{\log 7}{\log 5} \approx 1.21$

16. a. $f(x) = 3^x$ and $f^{-1}(x) = \log_3 x$

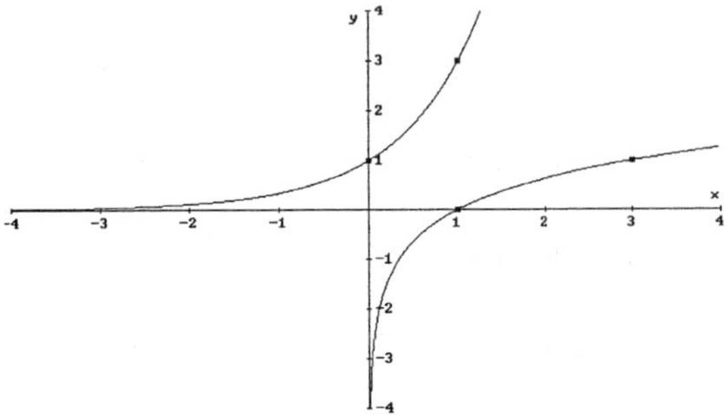

b. $f(x) = \left(\frac{1}{2}\right)^x$ and $f^{-1}(x) = \log_{\frac{1}{2}} x$

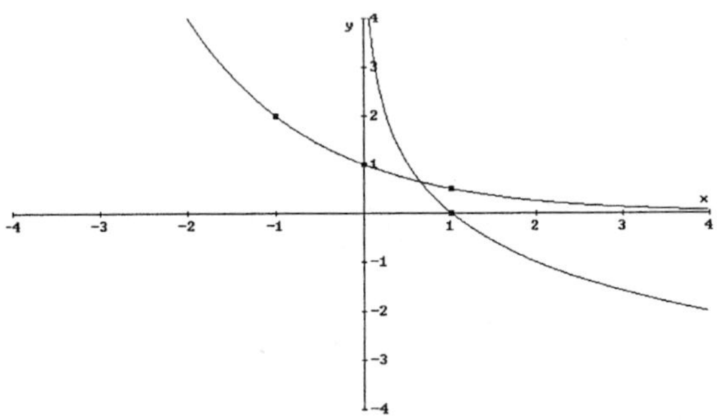

c. $f(x) = \left(\frac{2}{5}\right)^x$ and $f^{-1}(x) = \log_{\frac{2}{5}} x$

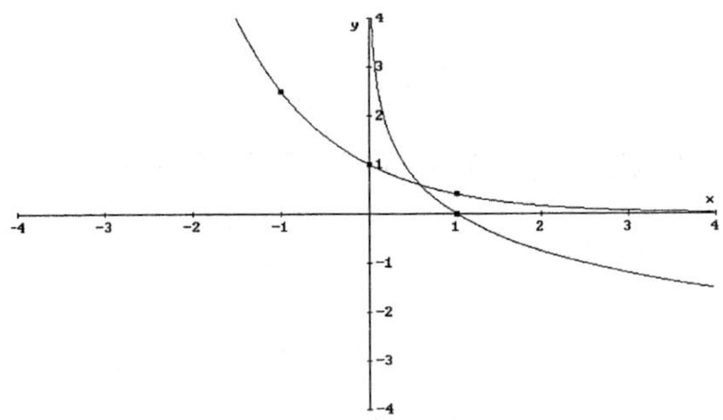

17. pH = – log[H$^+$] so 3.1 = – log[H$^+$] or -3.1 = log[H$^+$] so that H$^+ \approx .0008$

18. $\log_{10} \dfrac{1}{[H^+]} = \log_{10}[H^+]^{-1} = -\log_{10}[H^+]$

19. a. $\log_b \dfrac{x^2y^3}{z^4} = \log_b(x^2y^3) - \log_b z^4 =$

$\log_b x^2 + \log_b y^3 - \log_b z^4 =$

$2\log_b x + 3\log_b y - 4\log_b z$

b. $\log_b \sqrt{\dfrac{x}{yz^2}} = \log_b\left(\dfrac{x}{yz^2}\right)^{\frac{1}{2}} = \dfrac{1}{2}\log_b\left(\dfrac{x}{yz^2}\right) =$

$\dfrac{1}{2}\left[\log_b x - \log_b(yz^2)\right] =$

$\dfrac{1}{2}\left[\log_b x - \left(\log_b y + \log_b z^2\right)\right] =$

$\dfrac{1}{2}\left[\log_b x - \log_b y - 2\log_b z\right] =$

$\dfrac{1}{2}\log_b x - \dfrac{1}{2}\log_b y - \log_b z$

20. a. 3 $\log_b x$ – 5 $\log_b y$ + 7 $\log_b x$ = $\log_b x^3$ – $\log_b y^5$ + $\log_b x^7$ = $\log_b x^3x^7$ – $\log_b y^5$ = $\log_b \dfrac{x^{10}}{y^5}$.

b. $\dfrac{1}{2}\log_b x + 3\log_b y - 7\log_b z =$

$\log_b \sqrt{x} + \log_b y^3 - \log_b z^7 =$

$\log_b \sqrt{x}\, y^3 - \log_b z^7 = \log_b \dfrac{\sqrt{x}\, y^3}{z^7}$

21. a. $3^x = 7$ so that x log 3 = log 7 and x = $\dfrac{\log 7}{\log 3} \approx 1.77$.

b. $5^{x+2} = 625 = 5^4$ so that x + 2 = 4 or x = 2.

c. $2^x = 3^{x-1}$ so that x log 2 = (x – 1) log 3 = x log 3 – log 3 . Then x log 2 – x log 3 = x(log 2 – log 3) = – log 3 and x = $\dfrac{-\log 3}{\log 2 - \log 3} \approx 2.71$.

d. log x + log (29 – x) = 2 or log x(29 – x) = 2 and x(29 – x) = 10^2 = 100. Then $29x - x^2 - 100 = 0$ or $x^2 - 29x + 100 = (x-4)\cdot(x-25) = 0$ and x = 4 or x = 25.

e. $\frac{\log(7x-12)}{\log x} = 2$ or $\log(7x-12) =$

$2\log x = \log x^2$

Thus, $7x - 12 = x^2$ or $x^2 - 7x + 12 = (x-4)\cdot(x-3) = 0$ so that $x = 3$ or $x = 4$

f. $\log 3 - \log(x-1) = -1$ so that $\log\frac{3}{x-1} = -1$ and

$\frac{3}{x-1} = 10^{-1} = \frac{1}{10}$, so $x - 1 = 30$ or $x = 31$.

g. $\ln x = \ln(x-1)$ or $x = x - 1$. No solution.

h. $\ln x = \ln(x-1) + 1 = \ln(x-1) + \ln e$ so that $x = (x-1)e = xe - e$ and $x - xe = -e$.

Thus, $x(1-e) = -e$ and $x = \frac{-e}{1-e} = \frac{e}{e-1}$.

i. $\ln x = \log_{10} x = \frac{\ln x}{\ln 10}$ so that

$\ln x\left(1 - \frac{1}{\ln 10}\right) = 0$. Thus, $\ln x = 0$ or $x = 1$.

22. $A = A_0 2^{\frac{-t}{h}}$, $A = \frac{2}{3}A_0$, $h = 5700$

$\frac{2}{3}A_0 = A_0 2^{\frac{-t}{5700}}$ or $\frac{2}{3} = 2^{\frac{-t}{5700}}$

$\log\frac{2}{3} = \frac{-t}{5700}\log 2$ and $t = \frac{-5700\log\frac{2}{3}}{\log 2} \approx$

3334 or 3300 years.

23. $A = A_0 2^{\frac{-t}{h}}$, $A = \frac{2}{3}A_0$, $t = 20$

$\frac{2}{3}A_0 = A_0 2^{\frac{-20}{h}}$ or $\frac{2}{3} = 2^{\frac{-20}{h}}$. Then

$\log\frac{2}{3} = \frac{-20}{h}\log 2$ so $h = \frac{-20\log 2}{\log\frac{2}{3}} \approx 34.2$

year half-life.

Chapter 11 Test

In problems 1 - 6, $f(x) = 4x$, $g(x) = x - 1$

1. $(g \circ f)(1) = g(f(1)) = g(4) = 4 - 1 = 3$

2. $(f \circ g)(0) = f(g(0)) = f(-1) = 4(-1) = -4$

3. $(f \circ g)(-1) = f(g(-1)) = f(-2) = 4(-2) = -8$

4. $(g \circ f)(-2) = g(f(-2)) = g(-8) = -8 - 1 = -9$

5. $(f \circ g)(x) = f(g(x)) = f(x-1) = 4(x-1) = 4x - 4$.

6. $(g \circ f)(x) = g(f(x)) = g(4x) = 4x - 1$

7. $3x + 2y = 12$
   Interchange x and y: $3y + 2x = 12$
   Solve for y: $3y = -2x + 12$ or $y = \frac{-2}{3}x + 4$

8. $y = 3x^2 + 4$ ($x \le 0$)
   Interchange x and y: $x = 3y^2 + 4$ ($y \le 0$)
   Solve for y: $y^2 = \frac{x-4}{3}$ or $y = -\sqrt{\frac{x-4}{3}}$

9. $y = \frac{1}{4}x^2 - 3$

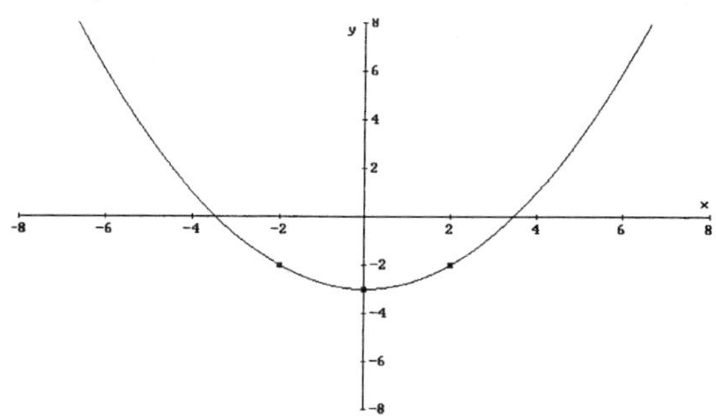

It is not one-to-one.

10. $y = 2^x + 1$

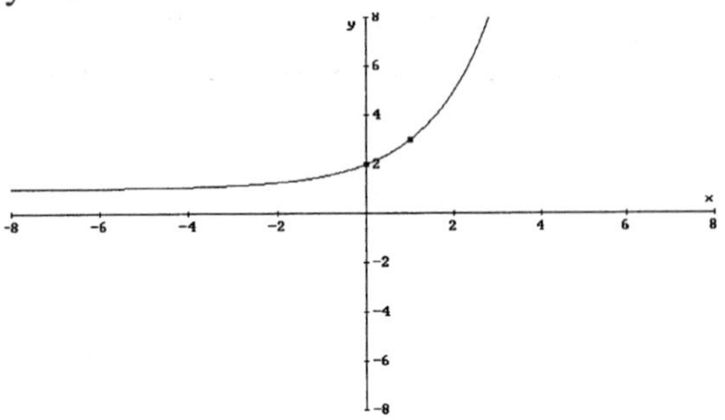

11. $y = 2^{-x}$

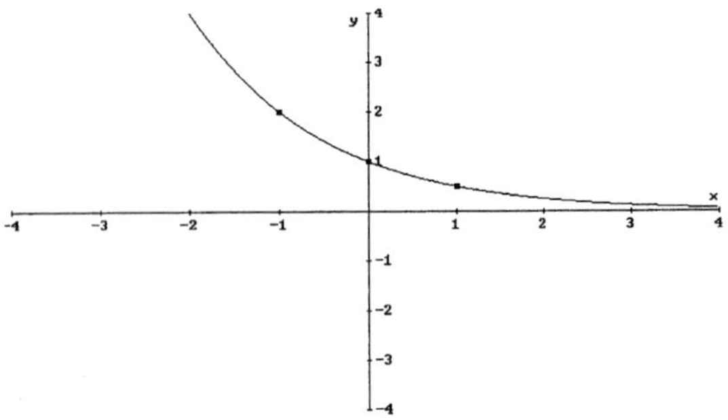

12. $A = A_0\, 2^{-t}$ with $A_0 = 3$, $t = 6$. $A = 3 \cdot 2^{-6} = \dfrac{3}{64}$ gram left in 6 years.

13. $A = Pe^{rt}$, $P = 2000$, $r = .08$, $t = 10$
$A = 2000e^{(.08)10} = \$4451.08$

14. $y = e^x$

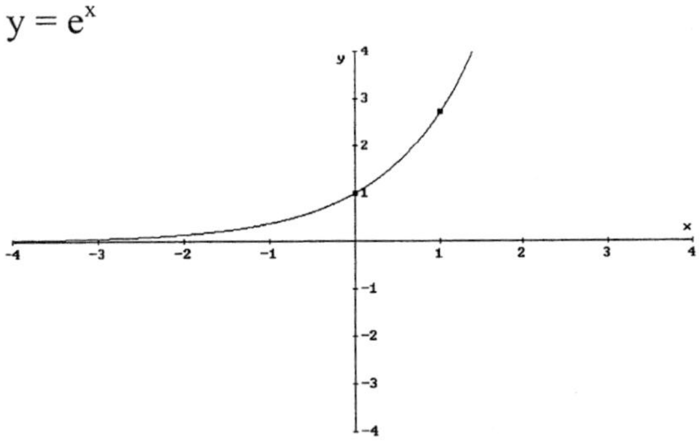

15. $\log_4 16 = x$ so that $4^x = 16 = 4^2$ and $x = 2$.

16. $\log_x 81 = 4$ so that $x^4 = 81 = 3^4$ and $x = 3$.

17. $\log_3 x = -3$ so that $3^{-3} = x$ or $x = \dfrac{1}{27}$.

18. $y = -\log_3 x$

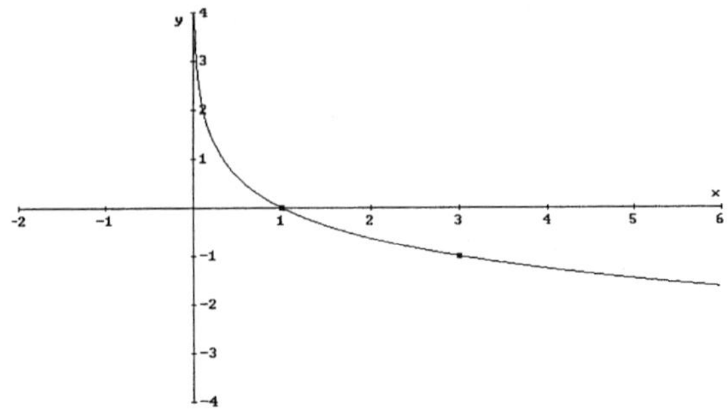

19. $y = \ln x$

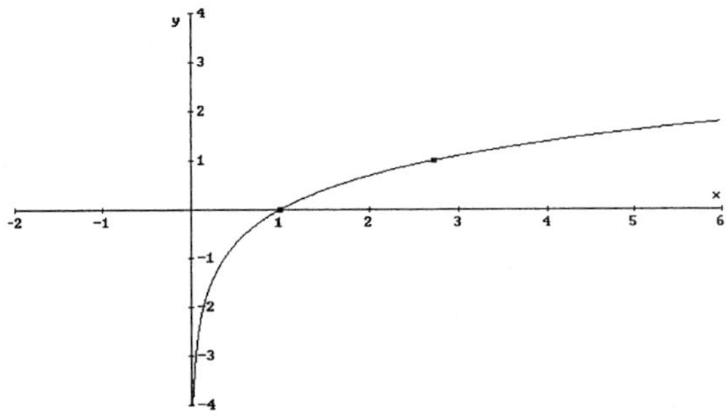

20. $\log a^2bc^3 = \log a^2 + \log b + \log c^3 = 2 \log a + \log b + 3 \log c$

21. $\ln\sqrt{\dfrac{a}{b^2c}} = \ln\left(\dfrac{a}{b^2c}\right)^{\frac{1}{2}} = \dfrac{1}{2}\ln\dfrac{a}{b^2c} =$

$\dfrac{1}{2}\left[\ln a - \ln(b^2c)\right] = \dfrac{1}{2}\left[\ln a - (2\ln b + \ln c)\right] =$

$\dfrac{1}{2}\ln a - \ln b - \dfrac{1}{2}\ln c$

22. $\dfrac{1}{2}\log(a+2) + \log b - 2\log c =$

$\log\sqrt{a+2} + \log b - \log c^2 =$

$\log b\sqrt{a+2} - \log c^2 = \log\dfrac{b\sqrt{a+2}}{c^2}$

23. $\frac{1}{3}(\log a - 2\log b) - \log c =$

$\frac{1}{3}(\log a - \log b^2) - \log c = \frac{1}{3}\log\frac{a}{b^2} - \log c =$

$\log\left(\frac{a}{b^2}\right)^{\frac{1}{3}} - \log c = \log\frac{\sqrt[3]{\frac{a}{b^2}}}{c}$

24. $\log_5 24 = \frac{\log 24}{\log 5}$

25. $\log_\pi e = \frac{\log e}{\log \pi}$

26. T

27. F; $\log a - \log b = \log\frac{a}{b}$

28. F; $\log\left(a^{-3}\right) = -3\log a$

29. F; At least one of -ln x and ln (-x) is undefined for real numbers.

30. $pH = -\log[H^+]$, $H^+ = 3.7 \times 10^{-7}$
Then $pH = -\log(3.7 \times 10^{-7}) \approx 6.4$

31. db gain $= 20 \log\left(\frac{E_O}{E_I}\right)$, $E_O = 60$, $E_I = 0.3$

Then dg gain $= 20 \log\left(\frac{60}{.3}\right) \approx 46$.

32. $5^x = 3$ so that $x \log 5 = \log 3$ and $x = \frac{\log 3}{\log 5}$.

33. $3^{x-1} = 100^x$ so that $(x - 1)\log 3 = x \log 100 = x \cdot 2 = 2x$. Then $x\log 3 - \log 3 = 2x$ and so $x(\log 3 - 2) = \log 3$. Thus, $x = \frac{\log 3}{\log 3 - 2}$.

34. $\log (5x + 2) = \log (2x + 5)$
$5x + 2 = 2x + 5$ or $3x = 3$ and $x = 1$

35. $\log x + \log (x - 9) = 1$ or $\log x(x - 9) = 1$ and so $x(x - 9) = 10^1 = 10$. Thus, $x^2 - 9x - 10 = (x - 10)(x + 1) = 0$ and $x = 10$ or -1, but -1 is extraneous, so 10 is the only solution.

Chapter 12 Systems of Equations and Inequalities

Section 12.1 Solving Systems of Two Equations in Two Variables

Vocabulary

1. system

3. inconsistent

5. dependent

Concepts

7. $\begin{cases} 2x - 3y = -7 \\ 4x - 5y = 25 \end{cases}$ (4,5) is not a solution of the system, since 4(4) – 5(5) = -9 ≠25.

9. $\begin{cases} 2x + y = 4 \\ 4x - 3y = 11 \end{cases}$ (½ ,3) is not a solution of the system since 4(½) – 3(3) = 2 – 9 = -7 ≠ 11

11. $\begin{cases} 5x - 4y = -6 \\ 8y = 10x + 12 \end{cases}$ $\left(\frac{-2}{5}, \frac{1}{4}\right)$ is not a solution of the system since $5(\frac{-2}{5}) - 4(\frac{1}{4})$ = -2 – 1 = -3 which is not equal to -6.

13. The solution set is {(4.5,0.5)}.

15. a. There is no solution. ϕ
    b. The equations are independent.
    c. The lines would be parallel.

Practice

17. $\begin{cases} x - y = 4 \\ 2x + y = 5 \end{cases}$ Solution: (3, -1)

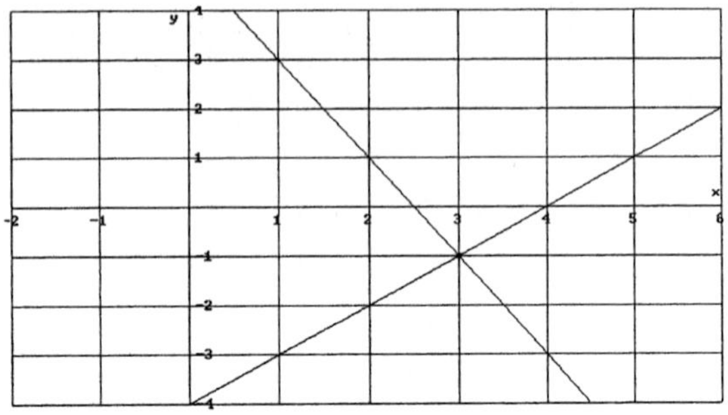

19. $\begin{cases} x = 13 - 4y \\ 3x = 4 + 2y \end{cases}$ Solution: (3, 2.5)

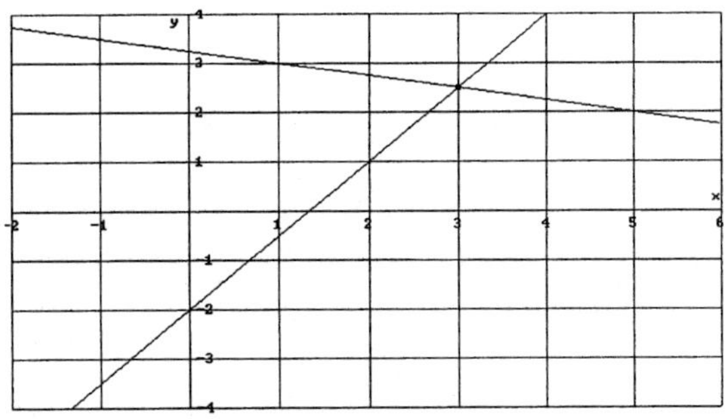

21. $\begin{cases} x = 3 - 2y \\ 2x + 4y = 6 \end{cases}$ Solution: infinitely many

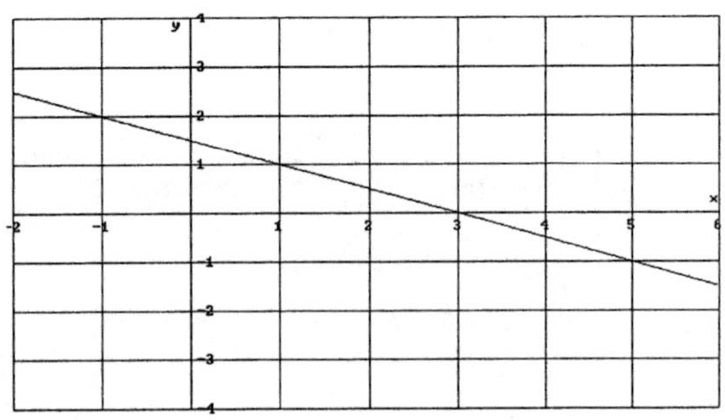

23. $\begin{cases} y = 3 \\ x = 2 \end{cases}$ Solution: (2,3)

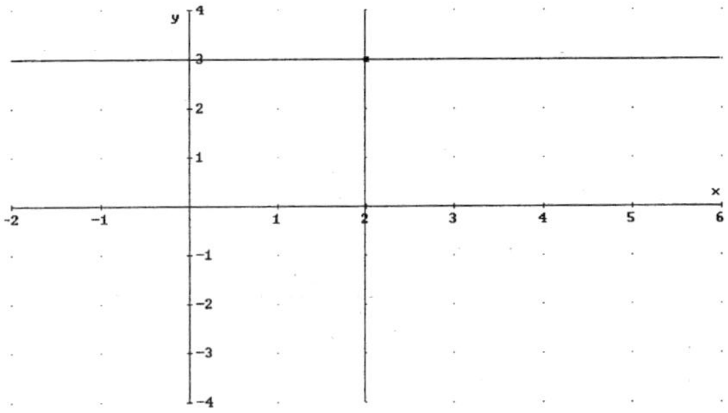

25. $\begin{cases} x = \dfrac{11-2y}{3} \\ y = \dfrac{11-6x}{4} \end{cases}$ Solution: ϕ

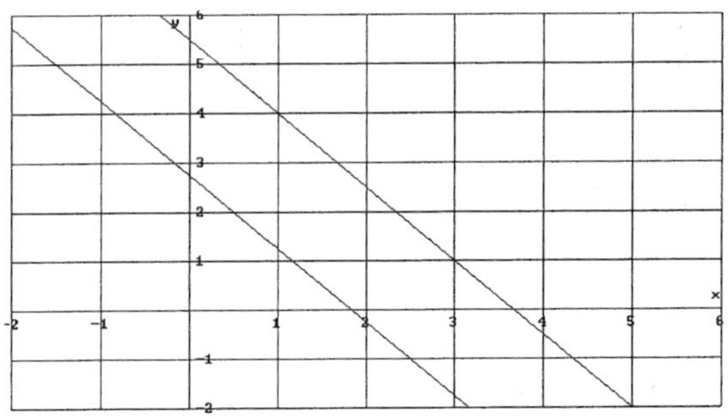

27. $\begin{cases} \dfrac{5}{2}x + y = \dfrac{1}{2} \\ 2x - \dfrac{3}{2}y = 5 \end{cases}$ Solution: (1,-2)

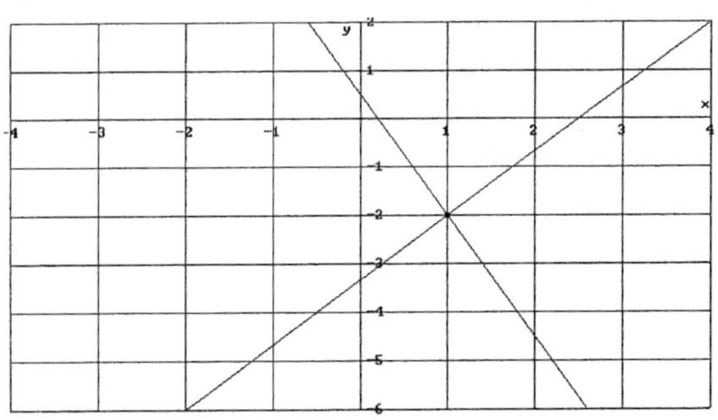

29. $\begin{cases} y = 3.2x - 1.5 \\ y = -2.7x - 3.7 \end{cases}$ Solution: (-.37, -2.69)

Use 'intersect' in [-5,5] by [-5,5].

31. $\begin{cases} 1.7x + 2.3y = 3.2 \\ y = 0.25x + 8.95 \end{cases}$ Solution: (-7.64, 7.04)

Use 'intersect' in [-10,10] by [-10,10]

33. $\begin{cases} y = x \\ x + y = 4 \end{cases}$ Solution: (2,2)

The first equation is solved for y in terms of x, so substitute x for y in the second equation to get $x + x = 4$, or $2x = 4$ so $x = 2$. Since $y = x$, $y = 2$ also.

35. $\begin{cases} x - y = 2 \\ 2x + y = 13 \end{cases}$ Solution: (5,3)

Solving the first equation for y gives y = x – 2. Substitute x – 2 for y in the second equation. $2x + (x - 2) = 13$

$$3x - 2 = 13$$
$$3x = 15$$
$$x = 5$$

Then y = x – 2 = 5 – 2 = 3.

37. $\begin{cases} x + 2y = 6 \\ 3x - y = -10 \end{cases}$ Solution: (-2,4)

Solving the first equation for x gives x = 6 – 2y. Substitute 6 – 2y for x in the second equation. $3(6 - 2y) - y = -10$

$$18 - 7y = -10$$
$$-7y = -28$$
$$y = 4$$

Then x = 6 – 2y = 6 – 2(4) = 6 – 8 = -2.

39. $\begin{cases} 3x = 2y - 4 \\ 6x - 4y = -4 \end{cases}$ Solution: ϕ

Solving the first equation for x gives x = $\dfrac{2y-4}{3}$. Substitute this value for x in the second equation. $6\left(\frac{2y-4}{3}\right)-4y=-4$

$$2(2y-4)-4y=-4$$
$$4y-8-4y=-4$$
$$8=-4, \text{ a contradiction}$$

Thus, the system is inconsistent.

41. $\begin{cases}3x-4y=9\\ x+2y=8\end{cases}$ Solution: (5,1.5)

Solving the second equation for x gives x = 8 – 2y. Substitute 8 – 2y for x in the first equation. $3(8-2y)-4y=9$

$$24-10y=9$$
$$-10y=-15$$
$$y=1.5$$

Then x = 8 – 2y = 8 – 2(1.5) = 8 – 3 = 5.

43. $\begin{cases}2x+2y=-1\\ 3x+4y=0\end{cases}$ Solution: (-2,1.5)

Solving the first equation for 2y gives 2y = -1 – 2x. Substitute -1 – 2x for 2y in the second equation. $3x+2(2y)=0$

$$3x+2(-1-2x)=0$$
$$-2-x=0$$
$$-2=x$$

Then 2y = -1 – 2x = -1 – 2(-2) = 3, or y =1.5.

45. Add to eliminate y.

$$\begin{cases}x-y=3\\ \underline{x+y=7}\end{cases}$$
$$2x \quad =10$$
$$x \quad =5$$

Replace x with 5 in equation one to find y. 5 – y = 3 or -y = -2 so y = 2. The solution is (5,2).

47. Add to eliminate y.

$$\begin{cases}2x+y=-10\\ \underline{2x-y=-6}\end{cases}$$
$$4x \quad = -16$$
$$x \quad = -4$$

Replace x with - 4 in equation one to find y. 2(- 4) + y = -10, or -8 + y = -10 and y = -2. The solution is (- 4, -2).

49. $\begin{cases}8x-4y=16\\ 2x-4=y\end{cases}$ Rewrite as $\begin{cases}8x-4y=16\\ 2x-y=4\end{cases}$.

Multiply the second equation by -4 and add to eliminate x. $\begin{cases}8x-4y=16\\ \underline{-8x+4y=-16}\end{cases}$

$$0=0$$

The system is dependent and there are infinitely many solutions.

51. $\begin{cases}x=\frac{3}{2}y+5\\ 2x-3y=8\end{cases}$ Rewrite as $\begin{cases}2x-3y=10\\ 2x-3y=8\end{cases}$.

The system is inconsistent since the number 2x – 3y cannot have two different values. No solution.

53. $\begin{cases}\frac{x}{2}+\frac{y}{2}=6\\ \frac{x}{2}-\frac{y}{2}=-2\end{cases}$ Rewrite as $\begin{cases}x+y=12\\ x-y=-4\end{cases}$. Add to eliminate y, giving 2x = 8 or x = 4. Replace x by 4 in equation one to find y. 4 + y = 12, or y = 8. The solution is (4,8).

55. $\begin{cases}\frac{3}{4}x+\frac{2}{3}y=7\\ \frac{3}{5}x-\frac{1}{2}y=18\end{cases}$ Multiply both sides of the top equation by 12, and both sides of the bottom equation by 10 to rewrite the system.

$\begin{cases}9x+8y=84\\ 6x-5y=180\end{cases}$ Now multiply the top equation by 5 and the bottom equation by 8 and then add to eliminate y.

$$\begin{cases} 45x + 40y = 420 \\ 48x - 40y = 1440 \end{cases}$$

$$\begin{aligned} 93x &= 1860 \\ x &= 20 \end{aligned}$$

Replace x by 20 in equation one to find y. $9(20) + 8y = 84$, or $8y = -96$ and $y = -12$. The solution is (20,-12).

57. $\begin{cases} y = 3x + 1 \\ 3x + 2y = 12 \end{cases}$ The best method is probably substitution since the first equation is already solved for y in terms of x.

59. David is correct since the equation involves only the one variable x.

Review

61. $2(4x + 5) - 3 = 23$

a. $8x + 10 - 3 = 23$

$$\begin{aligned} 8x + 7 &= 23 \\ 8x &= 16 \\ x &= 2 \end{aligned}$$

b. Use [-10,10] by [-10,25].

63. $3\sqrt{x} + 4 = 22$

a. $3\sqrt{x} = 18$

$$\begin{aligned} \sqrt{x} &= 6 \\ x &= 6^2 = 36 \end{aligned}$$

b. Use [-10,40] by [-10,25].

Section 12.2 Applications of Systems of Two Equations in Two Variables

1. Let x represent the smaller integer, y the larger. Then $\begin{cases} y = 2x \\ x + y = 96 \end{cases}$. Substitute 2x for y in the second equation and solve for x. $x + 2x = 96$, so $3x = 96$ and $x = 32$. Then $y = 2x = 2(32) = 64$. The integers are 32 and 64.

3. Let x and y represent the integers. Then $\begin{cases} 3x + y = 29 \\ x + 2y = 18 \end{cases}$. Multiply the top equation by -2 and then add the equations to eliminate y, giving $-5x = -40$ or $x = 8$. Then replace x by 8 in the bottom equation to find y. $8 + 2y = 18$, or $2y = 10$, so $y = 5$. The integers are 8 and 5.

5. Let c = number of cows, h = number of horses. Then $\begin{cases} c + h = 168 \\ c = 5h \end{cases}$. Substitute 5h for c in the first equation, giving $5h + h = 168$, or $6h = 168$ and so $h = 28$. Then $c = 5h = 5(28) = 140$. He has 140 head of cattle.

7. Let g = number of gallons of paint and b = number of brushes. Then $\begin{cases} 8g + 3b = 135 \\ 6g + 2b = 100 \end{cases}$. Multiply the top equation by 2 and the bottom equation by -3 and add to eliminate b.

$$\begin{cases} 16g + 6b = 270 \\ -18g - 6b = -300 \end{cases}$$

$$\begin{aligned} -2g &= -30 \\ g &= 15 \end{aligned}$$

Then $8(15) + 3b = 135$ so $3b = 15$ and $b = 5$. Each gallon of paint costs \$15, each brush \$5

9. Let x = cost of a bottle of lens cleaner and y = cost of a bottle of soaking solution. Then $\begin{cases} 2x + 3y = 29.40 \\ 3x + 2y = 28.60 \end{cases}$. Multiply the top equation by 2, the bottom one by -3 and add to eliminate y. $\begin{cases} 4x + 6y = 58.80 \\ -9x - 6y = -85.80 \end{cases}$

$$\begin{aligned} -5x &= -27 \\ x &= 5.40 \end{aligned}$$

Then $2(5.40) + 3y = 29.40$ so $3y = 18.60$ and $y = 6.20$. The lens cleaner cost \$5.40 and the soaking solution \$6.20.

11. Let x = length of the shorter piece and y = length of the longer piece. Then
$\begin{cases} x + y = 25 \\ y = x + 5 \end{cases}$. Substitute x + 5 for y in the top equation. $x + x + 5 = 25$ or $2x = 20$ so $x = 10$ and $y = 10 + 5 = 15$. The pieces are 10' and 15' long.

13. Let x = Susan's amount and y = Maria's amount. Then $\begin{cases} x + y = 250{,}000 \\ x = y + 50{,}000 \end{cases}$. Substitute y + 50,000 for x in the top equation, giving $y + 50{,}000 + y = 250{,}000$ or $2y = 200{,}000$ and so $y = 100{,}000$ and $x = y + 50{,}000 = 150{,}000$. Susan will get $150,000.

15. Let w = width of the rectangle and l = length of the rectangle. Then $\begin{cases} 2w + 2l = 110 \\ l = w + 5 \end{cases}$.
Substitute w + 5 for l in the top equation. $2w + 2(w + 5) = 110$ or $4w = 100$ and $w = 25$ so that $l = w + 5 = 25 + 5 = 30$. The rectangle is 25' by 30'.

17. Let w = width of the rectangle, l = length of the rectangle. Then $\begin{cases} 2w + 2l = 34 \\ l = 2w + 2 \end{cases}$.
Substitute 2w + 2 for l in the top equation. $2w + 2(2w + 2) = 34$ or $6w = 30$ and $w = 5$. Then $l = 2w + 2 = 2(5) + 2 = 12$. The area is $lw = 5(12) = 60$ ft.^{2}

19. Let y = number of years for the costs to be equal and c = common cost. Then
$\begin{cases} c = 2250 + 412y \text{ (90+ furnace)} \\ c = 1715 + 466y \text{ (80+ furnace)} \end{cases}$. Substitute 2250 + 412y for c in the second equation. $2250 + 412y = 1715 + 466y$ or $535 = 54y$ and $y \approx 9.9$. The break-even point is about 9.9 years.

21. In Exercise 19, letting y = 7 years gives $c = 2250 + 412(7) = 5134$ for the 90+ furnace and $c = 1715 + 466(7) = 4977$ for the 80+ furnace. You should choose the 80+ furnace.

23. Let x = amount Bill invested and y = amount Janette invested. Bill's interest = .05x(1) and Janette's interest = .07y(1). Thus,
$\begin{cases} x + y = 5000 \\ .05x + .07y = 310 \end{cases}$. Multiply the top equation by -7 and the bottom equation by 100 and add to eliminate y.

$$\begin{cases} -7x - 7y = -35000 \\ 5x + 7y = 31000 \end{cases}$$
$$-2x = -4000$$
$$x = 2000$$

Bill invested $2000.

25. Let x = number of student tickets sold and y = number of non-student tickets sold. Then
$\begin{cases} x + y = 350 \\ 1 \cdot x + 2 \cdot y = 450 \end{cases}$. Multiply the top equation by -2 and add to eliminate y.

$$\begin{cases} -2x - 2y = -700 \\ x + 2y = 450 \end{cases}$$
$$-x = -250$$
$$x = 250$$

There were 250 student tickets sold.

27. Let x = speed of the boat in still water, c = speed of the current. Then, using distance = rate · time, $\begin{cases} 24 = 2(x + c) \text{ (downstream)} \\ 24 = 3(x - c) \text{ (upstream)} \end{cases}$ or

$\begin{cases} 24 = 2x + 2c \\ 24 = 3x - 3c \end{cases}$. Multiply the top equation by 3, the bottom one by 2 and add to eliminate c

$$\begin{cases} 72 = 6x + 6c \\ 48 = 6x - 6c \end{cases}$$
$$120 = 12x$$
$$10 = x$$

The boat's speed is 10 mph in still water.

29. Let x = airspeed of the plane, w = wind speed Using distance = rate · time,

$$\begin{cases} 600 = 2(x + w) = 2x + 2w \text{ (downwind)} \\ 600 = 3(x - w) = 3x - 3w \text{ (upwind)} \end{cases}$$

Multiply the top equation by -3 and the bottom equation by 2 and add to eliminate x.

$$\begin{cases} -1800 = -6x - 6w \\ 1200 = 6x - 6w \end{cases}$$

$$-600 = -12w$$

$$50 = w$$

The wind speed is 50 mph.

31. Let x = number of liters of the 40% alcohol solution and y = number of liters of the 55% alcohol solution. Then

$$\begin{cases} x + y = 15 & \text{(total mixture)} \\ .40x + .55y = .50(15) & \text{(total alcohol)} \end{cases}$$

Multiply the top equation by - 40, the bottom one by 100, and add to eliminate x.

$$\begin{cases} -40x - 40y = -600 \\ 40x + 55y = 750 \end{cases}$$

$$15y = 150$$

$$y = 10$$

Then x + 10 = 15, so x = 5. She must use 5 l. of the 40% solution and 10 l. of the 55% solution.

33. Let x = number of pounds of peanuts and y = number of pounds of cashews. Then

$$\begin{cases} x + y = 48 & \text{(total mixture)} \\ 3x + 6y = 4(48) & \text{(total value of mixture)} \end{cases}$$

Multiply the top equation by -3 and then add to eliminate x. $\begin{cases} -3x - 3y = -144 \\ 3x + 6y = 192 \end{cases}$

$$3y = 48$$

$$y = 16$$

Then x + 16 = 48 so x = 32. He should use 32 lb. of peanuts, 16 lb. of cashews.

35. Let r = retail price and d = discount. Using the suggested equations, $\begin{cases} r - d = 384 \\ d = .40r \end{cases}$.

Substitute .40r for d in the first equation to get r – .40r = 384 or .60r = 384. Thus, r = 640. The original retail price was $640.

37. Let x = number of cheap radios sold, y = number of expensive radios sold. Then

$$\begin{cases} x + y = 25 & \text{(total radios)} \\ 87x + 119y = 2495 & \text{(total receipts)} \end{cases}$$

Multiply the top equation by -119 and add to eliminate y. $\begin{cases} -119x - 119y = -2975 \\ 87x + 119y = 2495 \end{cases}$

$$-32x = -480$$

$$x = 15$$

Thus, 15 cheap radios were sold.

39. Let x = lower rate, y = higher rate. Using I = Prt, the annual interest at rate x = 950x(1) and the annual interest at rate y = 1200y(1).

Then $\begin{cases} y = x + .01 \\ 950x + 1200y = 205.50 \end{cases}$. Substitute x + .01 for y in the bottom equation, giving

$$950x + 1200(x + .01) = 205.50$$

$$950x + 1200x + 12 = 205.50$$

$$2150x = 193.50$$

$$x = .09$$

The lower rate is 9%.

41. A solution method might be called 'easier' if a single equation in one variable could be obtained more easily.

Review

43. $|3x + 6| \geq 5$

$$3x + 6 \geq 5 \text{ or } 3x + 6 \leq -5$$

$$3x \geq -1 \text{ or } 3x \leq -11$$

$$x \geq -\frac{1}{3} \text{ or } x \leq -\frac{11}{3}$$

$$\text{Intervals}: (-\infty, -\frac{11}{3}] \cup [-\frac{1}{3}, \infty)$$

45. $|2x+5|-2<-6$

$|2x+5|<-4$

No solution

Section 12.3 Solving Systems of Three Equations in Three Variables

Vocabulary

1. plane

3. infinite

Practice

5. $\begin{cases} x+y+z=4 \\ 2x+y-z=1 \\ 2x-3y+z=1 \end{cases}$

Add equations one and two to eliminate z, add equations two and three to eliminate z.

$\begin{cases} x+y+z=4 \\ \underline{2x+y-x=1} \end{cases}$ $\quad$ $\begin{cases} 2x+y-z=1 \\ \underline{2x-3y+z=1} \end{cases}$

$3x+2y \quad =5 \qquad 4x-2y \quad =2$

Now solve $\begin{cases} 3x+2y=5 \\ 4x-2y=2 \end{cases}$ by adding to eliminate y, giving 7x = 7 or x = 1. Then 3(1) + 2y = 5, so 2y = 2 or y = 1. Finally, 1 + 1 + z = 4, so z = 2. Solution: (1,1,2)

7. $\begin{cases} 2x+2y+3z=10 \\ 3x+y-z=0 \\ x+y+2z=6 \end{cases}$

Multiply equation two by -2 and add to equation one to eliminate y.
Multiply equation two by -1 and add to equation three to eliminate y.

$\begin{cases} 2x+2y+3z=10 \\ \underline{-6x-2y+2z=0} \end{cases}$ $\quad$ $\begin{cases} -3x-y+z=0 \\ \underline{x+y+2x=6} \end{cases}$

$-4x \quad +5z=10 \qquad -2x \quad +3z=6$

Now solve $\begin{cases} -4x+5z=10 \\ -2x+3z=6 \end{cases}$. Multiply the second equation by -2 and add to eliminate x, giving -z = -2 or z = 2. Then - 4x + 5(2) = 10 or - 4x = 0 and so x = 0. Finally, using x + y + 2z = 6 gives 0 + y + 2(2) = 6 or y = 2. The solution is (0,2,2).

9. $\begin{cases} x+y+2z=7 \\ x+2y+z=8 \\ 2x+y+z=9 \end{cases}$

Multiply equation one by -1 and add to equation two to eliminate x.
Multiply equation one by -2 and add to equation three to eliminate x.

$\begin{cases} -x-y-2z=-7 \\ \underline{x+2y+z=8} \end{cases}$ $\quad$ $\begin{cases} -2x-2y-4z=-14 \\ \underline{2x+y+z=9} \end{cases}$

$y-z=1 \qquad -y-3z=-5$

Now solve $\begin{cases} y-z=1 \\ -y-3z=-5 \end{cases}$. Add to eliminate y, giving - 4z = - 4 or z = 1. Then y – z = 1 so y – 1 = 1 and y = 2. Finally, using x + y + 2z = 7, x + 2 + 2(1) = 7 so x = 7 – 4 = 3. The solution is (3,2,1).

11. $\begin{cases} 2x+y-z=1 \\ x+2y+2z=2 \\ 4x+5y+3z=3 \end{cases}$

Multiply equation one by 2 and add to equation two to eliminate z.
Multiply equation one by 3 and add to equation three to eliminate z.

$\begin{cases} 4x+2y-2z=2 \\ \underline{x+2y+2z=2} \end{cases}$ $\quad$ $\begin{cases} 6x+3y-3z=3 \\ \underline{4x+5y+3z=3} \end{cases}$ Now

$5x+4y \quad =4 \qquad 10x+8y \quad =6$

solve $\begin{cases} 5x+4y=4 \\ 10x+8y=6 \end{cases}$. Multiplying the first equation by -2 and add to eliminate y, giving 0 = -2. Since this is a contradiction, the system is inconsistent. Solution is ϕ.

13. $\begin{cases} 2x + 3y + 4z - 6 = 0 \\ 2x - 3y - 4z + 4 = 0 \\ 4x + 6y + 8z - 12 = 0 \end{cases}$

Since 2 · equation one = equation three, and equations one and two are independent, the system is dependent and has infinitely many solutions.

15. $\begin{cases} x + \frac{1}{3}y + z = 13 \\ \frac{1}{2}x - y + \frac{1}{3}z = -2 \\ x + \frac{1}{2}y - \frac{1}{3}z = 2 \end{cases}$

To eliminate fractions, multiply equation one by 3, equations two and three by 6.

$\begin{cases} 3x + y + 3z = 39 \\ 3x - 6y + 2z = -12 \\ 6x + 3y - 2z = 12 \end{cases}$

Multiply equation one by -1and add to equation two to eliminate x.
Multiply equation one by -2 and add to equation three to eliminate x.

$\begin{cases} -3x - y - 3x = -39 \\ \underline{3x - 6y + 2z = -12} \end{cases}$ $\begin{cases} -6x - 2y - 6z = -78 \\ \underline{6x + 3y - 2z = 12} \end{cases}$

$-7y - z = -51$ $y - 8z = -66$

Now solve $\begin{cases} -7y - z = -51 \\ y - 8z = -66 \end{cases}$. Multiply the second equation by 7 and add to eliminate y.

$\begin{cases} -7y - z = -51 \\ \underline{7y - 56z = -462} \end{cases}$

$-57z = -513$

$z = 9$

Then y – 8(9) = -66 or y = 6. Using x + ⅓y + z = 13, x + ⅓(6) + 9 = 13, so x = 2. The solution is (2,6,9).

Applications

17. Let x, y and z represent the three integers.

Then $\begin{cases} x + y + z = 18 \\ z = 4y \\ y = x + 6 \text{ or } y - 6 = x \end{cases}$

Replace z by 4y and x by y – 6 in equation one: (y – 6) + y + 4y = 18 Then 6y – 6 = 18 or 6y = 24 and y = 4. Then z = 4y = 4(4) = 16 and x = y – 6 = 4 – 6 = -2 and the integers are -2, 4 and 16.

19. Let a, b and c represent the measures of angles A, B and C respectively. Then

$\begin{cases} a + b + c = 180 \\ a = b + c - 100 \\ c = 2b - 40 \end{cases}$ Rewrite the system in

standard form: $\begin{cases} a + b + c = 180 \\ a - b - c = -100 \\ \quad - 2b + c = -40 \end{cases}$

Multiply equation one by -1 and add to equation two to eliminate a.

$\begin{cases} -a - b - c = -180 \\ \underline{a - b - c = -100} \end{cases}$

$-2b - 2c = -280$

or $2b + 2c = 280$

Now solve the system $\begin{cases} 2b + 2c = 280 \\ -2b + c = -40 \end{cases}$ by adding to eliminate b, giving 3c = 240 or c = 80. Then 2b +2(80) = 280 so 2b = 120 and b = 60. Finally, using a + b + c = 180, a + 60 + 80 = 180 or a = 40. The angles are A = 40°, B = 60° and C = 80°.

21. Let x, y and z represent the number of units to be used of foods A, B and C respectively.

Then $\begin{cases} 1 \cdot x + 2y + 2z = 11 & \text{(fat)} \\ 1 \cdot x + 1 \cdot y + 1 \cdot z = 6 & \text{(carbohydrate)} \\ 2x + 1 \cdot y + 2z = 10 & \text{(protein)} \end{cases}$

Multiply equation one by -1 and add to equation two to eliminate x.
Multiply equation one by -2 and add to equation three to eliminate x.

$$\begin{cases}-x-2y-2z=-11\\ x+y+z=6\end{cases} \qquad \begin{cases}-2x-4y-4z=-22\\ 2x+y+2z=10\end{cases}$$

$$-y-z=-5 \qquad -3y-2z=-12$$

or $y+z=5$

Now solve $\begin{cases}y+z=5\\ -3y-2z=-12\end{cases}$. Multiply equation one by 2 and add to eliminate z, giving -y = -2 or y = 2. Thus, 2 + z = 5 or z = 3. Finally, x + y + z = 6, so x + 2 + 3 = 6 and x = 1. You must use 1 unit of food A, 2 of food B and 3 of food C.

23. Let x, y and z represent, respectively, the number of $20, $12 and $9 statues to be produced. Then

$$\begin{cases}x+y+z=180 & \text{(total number of statues)}\\ 5x+4y+3z=650 & \text{(total cost)}\\ 20x+12y+9z=2100 & \text{(total revenue)}\end{cases}$$

Multiply equation one by -3 and add to equation two to eliminate z.

$$\begin{cases}-3x-3y-3z=-540\\ 5x+4y+3z=650\end{cases}$$

$$2x+y \qquad =110$$

Multiply equation one by -9 and add to equation three to eliminate z.

$$\begin{cases}-9x-9y-9z=-1620\\ 20x+12y+9z=2100\end{cases}$$

$$11x+3y \qquad =480$$

Now solve $\begin{cases}2x+y=110\\ 11x+3y=480\end{cases}$. Multiply equation one by -3 and add to equation three to eliminate y, giving 5x = 150 or x = 30. Then 2(30) + y = 110 or y = 50. Then,using x + y + z = 180, 30 +50 + z = 180 or z = 100. The artist should produce 30 of the $20 statues, 50 of the $12 ones and 100 or the $9 ones.

25. Let x, y and z represent, respectively, the number of $5, $3 and $2 tickets sold. Then

$$\begin{cases}x+y+z=750\\ 5x+3y+2z=2625\\ x=2z\end{cases}$$

Rewrite the system, replacing x by 2z.

$$\begin{cases}2z+y+z=750\\ 5(2z)+3y+2z=2625\end{cases} \text{ or } \begin{cases}y+3z=750\\ 3y+12z=2625\end{cases}$$

Multiply equation one by -3 and add to eliminate y, giving 3z = 375 or z = 125. Then y + 3z = y + 3(125) = 750 or y = 375. Also, x = 2z = 2(125) = 250. There were 250 $5 tickets sold, 375 $3 tickets and 125 $2 tickets

27. Let x, y and z represent, respectively, the number of totem poles, bears and deer to be produced. Then

$$\begin{cases}2x+2y+z=14 & \text{(carving)}\\ x+2y+2z=15 & \text{(sanding)}\\ 3x+2y+2z=21 & \text{(painting)}\end{cases}$$

Multiply equation two by -1 and add to equation one to eliminate y.
Multiply equation two by -1 and add to equation three to eliminate y.

$$\begin{cases}2x+2y+z=14\\ -x-2y-2z=-15\end{cases} \qquad \begin{cases}-x-2y-2z=-15\\ 3x+2y+2z=21\end{cases}$$

$$x \qquad -z=-1 \qquad 2x \qquad =6$$

$$\text{or } x \qquad =3$$

Thus, x – z = 3 – z = -1 and so z = 4. Finally, 2x + 2y + z = 6 + 2y + 4 = 14 so 2y = 4 or y = 2. Produce 3 totem poles, 2 bears, 4 deer.

Review

29. y = |x + 2| is a function, since each value of x results in a single value of y. Domain = ℝ and Range = [0, ∞).

31. slope = $\frac{3}{5}$, passing through (1,2) Using point-slope form, $y-2=\frac{3}{5}(x-1)$ or $y=\frac{3}{5}x+\frac{7}{5}$

33. $f(x) = -x^2 - 5x + 2$ so $f(-3) = -(-3)^2 - 5(-3) + 2 = -9 + 15 + 2 = 8$.

Section 12.4 Using Matrices to Solve Systems of Equations

Vocabulary

1. matrix

3. square

5. Gaussian elimination

Concepts

7. $\begin{bmatrix} 2 & 1 & 2 & | & 1 \\ 1 & -2 & 3 & | & 4 \\ 2 & -3 & 1 & | & 0 \end{bmatrix}$ is the augmented matrix.

9. The Multiplication Property of Equality allows you to multiply both sides of an equation by a non-zero number, and a row in the matrix represents an equation.

11. Substitute the values of the variables into the original equations to check the solution.

Practice

Note: R_i stands for 'row i', '▸' is read 'replaces' and '≈' means 'is row equivalent to'.

13. $\begin{cases} x + y = 2 \\ x - y = 0 \end{cases}$ Augmented matrix: $\begin{bmatrix} 1 & 1 & | & 2 \\ 1 & -1 & | & 0 \end{bmatrix}$

$(-1)R_1 + R_2 \blacktriangleright R_2$: $\begin{bmatrix} 1 & 1 & | & 2 \\ 0 & -2 & | & -2 \end{bmatrix}$

$(-.5)R_2 \blacktriangleright R_2$: $\begin{bmatrix} 1 & 1 & | & 2 \\ 0 & 1 & | & 1 \end{bmatrix}$

New system: $\begin{cases} x + y = 2 \\ y = 1 \end{cases}$. Then $x + 1 = 2$ or $x = 1$. The solution is (1,1).

15. $\begin{cases} x + 2y = -4 \\ 2x + y = 1 \end{cases}$

Augmented matrix: $\begin{bmatrix} 1 & 2 & | & -4 \\ 2 & 1 & | & 1 \end{bmatrix}$

$-2R_1 + R_2 \blacktriangleright R_2$: $\begin{bmatrix} 1 & 2 & | & -4 \\ 0 & -3 & | & 9 \end{bmatrix}$

$(-⅓)R_2 \blacktriangleright R_2$: $\begin{bmatrix} 1 & 2 & | & -4 \\ 0 & 1 & | & -3 \end{bmatrix}$ The new system is

$\begin{cases} x + 2y = -4 \\ y = -3 \end{cases}$, so $x + 2(-3) = -4$ or $x = 2$.
The solution is (2,-3).

17. $\begin{cases} 3x + 4y = -12 \\ 9x - 2y = 6 \end{cases}$

Augmented matrix: $\begin{bmatrix} 3 & 4 & | & -12 \\ 9 & -2 & | & 6 \end{bmatrix}$

$-3R_1 + R_2 \blacktriangleright R_2$: $\begin{bmatrix} 3 & 4 & | & -12 \\ 0 & -14 & | & 42 \end{bmatrix}$

$(-1/14)R_2 \blacktriangleright R_2$: $\begin{bmatrix} 3 & 4 & | & -12 \\ 0 & 1 & | & -3 \end{bmatrix}$

The new system is $\begin{cases} 3x + 4y = -12 \\ y = -3 \end{cases}$ and so $3x + 4(-3) = -12$ or $3x = 0$. Thus, $x = 0$ and the solution is (0,-3).

19. $\begin{cases} x + y + z = 6 \\ x + 2y + z = 8 \\ x + y + 2z = 9 \end{cases}$

Augmented matrix: $\begin{bmatrix} 1 & 1 & 1 & | & 6 \\ 1 & 2 & 1 & | & 8 \\ 1 & 1 & 2 & | & 9 \end{bmatrix}$

$\left.\begin{matrix} (-1)R_1 + R_2 \blacktriangleright R_2 \\ (-1)R_1 + R_3 \blacktriangleright R_3 \end{matrix}\right\}$: $\begin{bmatrix} 1 & 1 & 1 & | & 6 \\ 0 & 1 & 0 & | & 2 \\ 0 & 0 & 1 & | & 3 \end{bmatrix}$ The new system is $\begin{cases} x+y+z=6 \\ \quad y \quad\;\; =2 \\ \qquad\quad z=3 \end{cases}$ and so x + 2 + 3 = 6 and x = 1. The solution is (1,2,3).

21. $\begin{cases} 2x+y+3z=3 \\ -2x-y+z=5 \\ 4x-2y+2z=2 \end{cases}$

Augmented matrix: $\left[\begin{array}{ccc|c} 2 & 1 & 3 & 3 \\ -2 & -1 & 1 & 5 \\ 4 & -2 & 2 & 2 \end{array}\right]$

$\left.\begin{array}{l} 1R_1 + R_2 \blacktriangleright R_2 \\ -2R_1 + R_3 \blacktriangleright R_3 \end{array}\right\}: \left[\begin{array}{ccc|c} 2 & 1 & 3 & 3 \\ 0 & 0 & 4 & 8 \\ 0 & -4 & -4 & -4 \end{array}\right]$

$R_2 \blacktriangleleft\blacktriangleright R_3: \left[\begin{array}{ccc|c} 2 & 1 & 3 & 3 \\ 0 & -4 & -4 & -4 \\ 0 & 0 & 4 & 8 \end{array}\right]$ The new system is $\begin{cases} 2x+y+3z=3 \\ \quad -4y-4z=-4 \\ \qquad\quad 4z=8 \end{cases}$. Thus, z = 2 and - 4y – 4(2) = - 4 so that y = -1. Finally, 2x + (-1) + 3(2) = 3 so that 2x = -2 or x = -1. The solution is (-1,-1,2).

23. $\begin{cases} 3x-2y+4z=4 \\ x+y+z=3 \\ 6x-2y-3z=10 \end{cases}$

Augmented matrix: $\left[\begin{array}{ccc|c} 3 & -2 & 4 & 4 \\ 1 & 1 & 1 & 3 \\ 6 & -2 & -3 & 10 \end{array}\right]$

$\left.\begin{array}{l} -3R_1 + R_2 \blacktriangleright R_2 \\ -6R_1 + R_3 \blacktriangleright R_3 \end{array}\right\}: \left[\begin{array}{ccc|c} 1 & 1 & 1 & 3 \\ 0 & -5 & 1 & -5 \\ 0 & -8 & -9 & -8 \end{array}\right]$

$(-8/5)R_2 + R_3 \blacktriangleright R_3: \left[\begin{array}{ccc|c} 1 & 1 & 1 & 3 \\ 0 & -5 & 1 & -5 \\ 0 & 0 & -\frac{53}{5} & 0 \end{array}\right]$

The new system is $\begin{cases} x+y+z=3 \\ \; -5y+z=-5 \\ \qquad -\frac{53}{5}z=0 \end{cases}$ so that z = 0 and -5y + z = -5y + 0 = -5 or y = 1. Finally, x + y + z = x + 1 + 0 = 3 or x = 2. The solution is (2,1,0).

25. $\begin{cases} x+y=3 \\ 3x-y=1 \\ 2x+y=4 \end{cases}$

Augmented matrix: $\left[\begin{array}{cc|c} 1 & 1 & 3 \\ 3 & -1 & 1 \\ 2 & 1 & 4 \end{array}\right]$

$\left.\begin{array}{l} -3R_1 + R_2 \blacktriangleright R_2 \\ -2R_1 + R_3 \blacktriangleright R_3 \end{array}\right\}: \left[\begin{array}{cc|c} 1 & 1 & 3 \\ 0 & -4 & -8 \\ 0 & -1 & -2 \end{array}\right]$

$(-1/4)R_2 + R_3 \blacktriangleright R_3: \left[\begin{array}{cc|c} 1 & 1 & 3 \\ 0 & -4 & -8 \\ 0 & 0 & 0 \end{array}\right]$ Thus, the new system is $\begin{cases} x+y=3 \\ \; -4y=-8 \end{cases}$ and so y = 2 and x + y = x + 2 = 3, so x = 1. Solution: (1,2)

27. $\begin{cases} 2x-y=4 \\ x+3y=2 \\ -x-4y=-2 \end{cases}$

Augmented matrix: $\left[\begin{array}{cc|c} 2 & -1 & 4 \\ 1 & 3 & 2 \\ -1 & -4 & -2 \end{array}\right]$

$R_1 \leftrightarrow R_2$: $\begin{bmatrix} 1 & 3 & | & 2 \\ 2 & -1 & | & 4 \\ -1 & -4 & | & -2 \end{bmatrix}$

$\left.\begin{matrix} -2R_1 + R_2 \blacktriangleright R_2 \\ R_1 + R_3 \blacktriangleright R_3 \end{matrix}\right\}: \begin{bmatrix} 1 & 3 & | & 2 \\ 0 & -7 & | & 0 \\ 0 & -1 & | & 0 \end{bmatrix}$

(-1/7)$R_2 + R_3 \blacktriangleright R_3$: $\begin{bmatrix} 1 & 3 & | & 2 \\ 0 & -7 & | & 0 \\ 0 & 0 & | & 0 \end{bmatrix}$ The new

system is $\begin{cases} x + 3y = 2 \\ -7y = 0 \end{cases}$ and so y = 0 and

x + 3y = x + 3(0) = 2 and x = 2. The solution is (2,0).

29. $\begin{cases} 2x + y = 7 \\ x - y = 2 \\ -x + 3y = -2 \end{cases}$

Augmented matrix: $\begin{bmatrix} 2 & 1 & | & 7 \\ 1 & -1 & | & 2 \\ -1 & 3 & | & -2 \end{bmatrix}$

$R_1 \leftrightarrow R_2$: $\begin{bmatrix} 1 & -1 & | & 2 \\ 2 & 1 & | & 7 \\ -1 & 3 & | & -2 \end{bmatrix}$

$\left.\begin{matrix} -2R_1 + R_2 \blacktriangleright R_2 \\ 1 \cdot R_1 + R_3 \blacktriangleright R_3 \end{matrix}\right\}: \begin{bmatrix} 1 & -1 & | & 2 \\ 0 & 3 & | & 3 \\ 0 & 2 & | & 0 \end{bmatrix}$ The new

system is $\begin{cases} x - y = 2 \\ 3y = 3 \\ 2y = 0 \end{cases}$. Thus, y = 0 and y = 1,

a contradiction. No solution. ϕ

31. $\begin{cases} x + 3y = 7 \\ x + y = 3 \\ 3x + y = 5 \end{cases}$

Augmented matrix: $\begin{bmatrix} 1 & 3 & | & 7 \\ 0 & -2 & | & -4 \\ 0 & -8 & | & -16 \end{bmatrix}$

$\left.\begin{matrix} (-1)R_1 + R_2 \blacktriangleright R_2 \\ (-3)R_1 + R_3 \blacktriangleright R_3 \end{matrix}\right\}: \begin{bmatrix} 1 & 3 & | & 7 \\ 0 & -2 & | & -4 \\ 0 & -8 & | & -16 \end{bmatrix}$

(- 4)$R_2 + R_3 \blacktriangleright R_3$: $\begin{bmatrix} 1 & 3 & | & 7 \\ 0 & -2 & | & -4 \\ 0 & 0 & | & 0 \end{bmatrix}$ The new

system is $\begin{cases} x + 3y = 7 \\ -2y = -4 \end{cases}$,so that y = 2 and x =

3y = x + 6 = 7 and x = 1. Solution: (1,2)

33. $\begin{cases} x + 2y + 3z = -2 \\ -x - y - 2z = 4 \end{cases}$

Augmented matrix: $\begin{bmatrix} 1 & 2 & 3 & | & -2 \\ -1 & -1 & -2 & | & 4 \end{bmatrix}$

$R_1 + R_2 \blacktriangleright R_2$: $\begin{bmatrix} 1 & 2 & 3 & | & -2 \\ 0 & 1 & 1 & | & 2 \end{bmatrix}$ The new

system is $\begin{cases} x + 2y + 3z = -2 \\ y + z = 2 \end{cases}$. Solve equation two for y, y = 2 – z, and replace y by 2 – z in equation one, and solve for x .
x + 2(2 – z) + 3z = -2 or x + z = -6 and
x = -z – 6. There are infinitely many solutions of the form (-z – 6, 2 – z, z), one for each real value of z.

35. $\begin{cases} x - y \quad = 1 \\ \quad y + z = 1 \\ x \quad + z = 2 \end{cases}$

Augmented matrix: $\begin{bmatrix} 1 & -1 & 0 & | & 1 \\ 0 & 1 & 1 & | & 1 \\ 1 & 0 & 1 & | & 2 \end{bmatrix}$

$\left.\begin{matrix} -R_1 + R_3 \blacktriangleright R_3 \\ -R_2 + R_3 \blacktriangleright R_3 \end{matrix}\right\}: \begin{bmatrix} 1 & -1 & 0 & | & 1 \\ 0 & 1 & 1 & | & 1 \\ 0 & 0 & 0 & | & 0 \end{bmatrix}$ The new system is $\begin{cases} x - y = 1 \\ y + z = 1 \end{cases}$. Solve equation two for y in terms of z, y = 1 – z and replace y in equation one by 1 – z and solve for x in terms of z. x – y = x – (1 – z) = 1 or x + z – 1 = 1 and x = 2 – z. There are infinitely many solutions of the form (2 – z, 1 – z, z), one for each real value of z.

Applications

37. Let x and y be the atomic numbers of lead and iron, respectively. Then $\begin{cases} x = 3y + 4 \\ x - 2y = 30 \end{cases}$.

Rewrite the system. $\begin{cases} x - 3y = 4 \\ x - 2y = 30 \end{cases}$. The augmented matrix is $\left[\begin{array}{cc|c} 1 & -3 & 4 \\ 1 & -2 & 30 \end{array}\right]$.

$-R_1 + R_2 \blacktriangleright R_2$: $\left[\begin{array}{cc|c} 1 & -3 & 4 \\ 0 & 1 & 26 \end{array}\right]$ The new system is $\begin{cases} x - 3y = 4 \\ y = 26 \end{cases}$. Thus, x – 3(26) = 4 or x = 82. The atomic number of lead is 82, that of iron is 26.

Review

39. $f(x) = 3x^2 - 2x$, $g(x) = 4x - 2$

a. $(f \circ g)(x) = f(g(x)) = f(4x - 2) =$
$3(4x - 2)^2 - 2(4x - 2) =$
$3(16x^2 - 16x + 4) - 8x + 4 =$
$48x^2 - 48x + 12 - 8x + 4 = 48x^2 - 56x + 16$.

b. $(fg)(x) = (3x^2 - 2x)(4x - 2) =$
$12x^3 - 8x^2 - 6x^2 + 4x = 12x^3 - 14x^2 + 4x$.

c. $(f - g)(x) = (3x^2 - 2x) - (4x - 2) =$
$3x^2 - 2x - 4x + 2 = 3x^2 - 6x + 2$

41. The solutions of $3x^2 - 2x = 0$ are values of x: 0 and ⅔. The x-intercepts of f(x) are ordered pairs: (0,0) and (⅔,0).

Section 12.5 Using Determinants to Solve Systems of Equations

Vocabulary

1. determinant

Practice

3. $\begin{vmatrix} 2 & 3 \\ -2 & 1 \end{vmatrix} = 2 \cdot 1 - (-2) \cdot 3 = 2 - (-6) = 8$

5. $\begin{vmatrix} -1 & 2 \\ 3 & -4 \end{vmatrix} = (-1)(-4) - 3 \cdot 2 = 4 - 6 = -2$

7. $\begin{vmatrix} x & y \\ y & x \end{vmatrix} = x \cdot x - y \cdot y = x^2 - y^2$

9. $\begin{vmatrix} 1 & 0 & 1 \\ 0 & 1 & 0 \\ 1 & 1 & 1 \end{vmatrix} = 1\begin{vmatrix} 1 & 0 \\ 1 & 1 \end{vmatrix} - 0\begin{vmatrix} 0 & 0 \\ 1 & 1 \end{vmatrix} + 1\begin{vmatrix} 0 & 1 \\ 1 & 1 \end{vmatrix} =$
$1(1 - 0) + 1(0 - 1) = 1 - 1 = 0$

11. $\begin{vmatrix} -1 & 2 & 1 \\ 2 & 1 & -3 \\ 1 & 1 & 1 \end{vmatrix} = 1\begin{vmatrix} 2 & 1 \\ 1 & -3 \end{vmatrix} - 1\begin{vmatrix} -1 & 1 \\ 2 & -3 \end{vmatrix} + 1\begin{vmatrix} -1 & 2 \\ 2 & 1 \end{vmatrix} =$
$1(-6-1) - 1(3-2) + 1(-1-4) = -7 - 1 - 5 = -13$

13. $\begin{vmatrix} 1 & -2 & 3 \\ -2 & 1 & 1 \\ -3 & -2 & 1 \end{vmatrix} = -2\begin{vmatrix} -2 & 3 \\ -2 & 1 \end{vmatrix} - 1\begin{vmatrix} 1 & 3 \\ -3 & 1 \end{vmatrix} + 1\begin{vmatrix} 1 & -2 \\ -3 & -2 \end{vmatrix}$
$= -2(-2 - (-6)) - 1(1 - (-9)) + 1(-2 - 6) =$
$-2(4) - 1(10) + 1(-8) = -8 - 10 - 8 = -26$

15. $\begin{vmatrix} 1 & 2 & 3 \\ 4 & 5 & 6 \\ 7 & 8 & 9 \end{vmatrix} = 1\begin{vmatrix} 5 & 6 \\ 8 & 9 \end{vmatrix} - 2\begin{vmatrix} 4 & 6 \\ 7 & 9 \end{vmatrix} + 3\begin{vmatrix} 4 & 5 \\ 7 & 8 \end{vmatrix} =$
$1(45 - 48) - 2(36 - 42) + 3(32 - 35) =$
$-3 - 2(-6) + 3(-3) = -3 + 12 - 9 = 0$

17. $\begin{vmatrix} a & 2a & -a \\ 2 & -1 & 3 \\ 1 & 2 & -3 \end{vmatrix} = a\begin{vmatrix} -1 & 3 \\ 2 & -3 \end{vmatrix} - 2a\begin{vmatrix} 2 & 3 \\ 1 & -3 \end{vmatrix} - a\begin{vmatrix} 2 & -1 \\ 1 & 2 \end{vmatrix}$
$= a(3-6) - 2a(-6-3) - a(4-(-1)) =$
$-3a - 2a(-9) - a(5) = -3a + 18a - 5a = 10a$

19. $\begin{vmatrix} 1 & a & b \\ 1 & 2a & 2b \\ 1 & 3a & 3b \end{vmatrix} = 1\begin{vmatrix} 2a & 2b \\ 3a & 3b \end{vmatrix} - 1\begin{vmatrix} a & b \\ 3a & 3b \end{vmatrix} + 1\begin{vmatrix} a & b \\ 2a & 2b \end{vmatrix}$
$= 1(6ab - 6ab) - 1(3ab - 3ab) + 1(2ab - 2ab) = 0$

21. $\begin{cases} x + y = 6 \\ x - y = 2 \end{cases}$ $D = \begin{vmatrix} 1 & 1 \\ 1 & -1 \end{vmatrix} = -1 - 1 = -2$
$D_x = \begin{vmatrix} 6 & 1 \\ 2 & -1 \end{vmatrix} = -6 - 2 = -8$ and $D_y = \begin{vmatrix} 1 & 6 \\ 1 & 2 \end{vmatrix} =$
2– 6 = -4. Then $x = \dfrac{D_x}{D} = \dfrac{-8}{-2} = 4$ and $y =$
$\dfrac{D_y}{D} = \dfrac{-4}{-2} = 2$. The solution is (4,2).

23. $\begin{cases} 2x + y = 1 \\ x - 2y = -7 \end{cases}$ $D = \begin{vmatrix} 2 & 1 \\ 1 & -2 \end{vmatrix} = -4 - 1 = -5$,
$D_x = \begin{vmatrix} 1 & 1 \\ -7 & -2 \end{vmatrix} = -2 - (-7) = 5$,
$D_y = \begin{vmatrix} 2 & 1 \\ 1 & -7 \end{vmatrix} = -14 - 1 = -15$
Then $x = D_x / D$ = 5/-5 = -1 and $y = D_y / D$ = -15/-5 = 3. The solution is (-1,3).

25. $\begin{cases} 4x - 3y = -1 \\ 8x + 3y = 4 \end{cases}$ $D = \begin{vmatrix} 4 & -3 \\ 8 & 3 \end{vmatrix} = 12 - (-24) = 36$
$D_x = \begin{vmatrix} -1 & -3 \\ 4 & 3 \end{vmatrix} = -3 - (-12) = 9$,
$D_y = \begin{vmatrix} 4 & -1 \\ 8 & 4 \end{vmatrix} = 16 - (-8) = 24$
Then $x = D_x /D$ = 9/36 = ¼ and $y = D_y /D$ = 24/36 = ⅔. The solution is (¼, ⅔).

27. $\begin{cases} y = \dfrac{11-3x}{2} \\ x = \dfrac{11-4y}{6} \end{cases}$ Rewrite the system in standard form as $\begin{cases} 3x + 2y = 11 \\ 6x + 4y = 11 \end{cases}$. $D = \begin{vmatrix} 3 & 2 \\ 6 & 4 \end{vmatrix} = 12 - 12$
$= 0$ and $D_x = \begin{vmatrix} 11 & 2 \\ 11 & 4 \end{vmatrix} = 44 - 22 = 22$. Since D
= 0 and $D_x \neq 0$, there is no solution.

29. $\begin{cases} y = \dfrac{-2x+1}{3} \\ 3x - 2y = 8 \end{cases}$ Rewrite the system in standard form as $\begin{cases} 2x + 3y = 1 \\ 3x - 2y = 8 \end{cases}$. $D = \begin{vmatrix} 2 & 3 \\ 3 & -2 \end{vmatrix} = -4 - 9$
= -13, $D_x = \begin{vmatrix} 1 & 3 \\ 8 & -2 \end{vmatrix} = -2 - 24 = -26$ and $D_y =$
$\begin{vmatrix} 2 & 1 \\ 3 & 8 \end{vmatrix} = 16 - 3 = 13$. Then $x = D_x /D$ =
-26/-13 = 2 and $y = D_y /D$ = 13/-13 = -1. The solution is (2,-1).

31. $\begin{cases} x = \dfrac{5y-4}{2} \\ y = \dfrac{3x-1}{5} \end{cases}$ Rewrite the system in standard form as $\begin{cases} 2x - 5y = -4 \\ -3x + 5y = -1 \end{cases}$. Then $D = \begin{vmatrix} 2 & -5 \\ -3 & 5 \end{vmatrix}$
= 10 – 15 = -5, $D_x = \begin{vmatrix} -4 & -5 \\ -1 & 5 \end{vmatrix} = -20 - 5$ = -25,
$D_y = \begin{vmatrix} 2 & -4 \\ -3 & -1 \end{vmatrix} = -2 - 12 = -14$. Thus, $x =$
D_x /D = -25/-5 = 5, $y = D_y /D$ = -14/-5 = 14/5. The solution is (5, 14/5).

33. $\begin{cases} x + y + z = 4 \\ x + y - z = 0 \\ x - y + z = 2 \end{cases}$

$D =$

$$\begin{vmatrix} 1 & 1 & 1 \\ 1 & 1 & -1 \\ 1 & -1 & 1 \end{vmatrix} = 1\begin{vmatrix} 1 & -1 \\ -1 & 1 \end{vmatrix} - 1\begin{vmatrix} 1 & -1 \\ 1 & 1 \end{vmatrix} + 1\begin{vmatrix} 1 & 1 \\ 1 & -1 \end{vmatrix} =$$

$$(1-1)-(1-(-1))+(-1-1)=0-2-2=-4$$

$$D_x = \begin{vmatrix} 4 & 1 & 1 \\ 0 & 1 & -1 \\ 2 & -1 & 1 \end{vmatrix} = 4\begin{vmatrix} 1 & -1 \\ -1 & 1 \end{vmatrix} - 0 + 2\begin{vmatrix} 1 & 1 \\ 1 & -1 \end{vmatrix} =$$

$$4(0)-0+2(-2)=-4$$

$$D_y = \begin{vmatrix} 1 & 4 & 1 \\ 1 & 0 & -1 \\ 1 & 2 & 1 \end{vmatrix} = -4\begin{vmatrix} 1 & -1 \\ 1 & 1 \end{vmatrix} + 0 - 2\begin{vmatrix} 1 & 1 \\ 1 & -1 \end{vmatrix} =$$

$$-4(1-(-1))-2(-1-1)=-8+4=-4$$

$$D_z = \begin{vmatrix} 1 & 1 & 4 \\ 1 & 1 & 0 \\ 1 & -1 & 2 \end{vmatrix} = 4\begin{vmatrix} 1 & 1 \\ 1 & -1 \end{vmatrix} - 0 + 2\begin{vmatrix} 1 & 1 \\ 1 & 1 \end{vmatrix} =$$

$$4(-1-1)+2(1-1)=-8$$

Then $x = D_x/D$ = -4/-4=1, $y = D_y/D$ = -4/-4 = 1 and $z = D_z/D$ = -8/-4 = 2. Solution: (1,1,2)

35. $\begin{cases} x+y+2z=7 \\ x+2y+z=8 \\ 2x+y+z=9 \end{cases}$

$$D = \begin{vmatrix} 1 & 1 & 2 \\ 1 & 2 & 1 \\ 2 & 1 & 1 \end{vmatrix} = -4,\ D_x = \begin{vmatrix} 7 & 1 & 2 \\ 8 & 2 & 1 \\ 9 & 1 & 1 \end{vmatrix} = -12$$

$$D_y = \begin{vmatrix} 1 & 7 & 2 \\ 1 & 8 & 1 \\ 2 & 9 & 1 \end{vmatrix} = -8,\ D_z = \begin{vmatrix} 1 & 1 & 7 \\ 1 & 2 & 8 \\ 2 & 1 & 9 \end{vmatrix} = -4$$

Then $x = D_x/D$ = -12/-4 = 3, $y = D_y/D$ = -8/-4 = 2 and $z = D_z/D$ = -4/-4 = 1. The solution is (3,2,1).

37. $\begin{cases} 2x+y-z=1 \\ x+2y+2z=2 \\ 4x+5y+3z=3 \end{cases}$

$$D = \begin{vmatrix} 2 & 1 & -1 \\ 1 & 2 & 2 \\ 4 & 5 & 3 \end{vmatrix} = 0,\ D_x = \begin{vmatrix} 1 & 1 & -1 \\ 2 & 2 & 2 \\ 3 & 5 & 3 \end{vmatrix} = -8$$

Since $D = 0$ and $D_x \neq 0$, there is no solution.

39. $\begin{cases} 2x+y+z=5 \\ x-2y+3z=10 \\ x+y-4z=-3 \end{cases}$

$$D = \begin{vmatrix} 2 & 1 & 1 \\ 1 & -2 & 3 \\ 1 & 1 & -4 \end{vmatrix} = 20,$$

$$D_x = \begin{vmatrix} 5 & 1 & 1 \\ 10 & -2 & 3 \\ -3 & 1 & -4 \end{vmatrix} = 60$$

$$D_y = \begin{vmatrix} 2 & 5 & 1 \\ 1 & 10 & 3 \\ 1 & -3 & -4 \end{vmatrix} = -40$$

$$D_z = \begin{vmatrix} 2 & 1 & 5 \\ 1 & -2 & 10 \\ 1 & 1 & -3 \end{vmatrix} = 20$$

Then $x = D_x/D$ = 60/20= 3, $y = D_y/D$ = - 40/20 = -2 and $z = D_z/D$ = 20/20 = 1. The solution is (3,-2,1)

41. $\begin{cases} 4x \quad\quad +3z=4 \\ \quad 2y-6z=-1 \\ 8x+4y+3z=9 \end{cases}$

$$D = \begin{vmatrix} 4 & 0 & 3 \\ 0 & 2 & -6 \\ 8 & 4 & 3 \end{vmatrix} = 72,\ D_x = \begin{vmatrix} 4 & 0 & 3 \\ -1 & 2 & -6 \\ 9 & 4 & 3 \end{vmatrix} = 54$$

$$D_y = \begin{vmatrix} 4 & 4 & 3 \\ 0 & -1 & -6 \\ 8 & 9 & 3 \end{vmatrix} = 36,\ D_z = \begin{vmatrix} 4 & 0 & 4 \\ 0 & 2 & -1 \\ 8 & 4 & 9 \end{vmatrix} = 24$$

Then $x = D_x/D$ = 54/72 = 3/4, $y = D_y/D$ = 36/72 = ½ and $z = D_z/D$ = 24/72 = 1/3. The solution is (¾, ½ ,⅓).

43. $\begin{cases} x+y \quad\quad = 1 \\ \frac{1}{2}y + z = \frac{5}{2} \\ x \quad\quad - z = -3 \end{cases}$ Rewrite the system as:

$$\begin{cases} x+y \quad\quad = 1 \\ y + 2z = 5 \\ x \quad\quad - z = -3 \end{cases}$$

$$D = \begin{vmatrix} 1 & 1 & 0 \\ 0 & 1 & 2 \\ 1 & 0 & -1 \end{vmatrix} = 1,\ D_x = \begin{vmatrix} 1 & 1 & 0 \\ 5 & 1 & 2 \\ -3 & 0 & -1 \end{vmatrix} = -2$$

$$D_y = \begin{vmatrix} 1 & 1 & 0 \\ 0 & 5 & 2 \\ 1 & -3 & -1 \end{vmatrix} = 3,\ D_z = \begin{vmatrix} 1 & 1 & 1 \\ 0 & 1 & 5 \\ 1 & 0 & -3 \end{vmatrix} = 1$$

Then $x = D_x/D = -2/1 = -2$, $y = D_y/D = 3/1 = 3$ and $z = D_z/D = 1/1 = 1$. Solution: (-2,3,1).

45. $\begin{cases} 2x - y + 4z + 2 = 0 \\ 5x + 8y + 7z = -8 \\ x + 3y + z + 3 = 0 \end{cases}$ Rewrite the system as:

$$\begin{cases} 2x - y + 4z = -2 \\ 5x + 8y + 7z = -8 \\ x + 3y + z = -3 \end{cases}$$

$$D = \begin{vmatrix} 2 & -1 & 4 \\ 5 & 8 & 7 \\ 1 & 3 & 1 \end{vmatrix} = 0,\ D_x = \begin{vmatrix} -2 & -1 & 4 \\ -8 & 8 & 7 \\ -3 & 3 & 1 \end{vmatrix} = 41$$

Since $D = 0$, but $D_x \neq 0$, the system is inconsistent and has no solutions.

47. $\begin{vmatrix} x & 1 \\ 3 & 2 \end{vmatrix} = 1$

$2x - 3 = 1$

$2x = 4$ or $x = 2$

49. $\begin{vmatrix} x & -2 \\ 3 & 1 \end{vmatrix} = \begin{vmatrix} 4 & 2 \\ x & 3 \end{vmatrix}$

$x - (-6) = 12 - 2x$

$x + 6 = 12 - 2x$

$3x = 6$ or $x = 2$

Review

51. $\begin{cases} 2x + y = 1 \\ x - 2y = 7 \end{cases}$

a. Multiply the top equation by 2 and add to the bottom equation to eliminate y.

$$\begin{cases} 4x + 2y = 2 \\ x - 2y = 7 \end{cases}$$

$5x \quad = 9$

or $x = \frac{9}{5} = 1.8$

Then $1.8 - 2y = 7$ or $-2y = 5.2$ and $y = -2.6$. The solution is (1.8, -2.6)

b. Solve equation one for y, giving $y = 1 - 2x$ and replace y in equation by $1 - 2x$. Thus, $x - 2(1 - 2x) = 7$ or $5x - 2 = 7$ so $5x = 9$ and $x = 1.8$. Then $y = 1 - 2(1.8) = -2.6$

c. Use ‘intersect’ in [-5,5] by [-5,5].

53. a. $f(x) = \left(\frac{4}{5}\right)^x$

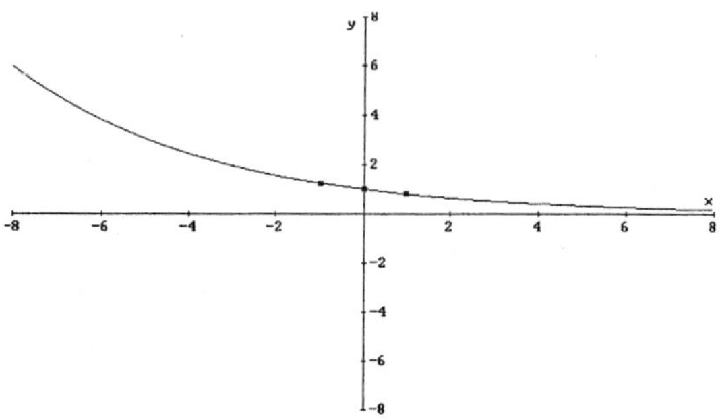

b. $g(x) = \log_{\frac{4}{5}} x$

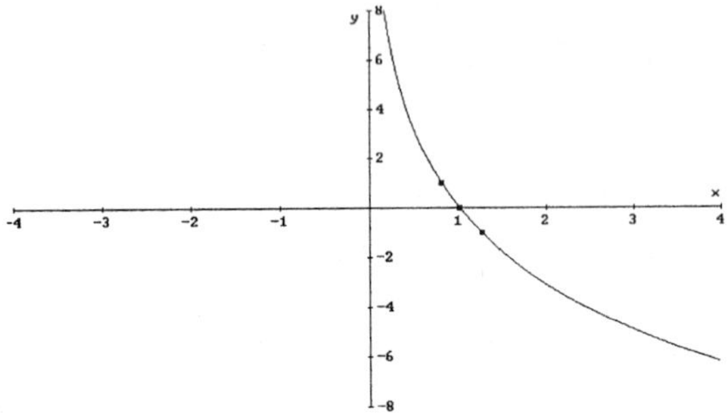

Section 12.6 Systems of Inequalities

1. $\begin{cases} x + 2y \le 3 \\ 2x - y \ge 1 \end{cases}$

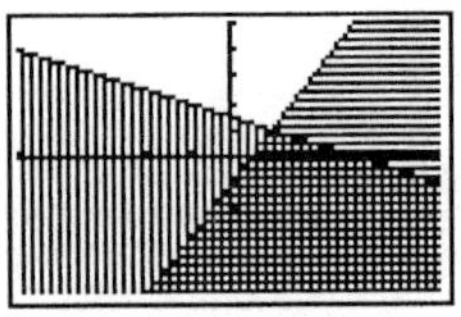

The solution set is the doubly-shaded region together with the two lines x + 2y = 3 and 2x – y = 1.

3. $\begin{cases} x + y < -1 \\ x - y > -1 \end{cases}$

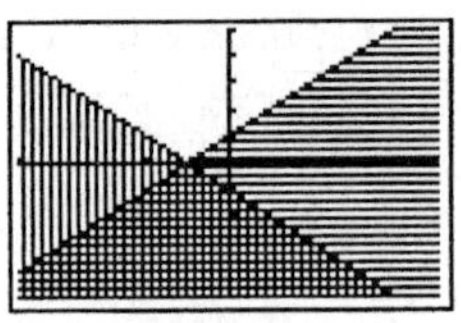

The solution set is the doubly-shaded region, not including either of the lines x + y = -1 or x – y = -1.

5. $\begin{cases} 2x - y < 4 \\ x + y \ge -1 \end{cases}$

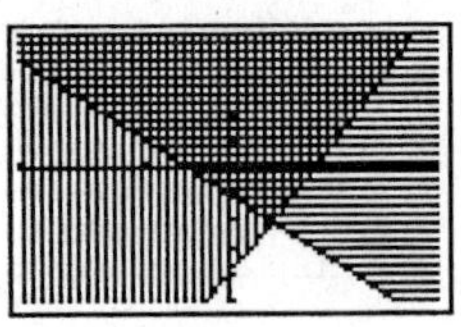

The solution set is the doubly-shaded region including the line x + y = -1, but not the line 2x – y = 4.

7. $\begin{cases} x > 2 \\ y \le 3 \end{cases}$

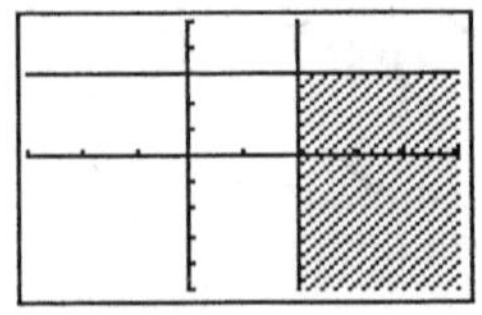

The solution set is the shaded region including the line y = 3, but not the line x = 2

9. $\begin{cases} 2x - 3y < 0 \\ y > x - 1 \end{cases}$

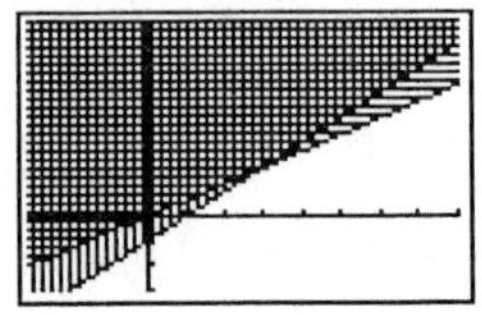

The solution set is the shaded region, not including the line 2x – 3y = 0 or the line y = x – 1.

11. $\begin{cases} x + y < 1 \\ x + y > 3 \end{cases}$

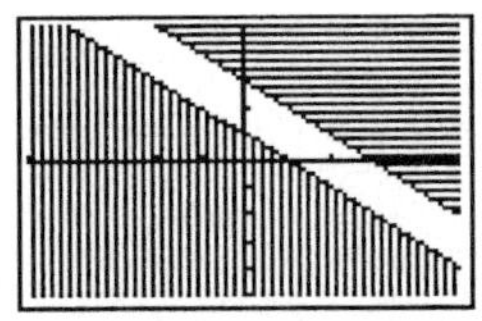

The solution set is empty since neither line is included and there is no doubly-shaded region.

13. $\begin{cases} x > 0 \\ y > 0 \end{cases}$

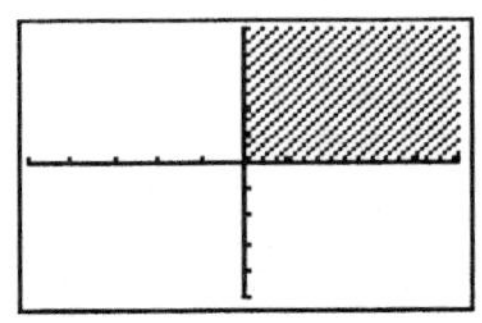

The solution set is the shaded region (Quadrant I), and does not include either the line x = 0 or the line y = 0.

15. $\begin{cases} 3x + 4y > -7 \\ 2x - 3y \ge 1 \end{cases}$

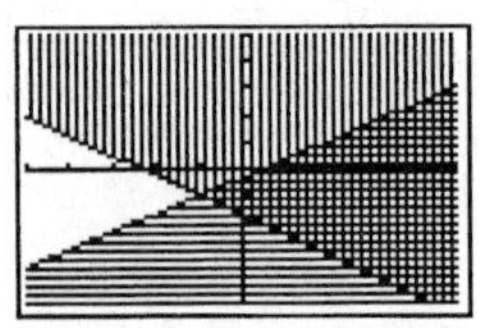

The solution set is the doubly-shaded region including the line 2x – 3y = 1, but not the line 3x + 4y = -7.

17. $\begin{cases} x < 3y - 1 \\ y \geq 2x - 3 \end{cases}$

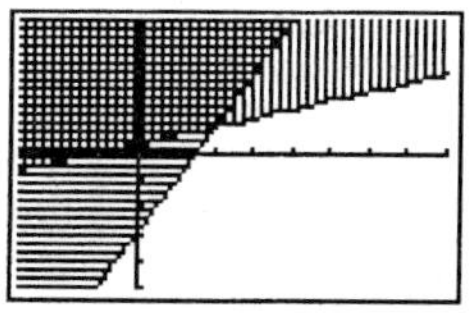

The solution set is the doubly-shaded region including the line y = 2x – 3, but not the line x = 3y – 1.

19. $\begin{cases} 2x + y < 7 \\ y > 2(1 - x) \end{cases}$

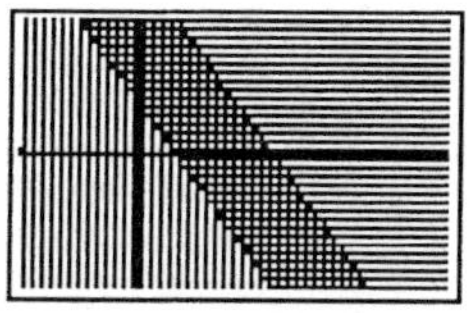

The solution set is the doubly-shaded region and does not include either the line 2x + y = 7 or the line y = 2(1 – x).

21. $\begin{cases} 2x - 4y > -6 \\ 3x + y \geq 5 \end{cases}$

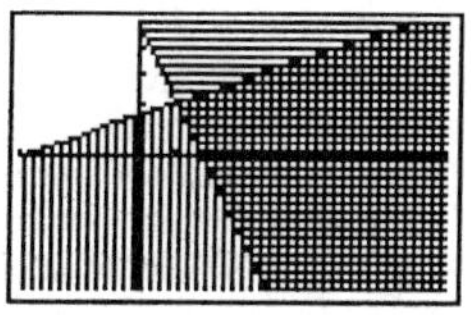

The solution set is the doubly-shaded region including the line 3x + y = 5, but not the line 2x – 4y = 6.

23. $\begin{cases} 3x - y \leq -4 \\ 3y > -2(x + 5) \end{cases}$

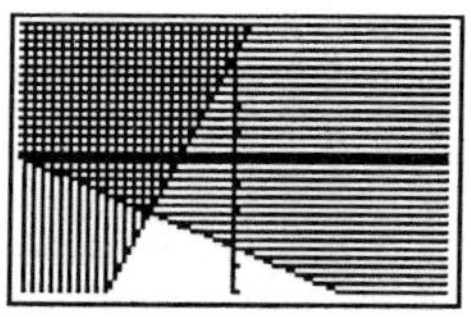

The solution set is the doubly-shaded region including the line 3x – y = -4, but not the line 3y = -2(x + 5).

25. $\begin{cases} \dfrac{x}{2} + \dfrac{y}{3} \geq 2 \\ \dfrac{x}{2} - \dfrac{y}{2} < -1 \end{cases}$ Rewrite as $\begin{cases} 3x + 2y \geq 12 \\ x - y < -2 \end{cases}$.

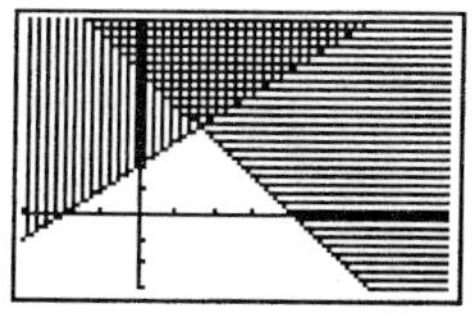

The solution set is the doubly-shaded region including the line 3x + 2y = 12, but not the line x – y = -2.

27. $\begin{cases} x \geq 0 \\ y \geq 0 \\ x + y \leq 3 \end{cases}$

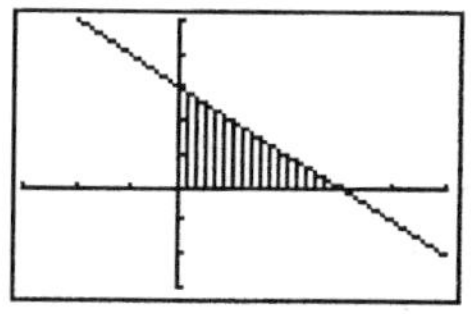

The solution set is the shaded region, including the lines x = 0, y = 0 and x + y = 3.

29. $\begin{cases} x \geq 0,\ y \geq 0 \\ x \leq 5 \\ y \leq x \end{cases}$

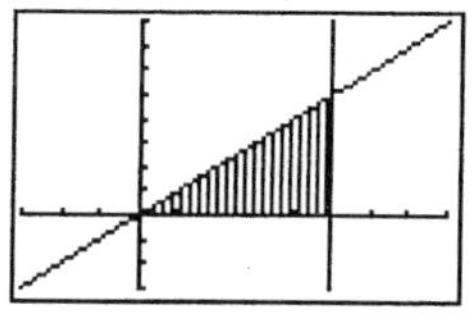

The solution set is the shaded area including the lines x = 0, y = 0, x = 5 and y = x.

31. $\begin{cases} x^2 - 6x - y \leq 5 \\ x^2 - 6x - y \geq -5 \end{cases}$

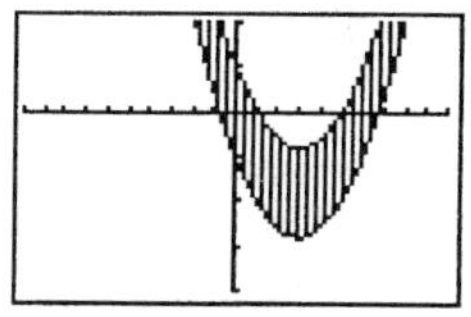

The solution set is the shaded region, including the curves $y = x^2 - 6x - 5$ and $y = x^2 - 6x + 5$.

33. $\begin{cases} x^2 - 6x - y < -5 \\ x^2 - 6x + y < -5 \end{cases}$

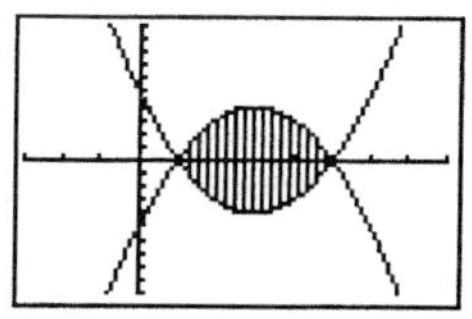

The solution set is the shaded area, not including either curve.

Applications

35. Let x = number of aluminum boats and y = number of fiberglass boats. Then

$$\begin{cases} x \geq 0,\ y \geq 0 \\ 800x + 600y \geq 2400 \\ 800x + 600y \leq 4800 \end{cases}$$

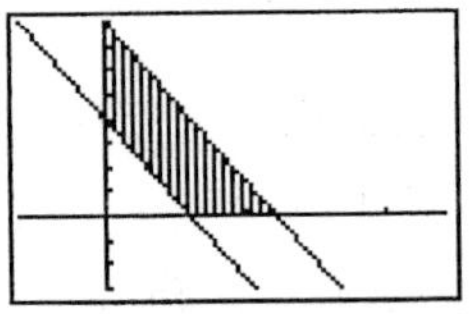

The solution set is the shaded area including the lines x = 0, y = 0, 4x + 3y = 12 and 4x + 3y = 24.

37. Let x = number of air cleaners and y = number of humidifiers. Then

$$\begin{cases} x \geq 0,\ y \geq 0 \\ 500x + 200y \leq 2000 \\ y > x \end{cases}$$

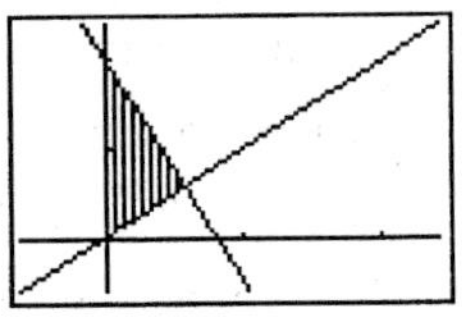

The solution set is the shaded area including the lines x = 0 and 5x + 2y = 20, but not y = x.

39. A system of inequalities will have no solution if the graphs of the individual inequalities have no points of intersection.

Review

41. $1-\sqrt{x} = \sqrt{x-3}$ Square both sides.

$1 - 2\sqrt{x} + x = x - 3$

$-2\sqrt{x} = -4$ or $\sqrt{x} = 2$

$x = 4$, but 4 is extraneous.

No solution.

43. $f(x) = \sqrt{x+2} - 4$

Domain = [-2, ∞) Range = [- 4, ∞)

Chapter 12 Summary

1. a. $\begin{cases} 3x - y = -2 \\ 2x + 3y = 17 \end{cases}$ The pair (1,5) is a solution to this system since 3(1) – 5 = 3 – 5 = -2 and 2(1) + 3(5) = 2 + 15 = 17.

b. $\begin{cases} 2x + 4y = 30 \\ \frac{x}{4} - y = 3 \end{cases}$ or $\begin{cases} 2x + 4y = 30 \\ x - 4y = 12 \end{cases}$. The pair (14, ½) is a solution to this system since 2(14) + 4(½) = 28 + 2 = 30 and 14 – 4(½) = 14 – 2 = 12.

2. a. $\begin{cases} x + y = 7 \\ 2x - y = 5 \end{cases}$

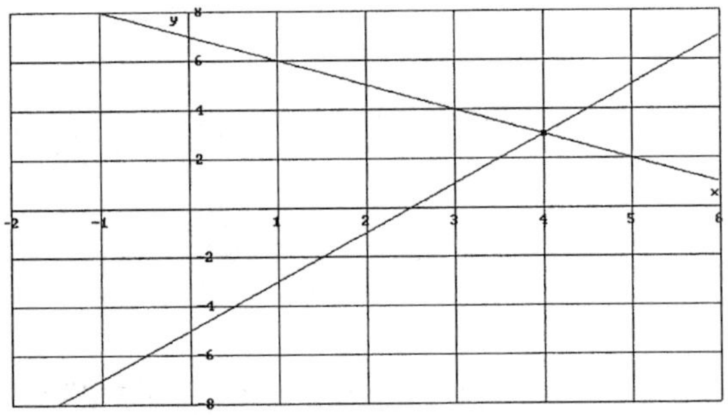

b. $\begin{cases} \frac{x}{3} + \frac{y}{5} = -1 \\ x - 3y = -3 \end{cases}$

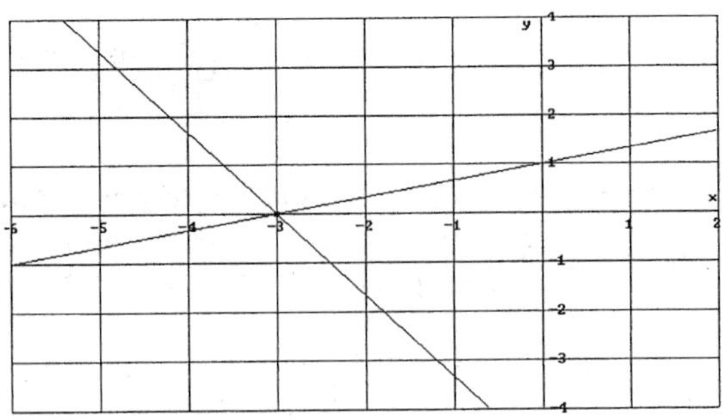

3. a. $\begin{cases} x = 3y + 5 \\ 5x - 4y = 3 \end{cases}$ The first equation is already solved for x, so replace x in equation two by

$3y + 5$. $5(3y + 5) - 4y = 3$ or $11y + 25 = 3$ and $11y = -22$ so $y = -2$. Then $x = 3(-2) + 5 = -1$. The solution is (-1,-2)

b. $\begin{cases} 6(x+2) = y - 1 \\ 5(y-1) = x + 2 \end{cases}$ The first equation is already solved for $y - 1$, so replace $y - 1$ in equation two by $6(x + 2)$. $5[6(x + 2)] = x + 2$ or $30x + 60 = x + 2$ and $29x = -58$ so $x = -2$. Then $y - 1 = 6(2 + -2) = 0$ so $y = 1$. The solution is (-2,1)

4. a. Add the equations to eliminate y.

$$\begin{cases} 2x + y = 1 \\ 5x - y = 20 \end{cases}$$
$$7x = 21$$
$$\text{or } x = 3$$

Then $2x + y = 2(3) + y = 1$ and $y = -5$. The solution is (3, -5).

b. $\begin{cases} 11x + 3y = 27 \\ 8x + 4y = 36 \end{cases}$

Multiply equation one by - 4 and equation two by 3 and add to eliminate y.

$$\begin{cases} -44x - 12y = -108 \\ 24x + 12y = 108 \end{cases}$$
$$-20x = 0$$
$$x = 0$$

Then $8x + 4y = 0 + 4y = 36$, so $y = 9$. The solution is (0,9)

5. Let x and y represent the two numbers. Then $\begin{cases} x + y = 18 \\ y = 5x \end{cases}$. Replace y by 5x in equation one to get $x + 5x = 18$ or $6x = 18$ and $x = 3$. Then $y = 5(3) = 15$. The numbers are 3, 15.

6. Let w = width of the rectangle, l = length of the rectangle. Then $\begin{cases} 2l + 2w = 24 \\ l = 3w \end{cases}$.

Replace l by 3w in equation one to get $2(3w) + 2w = 24$ or $8w = 24$ and $w = 3$. Then $l = 3w = 9$. The rectangle is 3' by 9'.

7. Let x = cost of a grapefruit and y = cost of an orange. Then $\begin{cases} x + y = 85 \\ x = y + 15 \end{cases}$. Replace x by y + 15 in equation one. $y + 15 + y = 85$ or $2y = 70$ and $y = 35$. Thus, $x = 35 + 15 = 50$. The cost of a grapefruit is 50¢.

8. Let g = amount of the gas bill and e = amount of the electric bill. Then $\begin{cases} g + e = 109 \\ e = g - 23 \end{cases}$. Replace e by $g - 23$ in equation one. $g + (g - 23) = 109$ or $2g - 23 = 109$ and $2g = 132$ so $g = 66$ and $e = g - 23 = 66 - 23 = 43$. His gas bill was \$66 and his electric bill was \$43.

9. Let x = amount invested at 10% and y = amount invested at 6%. Then

$$\begin{cases} x + y = 3000 \\ .10x + .06y = 270 \end{cases}$$

Multiply equation one by -10 and equation two by 100 and add to eliminate x.

$$\begin{cases} -10x - 10y = -30000 \\ 10x + 6y = 27000 \end{cases}$$
$$-4y = -3000$$
$$\text{or } y = 750$$

He invested \$750 at 6%.

10. $\begin{cases} x + y + z = 6 \\ x - y - z = -4 \\ -x + y - z = -2 \end{cases}$

Add equations one and two to eliminate z.

$$\begin{cases} x + y + z = 6 \\ x - y - z = -4 \end{cases}$$
$$2x = 2$$

Add equations one and three to eliminate z.

$$\begin{cases} x+y+z=6 \\ -x+y-z=-2 \end{cases}$$
$$2y \quad = 4$$

Then 2x = 2 or x = 1 and 2y = 4 or y = 2.
Finally, x + y + z = 1 + 2 + z = 6 so z = 3.
The solution is (1,2,3).

11. a. $\begin{cases} x+2y=4 \\ 2x-y=3 \end{cases}$

Augmented matrix: $\left[\begin{array}{cc|c} 1 & 2 & 4 \\ 2 & -1 & 3 \end{array}\right]$

$(-2)R_1 + R_2 \blacktriangleright R_2$: $\left[\begin{array}{cc|c} 1 & 2 & 4 \\ 0 & -5 & -5 \end{array}\right]$

New system: $\begin{cases} x+2y=4 \\ -5y=-5 \end{cases}$

Solve equation two for y, y = 1, and replace y by 1 in equation in equation one. x + 2(1) = 4 so x = 2. The solution is (2,1).

b. $\begin{cases} x+y+z=6 \\ 2x-y+z=1 \\ 4x+y-z=5 \end{cases}$

Augmented matrix: $\left[\begin{array}{ccc|c} 1 & 1 & 1 & 6 \\ 2 & -1 & 1 & 1 \\ 4 & 1 & -1 & 5 \end{array}\right]$

$\left.\begin{array}{l} -2R_1 + R_2 \blacktriangleright R_2 \\ -4R_1 + R_3 \blacktriangleright R_3 \end{array}\right\}: \left[\begin{array}{ccc|c} 1 & 1 & 1 & 6 \\ 0 & -3 & -1 & -11 \\ 0 & -3 & -5 & -19 \end{array}\right]$

$(-1)R_2 + R_3 \blacktriangleright R_3$: $\left[\begin{array}{ccc|c} 1 & 1 & 1 & 6 \\ 0 & -3 & -1 & -11 \\ 0 & 0 & -4 & -8 \end{array}\right]$

New system: $\begin{cases} x+y+z=6 \\ \quad -3y-z=-11 \\ \qquad -4z=-8 \end{cases}$

Solve equation three for z. z = 2. Then -3y – z = -3y – 2 = -11 so -3y = -9 or y = 3.
Finally x + y + z = x + 3 + 2 = 6 or x = 1.
The solution is (1,3,2).

12. a. $\begin{vmatrix} 2 & 3 \\ -4 & 3 \end{vmatrix} = 2\cdot 3-(-4)3 = 6+12 = 18$

b. $\begin{vmatrix} -1 & 2 & -1 \\ 2 & -1 & 3 \\ 1 & -2 & 2 \end{vmatrix} = (-1)\begin{vmatrix} -1 & 3 \\ -2 & 2 \end{vmatrix} - 2\begin{vmatrix} 2 & -1 \\ -2 & 2 \end{vmatrix} +$

$1\cdot\begin{vmatrix} 2 & -1 \\ -1 & 3 \end{vmatrix} = -(2-(-6)) - 2(4-2) +$

$(6-1) = -4-4+5 = -3$

13. a. $\begin{cases} 3x+4y=10 \\ 2x-3y=1 \end{cases}$ $D = \begin{vmatrix} 3 & 4 \\ 2 & -3 \end{vmatrix} = -9-8 = -17$

$D_x = \begin{vmatrix} 10 & 4 \\ 1 & -3 \end{vmatrix} = -30-4 = -34$

$D_y = \begin{vmatrix} 3 & 10 \\ 2 & 1 \end{vmatrix} = 3-20 = -17$. Then $x = D_x/D$ = -34/-17 = 2, and $y = D_y/D$ = -17/-17 = 1.
The solution is (2,1).

b. $\begin{cases} x+2y+z=0 \\ 2x+y+z=3 \\ x+y+2z=5 \end{cases}$

$D = \begin{vmatrix} 1 & 2 & 1 \\ 2 & 1 & 1 \\ 1 & 1 & 2 \end{vmatrix} = 1\begin{vmatrix} 1 & 1 \\ 1 & 2 \end{vmatrix} - 2\begin{vmatrix} 2 & 1 \\ 1 & 2 \end{vmatrix} + 1\begin{vmatrix} 2 & 1 \\ 1 & 1 \end{vmatrix} =$

$1-6+1=-4$

$D_x = \begin{vmatrix} 0 & 2 & 1 \\ 3 & 1 & 1 \\ 5 & 1 & 2 \end{vmatrix} = 0\begin{vmatrix} 1 & 1 \\ 1 & 2 \end{vmatrix} - 3\begin{vmatrix} 2 & 1 \\ 1 & 2 \end{vmatrix} + 5\begin{vmatrix} 2 & 1 \\ 1 & 1 \end{vmatrix} =$

$-9+5=-4$

$D_y = \begin{vmatrix} 1 & 0 & 1 \\ 2 & 3 & 1 \\ 1 & 5 & 2 \end{vmatrix} = 0\begin{vmatrix} 2 & 1 \\ 1 & 2 \end{vmatrix} + 3\begin{vmatrix} 1 & 1 \\ 1 & 2 \end{vmatrix} - 5\begin{vmatrix} 1 & 1 \\ 2 & 1 \end{vmatrix} =$

$3+5=8$

$$D_z = \begin{vmatrix} 1 & 2 & 0 \\ 2 & 1 & 3 \\ 1 & 1 & 5 \end{vmatrix} = 0\begin{vmatrix} 2 & 1 \\ 1 & 1 \end{vmatrix} - 3\begin{vmatrix} 1 & 2 \\ 1 & 1 \end{vmatrix} + 5\begin{vmatrix} 1 & 2 \\ 2 & 1 \end{vmatrix} =$$

$3 - 15 = -12$

Then x = D_x /D= -4/-4 = 1, y = D_y /D =

8/-4 = -2 and z = D_z /D = -12/-4 = 3. The solution is (1,-2,3).

14. a. $\begin{cases} y \ge x + 1 \\ 3x + 2y < 6 \end{cases}$

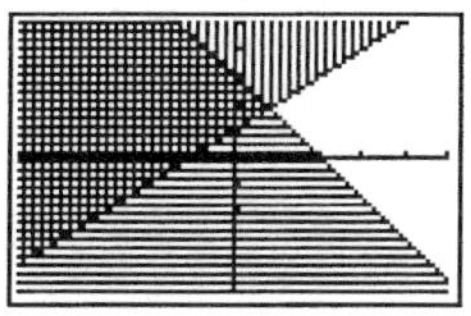

The solution set is the doubly-shaded area, not including either line.

b. $\begin{cases} y \ge x^2 - 4 \\ y < x + 3 \end{cases}$

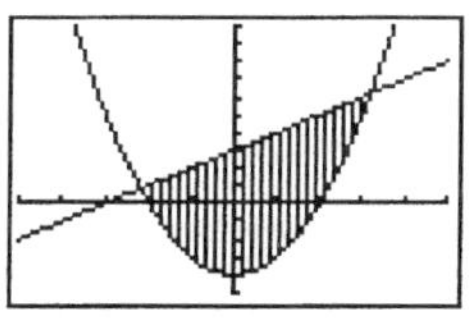

The solution set is the shaded area, including the curve $y = x^2 - 4$, but not the line y = x +3.

Chapter 12 Test

1. $\begin{cases} 3x + y = 7 \\ x - 2y = 0 \end{cases}$ Solution: (2,1)

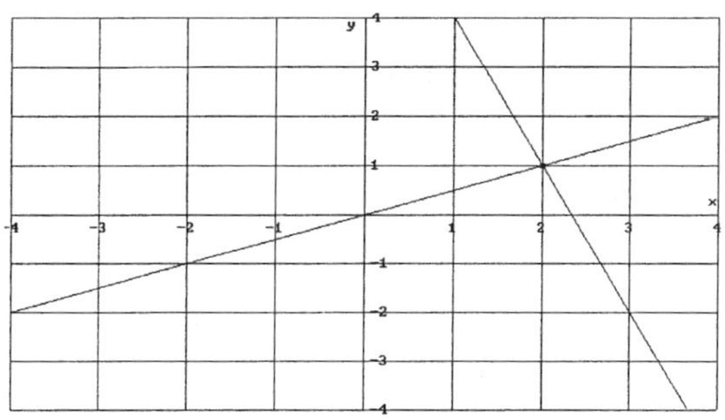

2. $\begin{cases} y = x - 1 \\ 2x + y = -7 \end{cases}$ Since equation one is solved for y already, replace y in equation two by x – 1. 2x + (x – 1) = -7 or 3x – 1 = -7 and so 3x = -6 or x = -2. Then y = x –1 = -2 – 1 = -3 and the solution is (-2, -3).

3. Add the two equations to eliminate y.

$$\begin{cases} 3x - y = 2 \\ 2x + y = 8 \end{cases}$$

$5x \quad = 10$

or $x \quad = 2$

Then 2(2) + y = 8 or y = 4. Solution: (2,4).

4. $\begin{cases} \dfrac{x}{6} + \dfrac{y}{10} = 3 \\ \dfrac{5x}{16} - \dfrac{3y}{16} = \dfrac{15}{8} \end{cases}$

Multiply the top equation by 30, the bottom one by 16 and add to eliminate y.

$$\begin{cases} 5x + 3y = 90 \\ 5x - 3y = 30 \end{cases}$$

$10x \quad = 120$

or $x \quad = 12$

Then 5x + 3y = 5(12) + 3y = 90, so 3y = 30 and y = 10. The solution is (12,10).

5. $\begin{cases} 4x + 3 = -3y \\ \dfrac{x}{7} + \dfrac{4y}{21} = 1 \end{cases}$ Rewrite as: $\begin{cases} 4x + 3y = -3 \\ -3x + 4y = 21 \end{cases}$

Multiply the top equation by 3, the bottom one by 4 and add to eliminate x.

$$\begin{cases} 12x + 9y = -9 \\ -12x + 16y = 84 \end{cases}$$

$25y = 75 \text{ or } y = 3$

Then 4x + 3y = 4x + 3(3) = -3 or 4x = -12 and x = -3. The solution is (-3,3).

6. $\begin{cases} 2x + 3y + z = -5 \\ -x + 2y - z = -6 \\ 3x + y + 2z = 4 \end{cases}$

Add equations one and two to eliminate z.

$$\begin{cases} 2x + 3y + z = -5 \\ -x + 2y - z = -6 \end{cases}$$

$$x + 5y \quad = -8$$

Multiply equation two by 2 and add to equation three to eliminate z.

$$\begin{cases} -2x + 4y - 2z = -12 \\ 3x + y + 2z = 4 \end{cases}$$

$$x + 5y \quad = -8$$

Now solve $\begin{cases} x + 5y = -11 \\ x + 5y = -8 \end{cases}$. Since this is impossible, the system has no solution.

7. $\begin{vmatrix} 2 & -3 \\ 4 & 5 \end{vmatrix} = 2(5) - 4(-3) = 10 - (-12) = 22$

8. $\begin{vmatrix} -3 & -4 \\ -2 & 3 \end{vmatrix} = -9 - 8 = -17$

9. $\begin{vmatrix} 1 & 2 & 0 \\ 2 & 0 & 3 \\ 1 & -2 & 2 \end{vmatrix} = 1\begin{vmatrix} 0 & 3 \\ -2 & 2 \end{vmatrix} - 2\begin{vmatrix} 2 & 3 \\ 1 & 2 \end{vmatrix} + 0\begin{vmatrix} 2 & 0 \\ 1 & -2 \end{vmatrix} =$

$(0 - (-6)) - 2(4 - 3) = 6 - 2 = 4$

10. $\begin{vmatrix} 2 & -1 & 1 \\ 3 & 1 & 0 \\ 0 & 1 & 2 \end{vmatrix} = 2\begin{vmatrix} 1 & 0 \\ 1 & 2 \end{vmatrix} - 3\begin{vmatrix} -1 & 1 \\ 1 & 2 \end{vmatrix} + 0\begin{vmatrix} -1 & 1 \\ 1 & 0 \end{vmatrix} =$

$2(2 - 0) - 3(-2 - 1) + 0 = 4 + 9 = 13$

For problems 11 - 14, $\begin{cases} x - y = -6 \\ 3x + y = -6 \end{cases}$

11. $D_x = \begin{vmatrix} -6 & -1 \\ -6 & 1 \end{vmatrix}$

12. $D_y = \begin{vmatrix} 1 & -6 \\ 3 & -6 \end{vmatrix}$

13. $D = \begin{vmatrix} 1 & -1 \\ 3 & 1 \end{vmatrix} = 1 - (-3) = 4,$

$D_x = \begin{vmatrix} -6 & -1 \\ -6 & 1 \end{vmatrix} = -6 - 6 = -12$

So $x = D_x / D = -12/4 = -3$.

14. $D = 4$ and $D_y = \begin{vmatrix} 1 & -6 \\ 3 & -6 \end{vmatrix} = -6 - (-18) = 12$ so that $y = D_y / D = 12/4 = 3$.

For problems 15 - 18. $\begin{cases} x + y + z = 4 \\ x + y - z = 6 \\ 2x - 3y + z = -1 \end{cases}$

15. Add equations one and two to eliminate z.

$$\begin{cases} x + y + z = 4 \\ x + y - z = 6 \end{cases}$$

$$2x + 2y \quad = 10$$

Add equations two and three to eliminate z.

$$\begin{cases} x + y - z = 6 \\ 2x - 3y + z = -1 \end{cases}$$

$$3x - 2y \quad = 5$$

Now eliminate y from: $\begin{cases} 2x + 2y = 10 \\ 3x - 2y = 5 \end{cases}$ by adding the two equations. Thus, $5x = 15$ or $x = 3$.

16. Since $x = 3$ and $2x + 2y = 10$, $2(3) + 2y = 10$ or $2y = 4$ and $y = 2$.

17. Since $x = 3$ and $y = 2$ and $x + y + z = 4$, $3 + 2 + z = 4$ or $z = -1$.

18. The augmented matrix for the system is

$$\left[\begin{array}{ccc|c} 1 & 1 & 1 & 4 \\ 1 & 1 & -1 & 6 \\ 2 & -3 & 1 & -1 \end{array}\right]$$

19. $\begin{cases} 2x - 3y \geq 6 \\ y \leq -x + 1 \end{cases}$

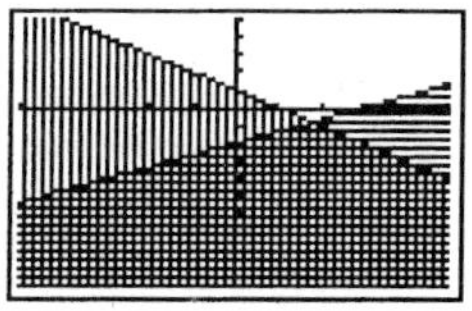

The solution set is the doubly-shaded area, including the lines $2x - 3y = 6$ and $y = -x + 1$.

20. $\begin{cases} y \geq x^2 \\ y < x + 3 \end{cases}$

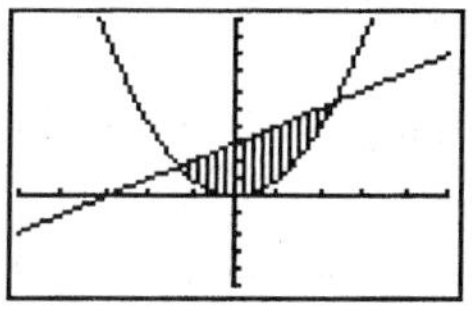

The solution set is the shaded area, including the curve $y = x^2$, but not the line $y = x + 3$.

Cumulative Review: Chapters 11 and 12

Vocabulary/Concepts

1. (b,a)
2. $f^{-1}(x)$
3. The inverse relation for {(1,3),(2,5),(-3,-2), (-4,-6)} is {(3,1),(5,2),(-2,-3),(-6,-4)}
4. not a function
5. not a one-to-one function
6. one-to-one
7. exponential
8. a. $f(x) = x^3$ is not an exponential function.
   b. $f(x) = 3^x$ is an exponential function.
   c. $f(x) = (\frac{1}{2})^x$ is an exponential function.
9. $\log_3 9 = x$ means $3^x = 9 = 3^2$ and so $x = 2$.
10. $-5\log_b x - 3\log_b y + \log_b z =$
    $-\log_b x^5 - \log_b y^3 + \log_b z =$
    $-(\log_b x^5 + \log_b y^3) + \log_b z =$
    $-\log_b(x^5y^3) + \log_b z = \log_b\left(\frac{z}{x^5y^3}\right)$
11. system
12. consistent
13. inconsistent
14. The solution is (-8,-7).
15. matrix
16. square
17. augmented; coefficient
18. coefficient matrix = $\begin{bmatrix} 1 & 2 & 2 \\ 2 & 1 & 2 \\ 2 & 2 & 1 \end{bmatrix}$
19. augmented matrix = $\left[\begin{array}{ccc|c} 1 & 2 & 2 & 10 \\ 2 & 1 & 2 & 9 \\ 2 & 2 & 1 & 11 \end{array}\right]$
20. Cramer's

Practice

For problems 21 - 24, $f(x) = 3x - 2$, $g(x) = x^2 + 1$

21. $(f \circ g)(3) = f(g(3))$ and $g(3) = 3^2 + 1 = 10$. So $f(g(3)) = f(10) = 3(10) - 2 = 30 - 2 = 28$.
22. $(g \circ f)(3) = g(f(3))$ and $f(3) = 3(3) - 2 = 7$. So $g(f(3)) = g(7) = 7^2 + 1 = 49 + 1 = 50$.
23. $(f \circ g)\left(\frac{1}{4}\right) = f(g\left(\frac{1}{4}\right))$ and $g\left(\frac{1}{4}\right) = \left(\frac{1}{4}\right)^2 + 1 =$
    $\frac{1}{16} + 1 = \frac{17}{16}$. Then $f\left(\frac{17}{16}\right) = 3\left(\frac{17}{16}\right) - 2 =$
    $\frac{51}{16} - \frac{32}{16} = \frac{19}{16}$
24. $(g \circ f)(h) = g(f(h)) = g(3h - 2) = (3h - 2)^2 + 1$
    $= 9h^2 - 12h + 5$.
25. $f(x) = x^2 + 5x + 6$, $g(x) = x + 3$
    a. $(f + g)(x) = f(x) + g(x) = (x^2 + 5x + 6) + (x + 3) = x^2 + 6x + 9$
    b. $(f - g)(x) = f(x) - g(x) = (x^2 + 5x + 6) - (x + 3) = x^2 + 4x + 3$
    c. $(f \cdot g)(x) = (x^2 + 5x + 6)(x + 3) = x^3 + 5x^2 + 6x + 3x^2 + 15x + 18 = x^3 + 8x^2 + 21x + 18$

d. $\left(\frac{f}{g}\right)(x)=\frac{f(x)}{g(x)}=\frac{x^2+5x+6}{x+3}=$
$\frac{(x+3)(x+2)}{x+3}=x+2,\ x\neq -3.$

26. $f(x) = x^2 - 4$, $g(x) = x + 2$
a. $(f + g)(x) = x^2 - 4 + x + 2 = x^2 + x - 2.$
Domain: $\mathbb{R}$

b. $(f - g)(x) = (x^2 - 4) - (x + 2) = x^2 - 4 - x - 2 = x^2 - x - 6.$ Domain: $\mathbb{R}$

c. $(fg)(x) = (x^2 - 4)(x + 2) = x^3 + 2x^2 - 4x - 8$
Domain: $\mathbb{R}$

d. $\left(\frac{f}{g}\right)(x)=\frac{x^2-4}{x+2}=x-2,\ x\neq -2$
Domain: $\{x|x \neq -2\}$

27. $x = 2y - 1$
a. Interchange x and y: $y = 2x - 1$. Since y is already isolated, $f^{-1}(x) = 2x - 1$ is the inverse function.

b. and c.

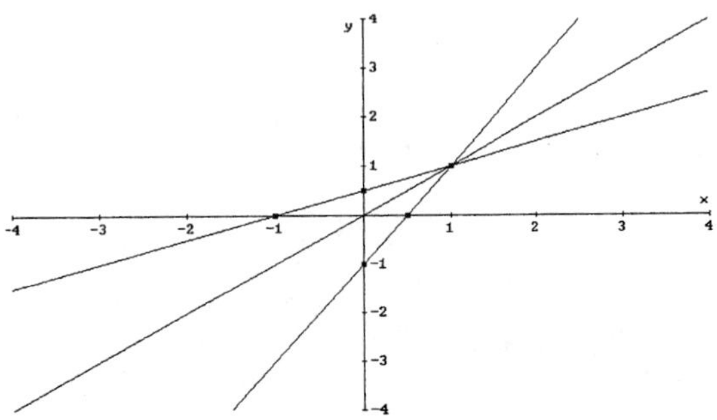

28. $3^x = 9^3 = (3^2)^3 = 3^6$ so $x = 6$.

29. $f(x) = \log_{\frac{1}{2}} x$.

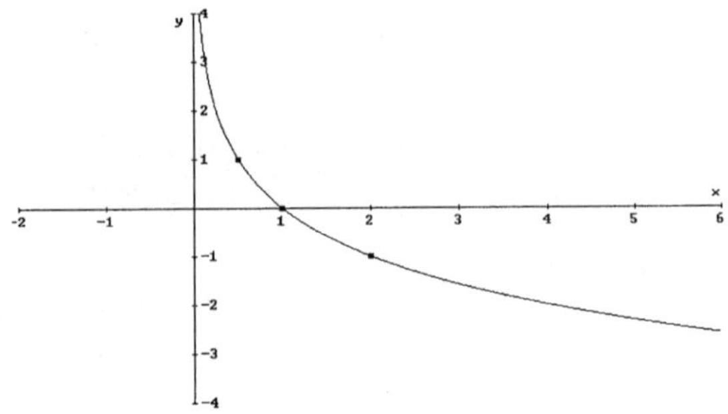

30. $125^{x-3} = 25^{2x+7}$ or $(5^3)^{x-3} = (5^2)^{2x+7}$ Thus,
$3(x-3) = 2(2x+7)$
$3x - 9 = 4x + 14$
$-23 = x$

31. $\begin{cases}2x - y = -21\\4x + 5y = 7\end{cases}$
Solve equation one for y, $y = 2x + 21$. Replace y by $2x + 21$ in equation two. Then $4x + 5(2x + 21) = 7$ or $4x + 10x + 105 = 7$. Thus, $14x = -98$ or $x = -7$. Now $2x - y = 2(-7) - y = -21$ or $-y = -7$ and $y = 7$. The solution is (-7,7),

32. $\begin{cases}2y - 3x = -13\\3x - 17 = 4y\end{cases}$ Rewrite as $\begin{cases}-3x + 2y = -13\\3x - 4y = 17\end{cases}$
Add to eliminate x, giving $-2y = 4$ or $y = -2$. Then $-3x + 2y = -3x + 2(-2) = -3x - 4 = -13$, and $-3x = -9$ or $x = 3$. Solution: (3, -2).

33. $\begin{cases}x + 2y + 2z = 10\\2x + y + 2z = 9\\2x + 2y + z = 11\end{cases}$
Add -2 times equation one to equation two to eliminate x. $\begin{cases}-2x - 4y - 4z = -20\\2x + y + 2z = 9\end{cases}$
$$-3y - 2z = -11$$
or $3y + 2z = 11$
Add -2 times equation one to equation three

to eliminate x. $\begin{cases} -2x-4y-4z=-20 \\ 2x+2y+z=11 \end{cases}$

$$-2y-3z=-9$$

or $\quad 2y+3z=9$

Now solve $\begin{cases} 3y+2z=11 \\ 2y+3z=9 \end{cases}$. Multiply the top equation by -2, the bottom equation by 3 and add to eliminate y. $\begin{cases} -6y-4z=-22 \\ 6y+9z=27 \end{cases}$. Thus,

$$5z=5$$

z = 1, and so 3y + 2z = 3y + 2(1) = 11 and 3y = 9 or y = 3. Finally, x + 2y + 2z = x + 2(3)+ 2(1) =10 or x = 2. The solution is (2,3,1).

34. $\begin{cases} 2x-3y=16 \\ -4x+y=-12 \end{cases}$

Augmented matrix: $\left[\begin{array}{cc|c} 2 & -3 & 16 \\ -4 & 1 & -12 \end{array}\right]$

$2R_1+R_2 \blacktriangleright R_2$: $\left[\begin{array}{cc|c} 2 & -3 & 16 \\ 0 & -5 & 20 \end{array}\right]$

The new system is $\begin{cases} 2x-3y=-16 \\ -5y=20 \end{cases}$. Thus, y = -4, and 2x – 3y = 2x – 3(- 4) = -16. Then 2x +12 = 16, 2x = 4 or x = 2. Solution: (2,-4)

35. $\begin{cases} 3x-2y+4z=4 \\ x+y+z=3 \\ 6x-2y-3z=10 \end{cases}$

Augmented matrix: $\left[\begin{array}{ccc|c} 3 & -2 & 4 & 4 \\ 1 & 1 & 1 & 3 \\ 6 & -2 & -3 & 10 \end{array}\right]$

$R_1 \blacktriangleleft\blacktriangleright R_2$: $\left[\begin{array}{ccc|c} 1 & 1 & 1 & 3 \\ 3 & -2 & 4 & 4 \\ 6 & -2 & -3 & 10 \end{array}\right]$

$\left.\begin{array}{l} -3R_1+R_2 \blacktriangleright R_2 \\ -6R_1+R_3 \blacktriangleright R_3 \end{array}\right\}$: $\left[\begin{array}{ccc|c} 1 & 1 & 1 & 3 \\ 0 & -5 & 1 & -5 \\ 0 & -8 & -9 & -8 \end{array}\right]$

$-\frac{1}{5}R_2 \blacktriangleright R_2$: $\left[\begin{array}{ccc|c} 1 & 1 & 1 & 3 \\ 0 & 1 & -\frac{1}{5} & 1 \\ 0 & -8 & -9 & -8 \end{array}\right]$

$8R_2+R_3 \blacktriangleright R_3$: $\left[\begin{array}{ccc|c} 1 & 1 & 1 & 3 \\ 0 & 1 & -\frac{1}{5} & 1 \\ 0 & 0 & -53/5 & 0 \end{array}\right]$

Thus, z = 0 and $y-\frac{1}{5}z=y-0=1$ or y = 1.

Finally, x + y + z = x + 1 + 0 = 3 or x = 2. The solution is (2,1,0).

36. $\begin{cases} 3x-y=-3 \\ 2x+y=-7 \end{cases}$

$$D=\begin{vmatrix} 3 & -1 \\ 2 & 1 \end{vmatrix}=3-(-2)=5$$

$$D_x=\begin{vmatrix} -3 & -1 \\ -7 & 1 \end{vmatrix}=-3-7=-10$$

$$D_y=\begin{vmatrix} 3 & -3 \\ 2 & -7 \end{vmatrix}=-21-(-6)=-15$$

Then x = D_x /D = -10/5 = -2 and y = D_y /D = -15/5 = -3. The solution is (-2,-3).

37. $\begin{cases} 2x+3y+4z=6 \\ 2x-3y-4z=-4 \\ 4x+6y+8z=12 \end{cases}$

$$D=\begin{vmatrix} 2 & 3 & 4 \\ 2 & -3 & -4 \\ 4 & 6 & 8 \end{vmatrix}=2\begin{vmatrix} -3 & -4 \\ 6 & 8 \end{vmatrix}-3\begin{vmatrix} 2 & -4 \\ 4 & 8 \end{vmatrix}+$$

$$4\begin{vmatrix} 2 & -3 \\ 4 & 6 \end{vmatrix}=2(0)-3(32)+4(24)=0$$

Also, $D_x = D_y = D_z = 0$ so the system is dependent and has infinitely many solutions.

38. $\begin{cases} 3x+y\le 1 \\ 4x-y>-8 \end{cases}$

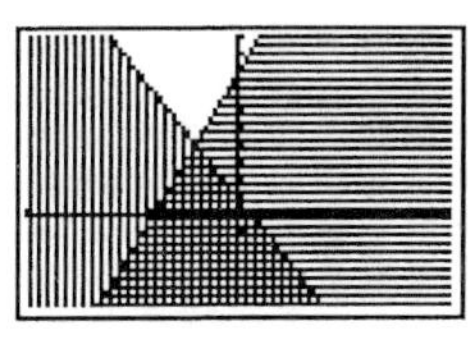

The solution set is the doubly-shaded region, including the line $3x + y = 1$, but not the line $4x - y = -8$.

39. $\begin{cases} x \geq 0 \\ y < 0 \end{cases}$

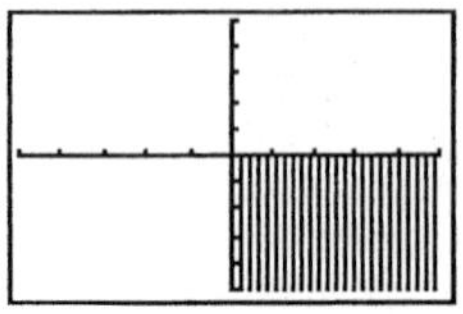

The solution set is the shaded area, including the line $x = 0$, but not the line $y = 0$.

40. $\begin{cases} y \geq x + 2 \\ x \leq y - 2 \end{cases}$

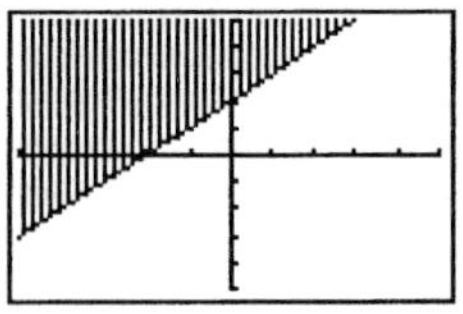

The solution set is the shaded area, including the line $y = x + 2$.

41. $\log x + \log (x + 3) = 1$ Use 'intersect' in the window [-5,5] by [-5,5].

42. $\log 3x = \log 9$
a. Thus, $3x = 9$ or $x = 3$.
b. Use 'intersect' in [-2,8] by [-3,3].

43. $2^{x^2-2x} = 8$

a. $2^{x^2-2x} = 8 = 2^3$ so $x^2 - 2x = 3$ or

$x^2 - 2x - 3 = 0$

$(x - 3)(x + 1) = 0$ so $x = 3$ or $x = -1$

b. Use 'intersect' in [-5,5] by [-5,10].

44. a. $\begin{cases} x^2 - 6x - y = 5 \\ x^2 - 6x + y = -5 \end{cases}$

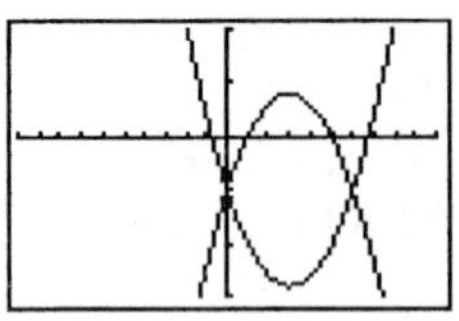

Using 'intersect' in [-10,10] by [-15,10], the solutions are found to be (0,-5) and (6,-5).

b. $\begin{cases} x^2 - x - y = 2 \\ 4x - 3y = 0 \end{cases}$

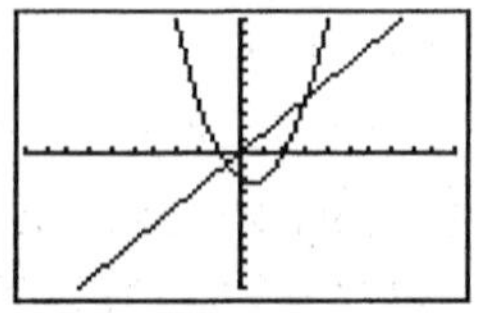

Using 'intersect' in [-10,10] by [-10,10] the solutions are found to be $(\frac{-2}{3}, \frac{-8}{9})$ and (3,4).

Applications

45. $y = 200\ e^{0.5x}$, x = days, y = number of ants

a. For $y = 1000$, $1000 = 200e^{0.5x}$ or $5 = e^{0.5x}$ so that $\ln 5 = .5x$ or $x = 2 \ln 5 \approx 3.2$ days.

b. After 10 days, $y = 200e^{0.5(10)} = 200\ e^5 \approx$ 29,683 ants, to the nearest ant.

46. $A = A_0\left(\frac{2}{3}\right)^t$, $A_0 = 5$g. and $t = 10$ years

Thus, $A = 5\left(\frac{2}{3}\right)^{10} \approx 0.08671$g.

47. Let x = the cost of a skirt and y = the cost of a sweater. Then $\begin{cases} 2x + 3y = 192 \\ 3x + y = 184 \end{cases}$.

Multiply equation two by -3 and add to equation one to eliminate y.

$$\begin{cases} 2x + 3y = 192 \\ -9x - 3y = -552 \end{cases}$$

$$-7x \quad = -360$$

or $x = \frac{360}{7} \approx 51.43$

A single skirt costs approximately \$51.43.

48. Let x = number of pounds of \$1.25 candy to be used and let y = number of pounds of \$2.50 candy to be used. Then

$$\begin{cases} x + y = 30 \\ 1.25x + 2.50y = 30(2.00) \end{cases}$$

Multiply the top equation by – 2.50 and add to the bottom equation to eliminate y.

$$\begin{cases} -2.50x - 2.50y = -75 \\ 1.25x + 2.50y = 60 \end{cases}$$

$$-1.25x \quad = -15$$

$$\text{or} \quad x \quad = 12$$

Then $x + y = 12 + y = 30$ so $y = 18$. He should use 12 pounds of the $1.25 per pound candy and 18 pounds of the $2.50 per pound candy.

Appendix A Conic Sections

Vocabulary

1. conic sections
2. circle
3. $(x-h)^2+(y-k)^2=r^2$
4. (0,0)
5. origin; vertical
6. origin; y
7. origin; x

Concepts

8. a
9. e
10. b
11. c
12. f
13. g
14. d

Practice

15. $x^2+y^2=9=3^2$
    center: (0,0) radius: 3

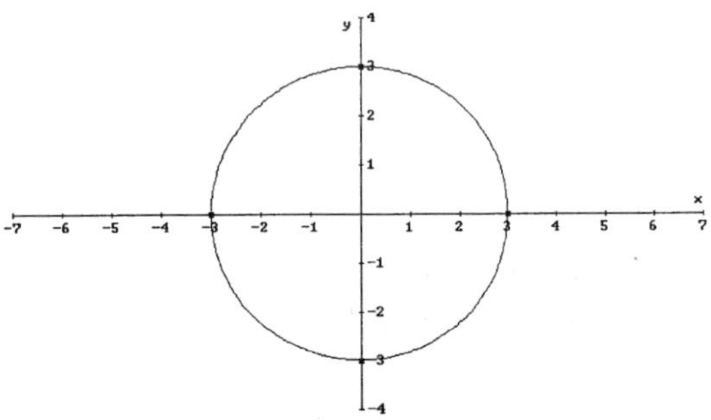

16. $x^2+y^2=16=4^2$
    center: (0,0) radius: 4

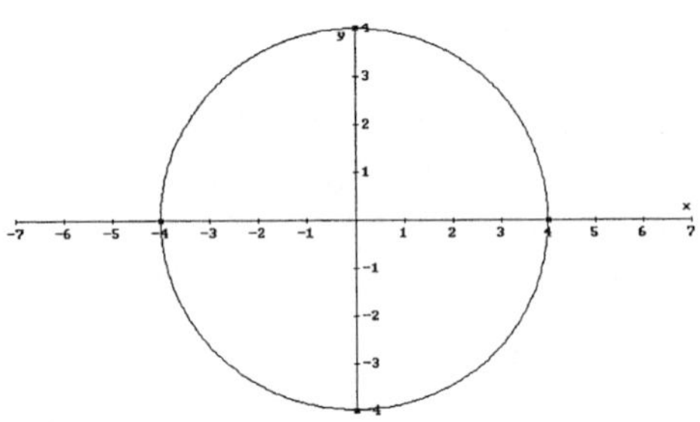

17. $(x-2)^2+(y-4)^2=4=2^2$
    center:(2,4) radius: 2

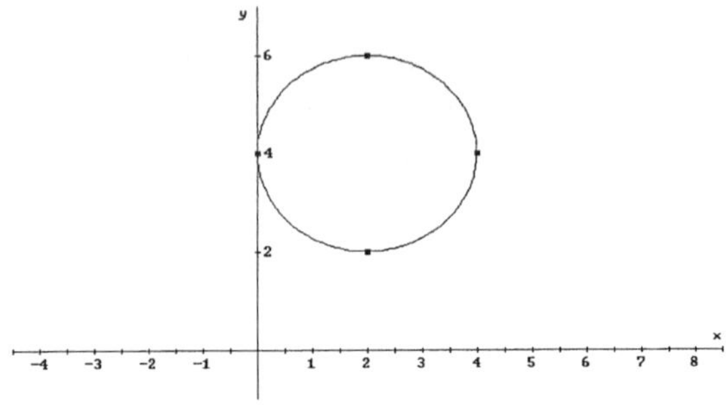

18. $(x-3)^2+(y-2)^2=4=2^2$
    center: (3,2) radius: 2

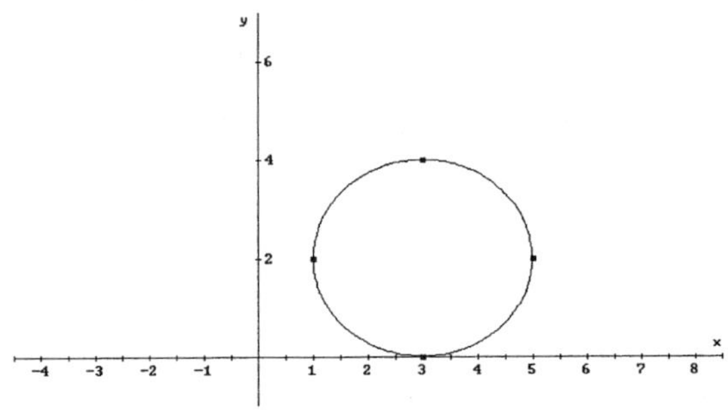

19. $x^2+y^2-2x+4y=-1$ Complete the square in both x and y to find the center and radius.
    $(x^2-2x+1)+(y^2+4y+4)=-1+1+4=4$
    or $(x-1)^2+(y+2)^2=2^2$
    center: (1, -2) radius: 2

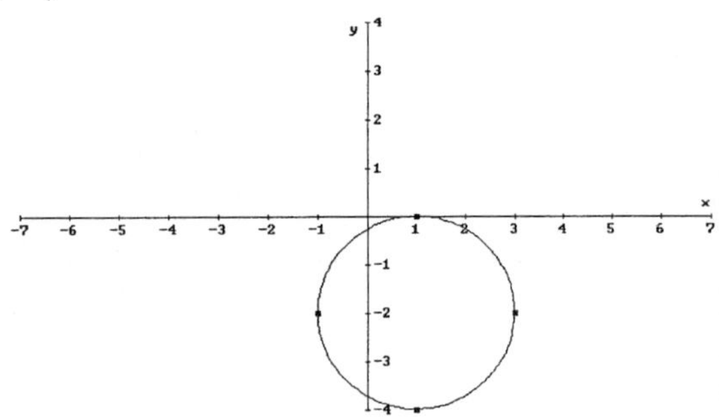

20. $x=y^2$
    This can be written as $x=(y-0)^2+0$
    Vertex: (0,0)

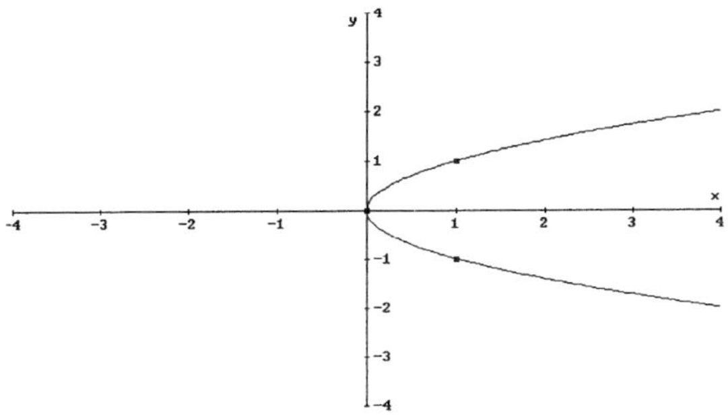

21. $x = 4y^2$

This can be written as $x = 4(y - 0)^2 + 0$

Vertex: (0,0)

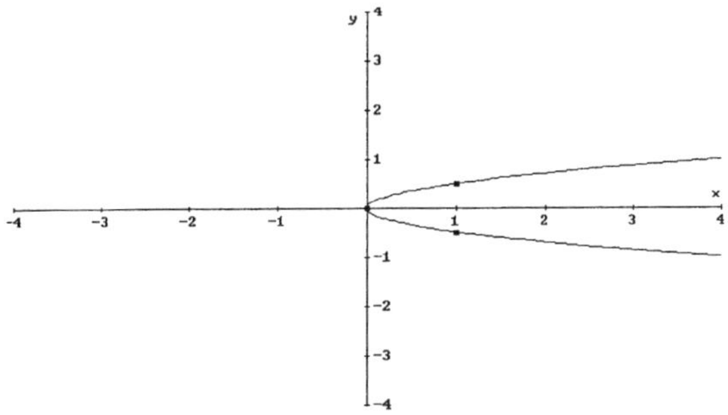

22. $x = -3(y + 2)^2 - 2$ Vertex: (-2,-2)

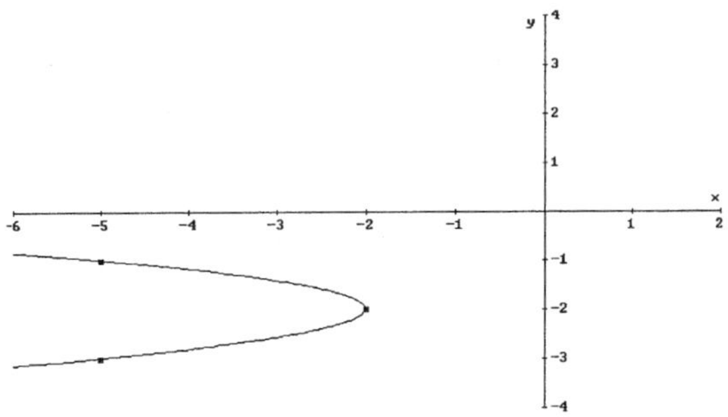

23. $x = 2(y - 3)^2 - 4$ Vertex: (- 4, 3)

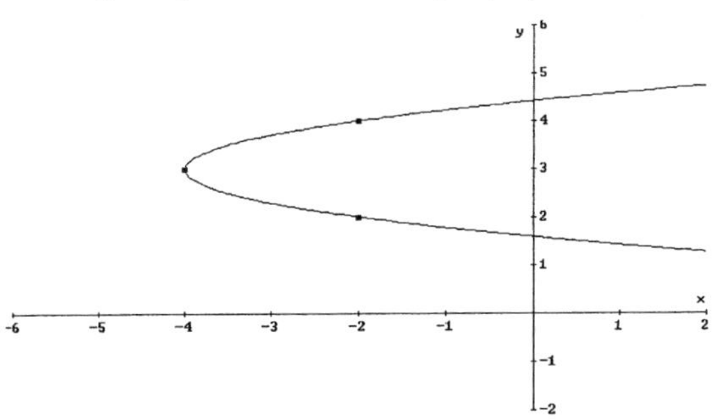

24. $x = -y^2 - 2y + 3$

Complete the square in y to find the vertex.

$x = -(y^2 + 2y + 1) + (3 + 1) = -(y + 1)^2 + 4$

Vertex: (4,-1)

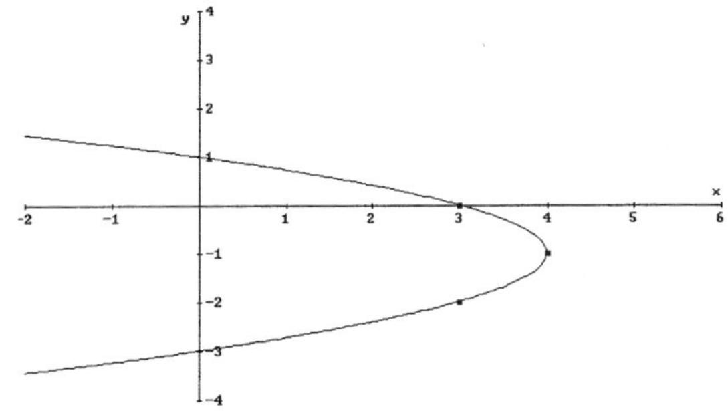

25. $x = y^2 + 4y + 5$

Complete the square in y to find the vertex.

$x = (y^2 + 4y + 4) + (5 - 4) = (y + 2)^2 + 1$

Vertex: (1,-2)

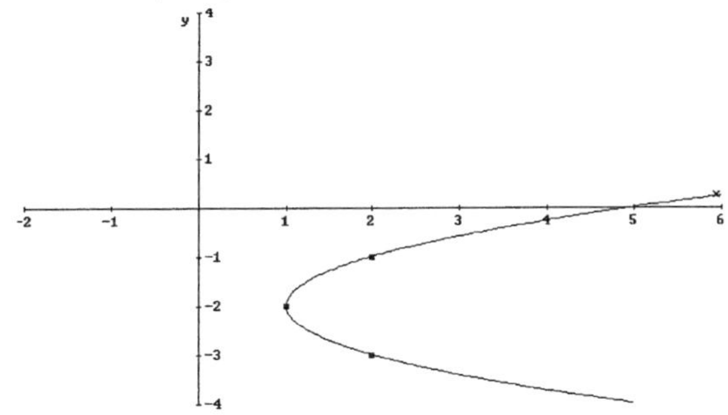

26. $\dfrac{x^2}{4} + \dfrac{y^2}{9} = 1$

x-intercepts: (2,0), (-2,0)

y-intercepts: (0,3), (0,-3)

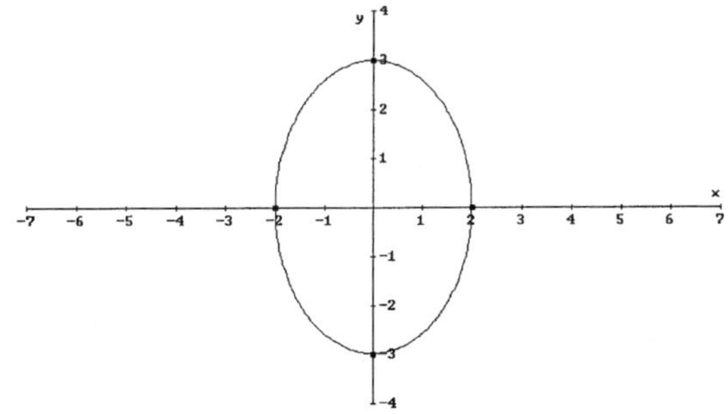

27. $x^2 + \dfrac{y^2}{9} = 1$

x-intercepts: (1,0), (-1,0)
y-intercepts: (0,3), (0,-3)

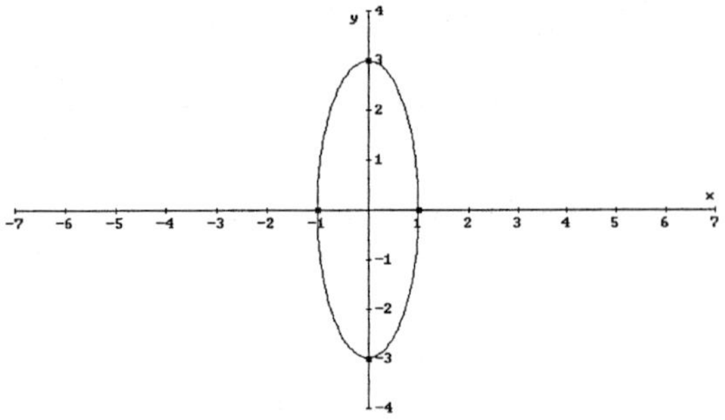

28. $25x^2 + 9y^2 = 225$
x-intercepts: (3,0), (-3,0)
y-intercepts: (0,5), (0,-5)

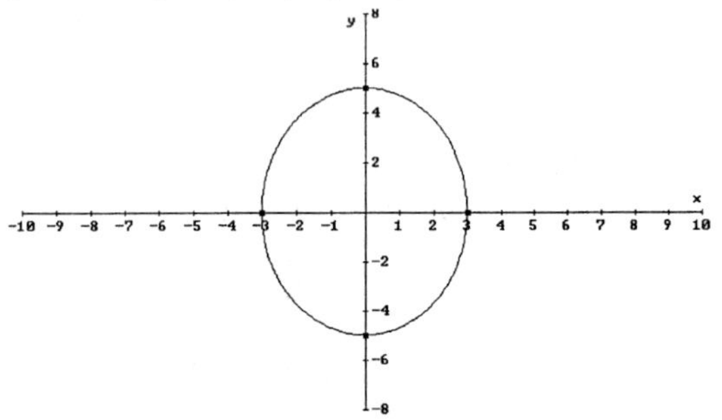

29. $4x^2 + 16y^2 = 64$
x-intercepts: (4,0), (- 4,0)
y-intercepts: (0,2), (0,-2)

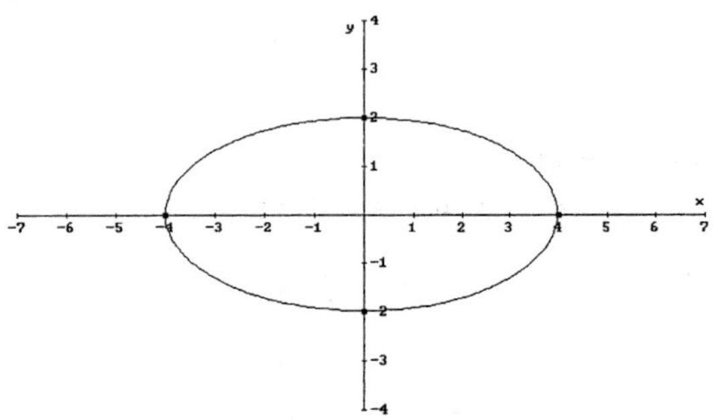

30. $\frac{x^2}{9} - \frac{y^2}{4} = 1$
Vertices: (3,0), (-3,0). To find the asymptotes

solve $\frac{x^2}{9} - \frac{y^2}{4} = 0$ or $4x^2 = 9y^2$

$$y = \pm\frac{4}{9}x = \pm\frac{2}{3}x$$

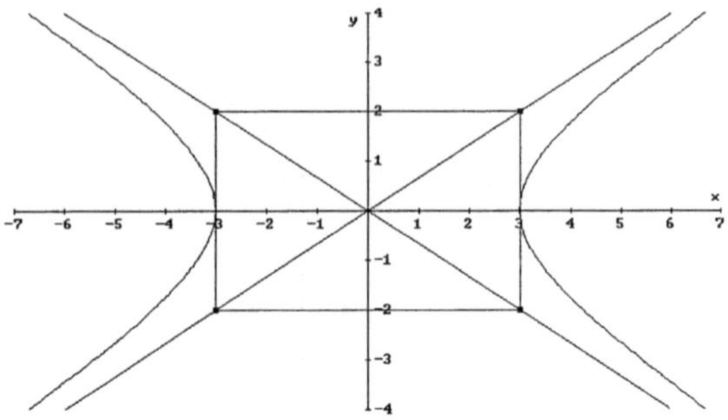

31. $\frac{y^2}{4} - \frac{x^2}{9} = 1$
Vertices: (0,2), (0,-2)
To find the asymptotes, solve

$\frac{y^2}{4} - \frac{x^2}{9} = 0$ or $9y^2 = 4x^2$

$$y = \pm\frac{4}{9}x = \pm\frac{2}{3}x$$

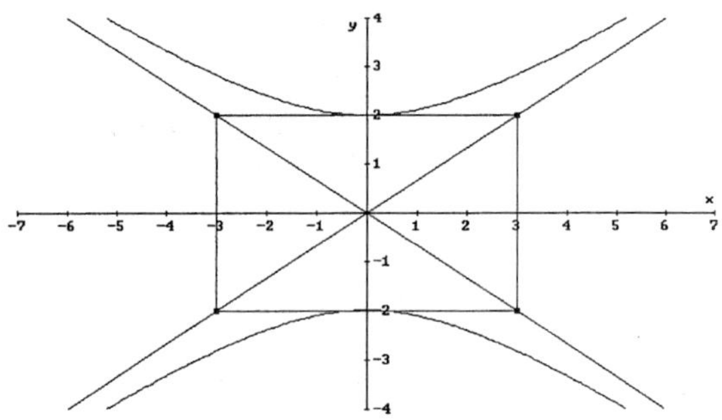

32. $\frac{y^2}{25} - \frac{x^2}{4} = 1$
Vertices: (0,5), (0,-5)

To find the asymptotes, solve $\frac{y^2}{25} - \frac{x^2}{4} = 0$.

Then $4y^2 = 25x^2$ or $y = \pm\frac{5}{2}x$.

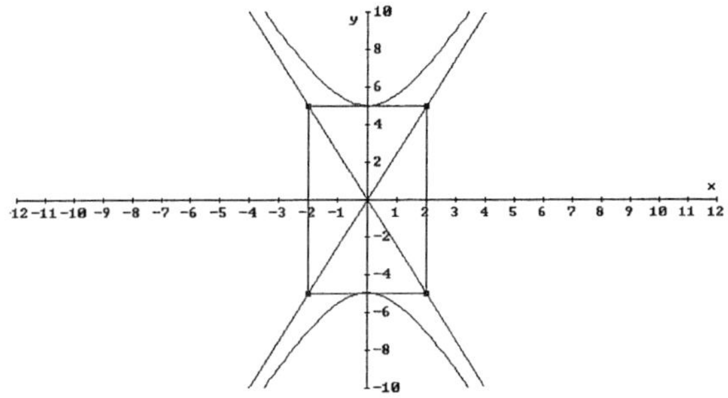

33. $\frac{x^2}{4} - \frac{y^2}{25} = 1$

Vertices: (2,0), (-2,0)

To find the asymptotes, solve $\frac{x^2}{4} - \frac{y^2}{25} = 0$.

Then $25x^2 = 4y^2$ or $y = \pm\frac{5}{2}x$.

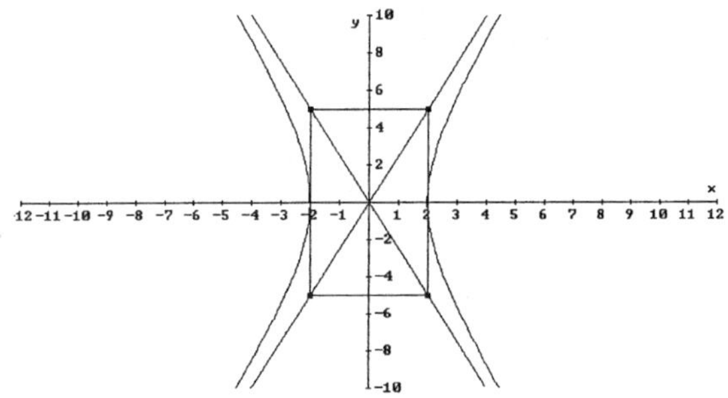

34. $25x^2 - y^2 = 25$ Divide by 25 to put in standard form. $x^2 - \frac{y^2}{25} = 1$.

Vertices: (1,0), (-1,0)

To find the asymptotes, solve $x^2 - \frac{y^2}{25} = 0$.

Then $y^2 = 25x^2$ or $y = \pm 5x$.

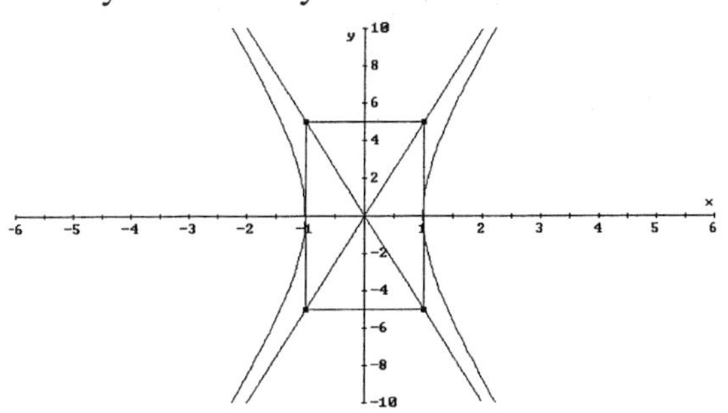

35. $9x^2 - 4y^2 = 36$ Divide by 36 to put into standard form. $\frac{x^2}{4} - \frac{y^2}{9} = 1$.

Vertices: (2,0), (-2,0)

To find the asymptotes, solve $\frac{x^2}{4} - \frac{y^2}{9} = 0$.

Then $4y^2 = 9x^2$ or $y = \pm\frac{3}{2}x$.

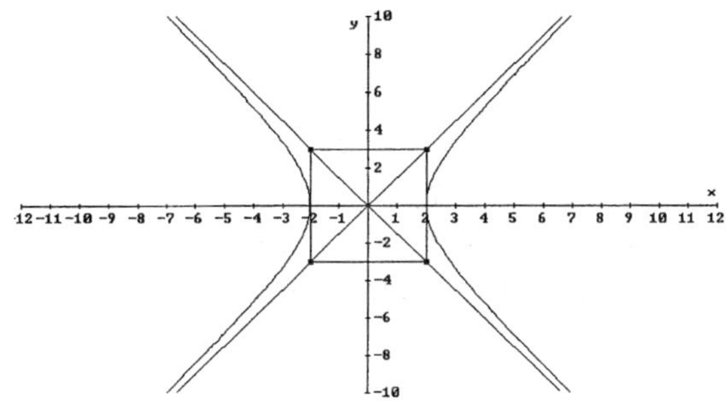

36. $xy = 8$

The asymptotes are the axes.

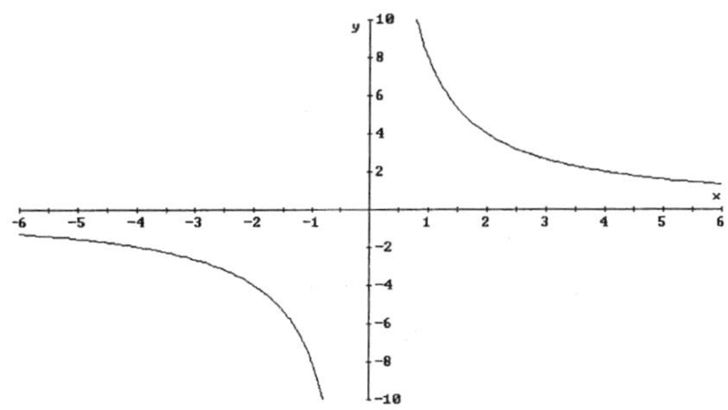

37. $xy = -10$

The asymptotes are the axes.

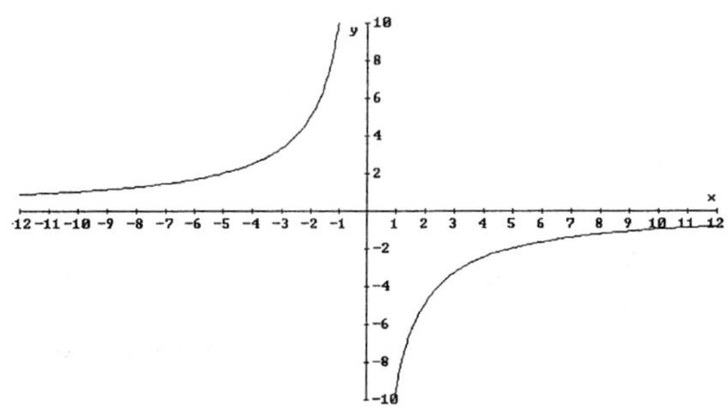

Applications

38. The larger circle is centered at (0,0) with radius 4. Thus, the smaller circle has radius $7 - 4 = 3$, and equation $(x - 7)^2 + y^2 = 3^2$.

39. The larger circle is centered at (0,0) with radius 50' and the smaller circle is centered at (10,0) with radius 30'. Thus, the x-intercepts of the larger circle are (±50,0) and of the smaller circle are (40,0) and (-20,0). The smallest width is 50 – 40 = 10' and the largest is -20 – (-50) = 30'.

40. Put each circle in standard form by completing the square in x and y.
$x^2 + y^2 - 8x - 20y + 16 = 0$ or
$(x^2 - 8x + 16) + (y^2 - 20y + 100) = -16 + 16 + 100 = 10^2$ Thus, $(x - 4)^2 + (y - 10)^2 = 10^2$ and the center is (4,10), the radius r = 10.
$x^2 + y^2 + 2x + 4y - 11 = 0$ or
$(x^2 + 2x + 1) + (y^2 + 4y + 4) = 11 + 1 + 4 = 16$ and $(x + 1)^2 + (y + 2)^2 = 4^2$. Thus, the center is (-1,-2) and the radius r = 4. The distance between the centers P(4,10) and Q(-1,-2) is found by the distance formula to be 13. Since the sum of the radii is 10 + 4 = 14, the circles overlap and they may not be licensed for the same frequency.

41. Complete the square in x and y to find the center and radius of the circle.
$x^2 + y^2 - 16x - 20y + 155 = 0$ or
$(x^2 - 16x + 64) + (y^2 - 20y + 100) = -155 + 64 + 100 = 9$. Thus, $(x - 8)^2 + (y - 10)^2 = 3^2$ and the center is (8,10) and the radius r = 3. From the center of town north to the intersection of the highway and Main will be y = 10 + 3 = 13 km. From the center of town east to the intersection of the highway and State will be 8 + 3 = 11 km. Thus, the locations are (0,13) and (11,0), respectively.

42. The parabola is $2y^2 - 9x = 18$. Writing it in standard form: $2y^2 - 18 = 9x$

$$\frac{2}{9}y^2 - 2 = x$$

$$\frac{2}{9}(y - 0)^2 - 2 = x$$

The vertex is at (-2, 0), a distance of 2 AU from the sun, which is located at (0,0).

43. Given $x = \frac{1}{16}y^2$. If the dish is 8 feet wide vertically, then the top (and bottom) is 4 feet from the x-axis or we have y = 4 (or y = -4 for the bottom). Thus, the depth $x = \frac{1}{16}(4^2)$ = 1 foot.

44. The arch is the top half of an ellipse with center at (0,0), vertices at (±20,0) and y-intercept at (0,10). Thus, the ellipse is $\frac{x^2}{20^2} + \frac{y^2}{10^2} = 1$. Solving for y gives

$$\frac{y^2}{100} = 1 - \frac{x^2}{400}$$

$$y^2 = 100 - \frac{x^2}{4}$$

$$y = \sqrt{100 - \frac{x^2}{4}} = \frac{\sqrt{400 - x^2}}{2}$$

where the principal root was chosen since y must be positive.

45. At a point 10' from the center of the roadway in Exercise 44, x = 10, so that y =

$$\frac{\sqrt{400 - 10^2}}{2} = \frac{\sqrt{300}}{2} = \frac{10\sqrt{3}}{2} = 5\sqrt{3} \text{ feet.}$$

46. Putting the ellipse $9x^2 + 16y^2 = 144$ in standard form gives $\frac{x^2}{16} + \frac{y^2}{9} = 1$, so that a = 4 and b = 3. Thus, the area $A = \pi \cdot 4 \cdot 3 = 12\pi$ square units.

47. Put each ellipse in standard form. The

ellipse $4x^2 + 9y^2 = 576$ becomes
$\frac{x^2}{144} + \frac{y^2}{64} = 1$ or $\frac{x^2}{12^2} + \frac{y^2}{8^2} = 1$. The ellipse

$9x^2 + 25y^2$ becomes $\frac{x^2}{100} + \frac{y^2}{36} = 1$ or

$\frac{x^2}{10^2} + \frac{y^2}{6^2} = 1$. The area of the larger ellipse is $\pi \cdot 12 \cdot 8 = 96\pi$ and the area of the smaller ellipse in $\pi \cdot 10 \cdot 6 = 60\pi$. The difference, $96\pi - 60\pi = 36\pi$ is the area of the track.

48. Solving $9y^2 - x^2 = 81$ for y when $x = 0$ gives:
$9y^2 - 0 = 81$ The closest the particle
$y^2 = 9$ or $y = \pm 3$
comes to the nucleus is 3 units, the distance from (0,0) to (0,3).

49. The ship is on the curve $x^2 - 4y^2 = 576$. If $y = 5$, $x^2 - 4(5^2) = 576$ or $x^2 - 100 = 576$. Then $x^2 = 676$ or $x = 26$. It is located at (26,5).

50. The vertex of the hyperbola $y^2 - x^2 = 25$ is found by setting $x = 0$. Thus, $y^2 = 25$ or $y = 5$ is the vertex. At 5 miles from the vertex, along the y-axis, $y = 10$. Thus, $10^2 - x^2 = 25$ and so $100 - 25 = x^2$. Then $x^2 = 75$ and $x = 5\sqrt{3}$. The width at the point where $y = 10$ miles is then $2(5\sqrt{3}) = 10\sqrt{3}$ miles.

51. The minimum distance between the particles occurs when they are at the vertices. To find the vertices of $x^2 - 4y^2 = 4$, set $y = 0$. Then $x^2 - 4(0^2) = 4$ or $x^2 = 4$, so $x = \pm 2$. The vertices are at (2,0) and (-2,0) and so the minimum distance is 4 units.

APPENDIX B: BINOMIAL THEOREM

Vocabulary

1. Pascal's triangle

2. factorial; n(n-1)·...·2·1

3. 1

Concepts

4. The first row has a single 1 and the second row has two ones. Each succeeding row begins and ends with a 1, with each other entry being the sum of the two numbers to its right and left in the row directly above.

5. For the first nine rows of Pascal's triangle, see the text. The tenth row is:
1 9 36 84 126 126 84 36 9 1

6. The sums are 1, 2, 4, 8, 16, 32, 64, 128, 256 and 512. The sum of the entries in row k is 2 to the k - 1 power.

7. The diagonal sums are 1, 1, 2, 3, 5, 8, 13, 21. These are the first 8 Fibonacci numbers. The sequence begins with two 1's and each subsequent entry is the sum of the previous two entries.

8. In $(r + s)^4$, the first term has r^4s^0 and in each succeeding term, the power of r decreases by 1 and the power of s increases by 1, ending with r^0s^4 in the 5th term.

Practice

9. $3! = 3{\cdot}2{\cdot}1 = 6$

10. $-5! = -5{\cdot}4{\cdot}3{\cdot}2{\cdot}1 = -120$

11. $11!(5) = 11{\cdot}10{\cdot}9{\cdot}8{\cdot}7{\cdot}6{\cdot}5{\cdot}4{\cdot}3{\cdot}2{\cdot}1{\cdot}5 =$
199,584,000

12. $8(7!) = 8! = 8{\cdot}7{\cdot}6{\cdot}5{\cdot}4{\cdot}3{\cdot}2{\cdot}1 = 40{,}320$

13. $\dfrac{5!}{3!(5-3)!} = \dfrac{5!}{3!2!} = \dfrac{120}{6(2)} = 10$

14. $\dfrac{6!}{4!(6-4)!} = \dfrac{720}{24(2)} = 15$

15. $(x + y)^3 = x^3 + \dfrac{3!}{1!2!}x^2y + \dfrac{3!}{2!1!}xy^2 + y^3$
$= x^3 + 3x^2y + 3xy^2 + y^3$

16. $(x - y)^3 = [x + (-y)]^3 =$
$x^3 + \dfrac{3!}{1!2!}x^2(-y) + \dfrac{3!}{2!1!}x(-y)^2 + (-y)^3$
$= x^3 - 3x^2y + 3xy^2 - y^3$

17. $(2x + y)^4 = [(2x) + y]^4 = (2x)^4 +$
$\dfrac{4!}{1!3!}(2x)^3y + \dfrac{4!}{2!2!}(2x)^2y^2 + \dfrac{4!}{3!1!}(2x)y^3 + y^4 =$
$16x^4 + 4(8x^3)y + 6(4x^2)y + 4(2x)y^3 + y^4 =$
$16x^4 + 32x^3y + 24x^2y^2 + 8xy^3 + y^4$

18. $(x + 2y)^4 = [x + (2y)]^4 =$
$x^4 + \dfrac{4!}{1!3!}x^3(2y)^1 + \dfrac{4!}{2!2!}x^2(2y)^2 + \dfrac{4!}{3!1!}x(2y)^3 + (2y)^4$
$= x^4 + 4x^3(2y) + 6x^2(4y^2) + 4x(8y^3) + 16y^4$
$= x^4 + 8x^3y + 24x^2y^2 + 32xy^3 + 16y^4$

19. $(x - 2y)^5 = [x + (-2y)]^5 =$
$x^5 + \dfrac{5!}{1!4!}x^4(-2y) + \dfrac{5!}{2!3!}x^3(-2y)^2 + \dfrac{5!}{3!2!}x^2(-2y)^3$
$+ \dfrac{5!}{4!1!}x(-2y)^4 + (-2y)^5 =$
$x^5 + 5x^4(-2y) + 10x^3(4y^2) + 10x^2(-8y^3) +$
$5x(16y^4) - 32y^5 =$
$x^5 - 10x^4y + 40x^3y^2 - 80x^2y^3 + 80xy^4 - 32y^5$

20. $(2x - y)^5 = [(2x) + (-y)]^5 =$
$(2x)^5 + \dfrac{5!}{1!4!}(2x)^4(-y)^1 + \dfrac{5!}{2!3!}(2x)^3(-y)^2 +$
$\dfrac{5!}{3!2!}(2x)^2(-y)^3 + \dfrac{5!}{4!1!}(2x)(-y)^4 + (-y)^5 =$
$32x^5 + 5(16x^4)(-y) + 10(8x^3)y^2 + 10(4x^2)(-y^3)$
$+ 5(2x)y^4 - y^5 =$
$32x^5 - 80x^4y + 80x^3y^2 - 40x^2y^3 + 10xy^4 - y^5$

21. $(2x + 3y)^4 = [(2x) + (3y)]^4 = (2x)^4 + \frac{4!}{1!3!}(2x)^3(3y) + \frac{4!}{2!2!}(2x)^2(3y)^2 + \frac{4!}{3!1!}(2x)(3y)^3 + (3y)^4 = 16x^4 + 4(8x^3)(3y) + 6(4x^2)(9y^2) + 4(2x)(27y^3) + 81y^4 = 16x^4 + 96x^3y + 216x^2y^2 + 216\,xy^3 + 81y^4$

22. $(3x - 2y)^4 = [(3x) + (-2y)]^4 = (3x)^4 + \frac{4!}{1!3!}(3x)^3(-2y) + \frac{4!}{2!2!}(3x)^2(-2y)^2 + \frac{4!}{3!1!}(3x)(-2y)^3 + (-2y)^4 = 81x^4 + 4(27x^3)(-2y) + 6(9x^2)(4y^2) + 4(3x)(-8y^3) + 16y^4 = 81x^4 - 216x^3y + 216x^2y^2 - 96xy^3 + 16y^4$

23. $\left(\frac{x}{2}-\frac{y}{2}\right)^4 = \left(\frac{x}{2}+\left(\frac{-y}{2}\right)\right)^4 = \left(\frac{x}{2}\right)^4 + \frac{4!}{1!3!}\left(\frac{x}{2}\right)^3\left(\frac{-y}{2}\right) + \frac{4!}{2!2!}\left(\frac{x}{2}\right)^2\left(\frac{-y}{2}\right)^2 + \frac{4!}{3!1!}\left(\frac{x}{2}\right)\left(\frac{-y}{2}\right)^3 + \left(\frac{-y}{2}\right)^4 = \frac{x^4}{16} + 4\left(\frac{x^3}{8}\right)\left(\frac{-y}{2}\right) + 6\left(\frac{x^2}{4}\right)\left(\frac{y^2}{4}\right) + 4\left(\frac{x}{2}\right)\left(\frac{-y^3}{8}\right) + \frac{y^4}{16}$

$= \frac{x^4}{16} - \frac{x^3y}{4} + \frac{3x^2y^2}{8} - \frac{xy^3}{4} + \frac{y^4}{16}$

24. $\left(\frac{x}{2}+\frac{y}{2}\right)^4 = \left(\frac{x}{2}\right)^4 + \frac{4!}{1!3!}\left(\frac{x}{2}\right)^3\left(\frac{y}{2}\right) + \frac{4!}{2!2!}\left(\frac{x}{2}\right)^2\left(\frac{y}{2}\right)^2 + \frac{4!}{3!1!}\left(\frac{x}{2}\right)\left(\frac{y}{2}\right)^3 + \left(\frac{y}{2}\right)^4 = \frac{x^4}{16} + 4\left(\frac{x^3}{8}\right)\left(\frac{y}{2}\right) + 6\left(\frac{x^2}{4}\right)\left(\frac{y^2}{4}\right) + 4\left(\frac{x}{2}\right)\left(\frac{y^3}{8}\right) + \frac{y^4}{16} =$

$\frac{x^4}{16} + \frac{x^3y}{4} + \frac{3x^2y^2}{8} + \frac{xy^3}{4} + \frac{y^4}{16}$

25. The second term of $(a + b)^4$ will have $b^{2-1} = b^1$ and hence a^3. The coefficient will be $\frac{4!}{1!3!} = 4$. The term is $4a^3b$.

26. The third term of $(a + b)^4$ will have $b^{3-1} = b^2$ and hence a^2. The coefficient will be $\frac{4!}{2!2!} = 6$. The term will be $6a^2b^2$.

27. The third term of $(x - 3y)^8 = [x + (-3y)]^8$ will have $(-3y)^{3-1} = (-3y)^2$ and hence x^6. The coefficient will be $\frac{8!}{2!6!} = 28$. The term is $28x^6(-3y)^2 = 28x^6(9y^2) = 252x^6y^2$.

28. The fifth term of $(x - 3y)^8 = [x + (-3y)]^8$ will have $(-3y)^{5-1} = (-3y)^4$ and hence x^4. The coefficient will be $\frac{8!}{4!4!} = 70$. The term is $70x^4(-3y)^4 = 5670x^4y^4$.

29. The sixth term of $(2x + 3)^6 = [(2x) + 3]^6$ will have $3^{6-1} = 3^5$ and hence $(2x)^1$. The coefficient will be $\frac{6!}{5!1!} = 6$. The term is $6(2x)(3^5) = 12x(243) = 2916x$.

30. The fourth term of $(2x + 3)^6 = [(2x) + 3]^6$ will have $3^{4-1} = 3^3$ and hence $(2x)^3$. The coefficient will be $\frac{6!}{3!3!} = 20$. The term is $20(2x)^3(3^3) = 20(8x^3)(27) = 4320x^3$.

APPENDIX C: ARITHMETIC AND GEOMETRIC SEQUENCES

Vocabulary

1. sequence
2. arithmetic
3. mean
4. $\sum$
5. geometric; common ratio

Concepts

6. a. 1,4,7,...
   The common difference d = 7 - 4 = 3.
   b.2,6,10...
   The common difference d = 10 - 6 = 4.

7. a. a = 3, d = 2
   3, 3 + 2, 3 + 2(2), 3 + 3(2), 3 + 4(2) or 3,5,7,9,11

   b. a = -5, d = -3
   -5, -5 + (-3), -5 + 2(-3), -5 + 3(-3), -5 + 4(-3) or -5,-8,-11,-14,-17

8. a. 3,6,12,... The common ratio is 12/6 = 2.
   b. 3,-6,12,... The common ratio is 12/-6 = -2.

9. a. a = 3, r = 2
   $3, 3\cdot2^1, 3\cdot2^2, 3\cdot2^3, 3\cdot2^4$ or 3,6,12,24,48
   b. a = -5, r = 1/5
   $-5, -5(1/5)^1, -5(1/5)^2, -5(1/5)^3, -5(1/5)^4$ or -5, -1, -1/5, -1/25, -1/125

10. Subtracting 5 is the same as adding -5, so the common difference is -5.

11. Dividing by 2 is the same as multiplying by ½ so the common ratio is ½ .

Practice

12. For a = 7 and d = 12, the 30th term of the arithmetic sequence (AS) is a + (n - 1)d = 7 + (30 - 1)(12) = 355

13. For a = -5 and d = 4, the 55th term of the arithmetic sequence (AS) is a + (n - 1)d = -5 + (55 - 1)(4) = -5 + 54·4 = 211.

14. For d = 11, the 27th term of the AS is a + (27 - 1)(11) = a + 286. Thus, a + 286 = 263 or a = -23 is the first term.

15. For a = -164, the 36th term of the AS is -164 + (36 - 1)d = -164 + 35d.
    Thus, -164 + 35d = -24 or 35d = 140 so that d = 4 is the common difference.

16. Given 2 and 11, find d so that 2, 2 + d, 2 + 2d 2 + 3d, 11 is an AS. Since 11 is the 5th term and a = 2, 2 + (5 - 1)d = 11. Hence, d = 9/4. The three means are 2 + 9/4= 17/4, 2 + 18/4 = 26/4 and 2 + 27/4 = 35/4.

17. Given 5 and 25, find d so that 5, 5 + d, 5 + 2d, 5 + 3d, 5 + 4d, 25 is an AS. Since 25 is the 6th term and a = 5, 5 + (6 - 1)d = 25 or 5 + 5d = 25. Thus, d = 4. The 4 means are 5 + 4 = 9, 5 + 8 = 13, 5 + 12 = 17 and 5 + 16 = 21.

18. Given 10 and 19, find d so that 10, 10 + d, 19 is an AS. Since 19 is the 3rd term and a = 10, 10 + (3 - 1)d = 19, so d = 9/2. The mean is 10 + 9/2 = 29/2

19. Given 5 and 23, find d so that 5, 5 + d, 23 is an AS. Since 23 is the 3rd term and a = 5, 5 + (3 - 1)d = 23 or 5 + 2d = 23. Thus, d = 9 and the mean is 5 + 9 = 14.

20. For 1,4,7,... a = 1 and d = 7 - 4 = 3. The sum of the first 30 terms is (30/2)[2(1) + (30 -1)3] = 15[2 + 29·3] = 15·89 = 1335.

21. For 2,6,10,... a = 2 and d = 10 - 6 = 4. The sum of the first 28 terms is $\frac{n}{2}[2a + (n-1)d]$ = (28/2)[2·2 + (28 - 1)4] = 14[4 + 108] = 14·112 = 1568.

22. f(n) = 2n + 1, so a = f(1) = 3. The 31st term is f(31) = 63. The sum of the first 31 terms is S = (31/2)[3 + 63] = 15.5(66) =1023.

23. $f(n) = 4n + 3$, so $a = f(1) = 4\cdot1 + 3 = 7$. The 23rd term is $f(23) = 4\cdot23 + 3 = 95$. The sum of the first 23 terms is $\frac{n}{2}[f(1)+f(n)] =$ $(23/2)[7 + 95] = (23/2)(102) = 1173$.

24. Here $a = 1$ and $l = 50$, so the sum $S = 50(1 + 50)/2 = 25\cdot51 = 1275$.

25. Here $a = 1$, $d = 1$, $n = 100$ and so the sum $S = \frac{n}{2}[2a+(n-1)d] = (100/2)[2(1) + (100 - 1)1]$ $= 50[101] = 5050$.

26. For $a = 7$, $r = 2$, the 10th term of the geometric sequence (GS) is $a\,r^{n-1} = 7\cdot2^{10-1}$ $= 7\cdot2^9 = 7(512) = 3584$.

27. For $a = 64$, $r = ½$ the 12th term of the geometric sequence (GS) is $a\,r^{n-1} = 64(½)^{12-1}$ $= 64(½)^{11} = 1/32$.

28. For $a = -8$ and common ratio r, the 6th term is $(-8)r^{6-1} = -8r^5$. Thus, $-8r^5 = -1944$ so $r^5 = 243$ and $r = 3$.

29. For $a = 12$ and common ratio r, the 6th term is $12r^{6-1} = 12\,r^5$. Thus, $12r^5 = 3/8$ so $r^5 = 3/96 =$ $1/32$. Thus $r = ½$.

30. For 2 and 162, find r so that 2, 2r, $2r^2$, $2r^3$, 162 is a GS. Since 162 is the 5th term, $2r^{5-1} =$ 162. Thus, $2r^4 = 162$, $r^4 = 81$ and $r = 3$. The means are $2\cdot3 = 6$, $2\cdot3^2 = 18$ and $2\cdot3^3 = 54$.

31. For 3 and 96, find r so that 3, 3r, $3r^2$, $3r^3$, $3r^4$, 96 is a GS. Since 96 is the 6th term, $96 = 3r^{6-1}$ $= 3r^5$. Thus, $32 = r^5$ and $2 = r$. The means are $3\cdot2 = 6$, $3\cdot2^2 = 12$, $3\cdot2^3 = 24$ and $3\cdot2^4 = 48$.

32. For 2 and 128, find $r < 0$ so that 2, 2r, 128 is a GS. Since 128 is the 3rd term, $128 = 2r^{3-1} =$ $2r^2$ .Thus, $r^2 = 64$ and $r = -8$. The mean is $2(-8) = -16$.

33. For 3 and 243, find $r > 0$ so that 3, 3r, 243 is a GS. Since 243 is the 3rd term, $243 = 3r^{3-1} =$ $3r^2$. Thus, $r^2 = 81$ and $r = 9$. The mean is $3\cdot9$ $= 27$.

34. For 2,6,18,... $a = 2$ and $r = 18/6 = 3$. The sum of the first 6 terms of the GS is $S =$

$$\frac{2(1-3^6)}{1-3} = \frac{2(-728)}{-2} = 728.$$

35. For 2, -6, 18,... $a = 2$ and $r = 18/-6 = -3$. The sum of the first 6 terms is $S = \frac{a(1-r^n)}{1-r} =$

$$\frac{2(1-(-3)^6)}{1-(-3)} = \frac{2(-728)}{4} = -364.$$

36. The 3rd term is -2 and the 4th term is 1, and so the common ratio is $1/-2 = -½$. Then the 3rd term $-2 = a(-½)^{3-1} = a(-½)^2 = a/4$. Thus, $a =$ -8. The sum of the first 6 terms is $S =$

$$\frac{a(1-r^n)}{1-r} = \frac{-8\left(1-\left(\frac{-1}{2}\right)^6\right)}{1-\left(\frac{-1}{2}\right)} = \frac{-8\left(\frac{63}{64}\right)}{\frac{3}{2}} = \frac{-21}{4}$$

37. The 3rd term is -3 and the 4th term is 1 so the common ratio is $1/-3$. Thus, the 3rd term $-3 =$ $a(-1/3)^{3-1} = a(-1/3)^2 = a(1/9)$. Thus, $a = -27$. Then the sum of the first 5 terms is $S =$

$$\frac{a(1-r^n)}{1-r} = \frac{-27\left(1-\left(\frac{-1}{3}\right)^5\right)}{1-\left(\frac{-1}{3}\right)} = \frac{-27\left(\frac{244}{243}\right)}{\frac{4}{3}} =$$

$-61/3$.

38. $\sum_{k=1}^{4} 6k = 6\bullet1 + 6\bullet2 + 6\bullet3 + 6\bullet4 = 6 + 12 + 18 + 24$ $= 60$

39. $\sum_{k=2}^{5} 3k = 3\bullet2 + 3\bullet3 + 3\bullet4 + 3\bullet5 = 6 + 9 + 12 + 15$ $= 42$

40. $\sum_{k=3}^{4}(k^2+3) = (3^2+3)+(4^2+3) = 12+19 = 31$

41. $\sum_{k=2}^{6}(k^2+1) = (2^2+1)+(3^2+1)+(4^2+1)+$ $(5^2+1)+(6^2+1) = 5+10+17+26+37 =$ 95

42. Here a = 60 and d = 50, so the first 6 terms of the AS are 60, 110, 160, 210, 260, 310. After 10 years Fred will have made 120 deposits and his savings will be the 121st term in the AS. The amount is $a + (n - 1)d = 60 + (121 - 1)50 = \$6060$.

43. Here a = 10000 and d = -275, so the first 6 terms are 10000, 9725, 9450, 9175, 8900, 8625. After 17 months the balance due will be 18th term of the AS. The balance due is $a + (n - 1)d = 10000 + (18 - 1)(-275) =$ \$5325.

44. Here a = 1, d = 1 and l = 150. Then 150 = a + (n - 1)d = 1 + (n - 1)·1, so 150 = n. The total number of bricks is $S = \frac{n}{2}(a+\ell) = \frac{150}{2}(1+150) = 11{,}325$.

45. After 1 second, the object has fallen 16 ft. and after 3 seconds it has fallen $16 \cdot 3^2 = 144$ ft. Thus, during the second and third seconds, it fell 144 - 16 = 128 ft.

46. During the 12th second the object in Exercise 45 falls $16(12^2) - 16(11^2) = 16(144 - 121) = 16 \cdot 23 = 368$ ft.

47. The sum of the ang;les is an AS with a = 180 and d = 180. For an octagon, n = 6, so the angle sum is 180 + (6 - 1)180 = 1080°. For a dodecagon, n = 10, so the angle sum is 180 + (10 - 1)180 = 1800°.

48. The current population is a = 500 and with a growth rate of 6%, r = 1.06. The population 5 years from now will be the 6th term in the GS = $a \cdot r^{n-1} = 500(1.06)^{6-1} = 500(1.06)^5 \approx$ 669 people.

49. The current population is a = 98, and with a growth rate of -10%, r = 1 - .10 = .9. The population 8 years from now will be the 9th term in the GS = $a \cdot r^{n-1} = 98(.9)^{9-1} = 98(.9)^8$ ≈42 people.

50. Here a = 10000, r = 1 - .12 = .88. The amount left in the box after 15 years will be the 16th term in the GS = $a \cdot r^{n-1} = 10000(.88)^{16-1} = 10000(.88)^{15} \approx \$1469.74$.

51. Here a = 5000 and the growth rate is 12%, so r = 1.12. The amount in Sally's account 10 years from now will be the 11th term of the GS = $a \cdot r^{n-1} = 5000(1.12)^{11-1} = 5000(1.12)^{10}$ ≈ \$15,529.24.

52. The value today is a = 70000, and the rate of appreciation is 6%, so r = 1.06. The value of the house in 12 years is the 13th term of the GS = $a \cdot r^{n-1} = 70000(1.06)^{13-1} = 70000(1.06)^{12} \approx \$140{,}853.75$.

53. The original cost a = 5000 and the rate of depreciation is 9%, so r = 1 - .09 = .91. The value of the boat in 5 years is the 6th term in the GS= $a \cdot r^{n-1} = 5000(.91)^{6-1} = 5000(.91)^5$ ≈ \$3120.16.

54. Here a = 1. By the Pythagorean Theorem, one side of the next smaller square is $\frac{1}{\sqrt{2}}$ and so its area is ½.. Thus, r = ½ and the area of the 12th square is the 12th term is the GS = $a \cdot r^{n-1} = 1(½)^{12-1} = (½)^{11} = 1/2048$ square units.

55. Each generation, the number of people doubles, so r = 2. Also, a = 1 and n = 10 generations. Thus, the number of names would be $S = \frac{a(1-r^n)}{1-r} = \frac{1(1-2^{10})}{1-2} = \frac{-1023}{-1} =$ 1023.

APPENDIX D: PERMUTATIONS AND COMBINATIONS

Vocabulary

1. event
2. multiplication; ab
3. permutation
4. combination

Concepts

5. a. The sets are combinations of 1, 2 and 3.
   b. The sets represent permutations of 1,2 and 3.

6. a. C(8,4) = 70
   b. C(8,5) = 56
   c. C(8,8) = 1

7. a. P(8,4) = 1680
   b. P(8,5) = 6720
   c. P(8,8) = 40,320

8. C(8,8) is only <u>one</u> <u>group</u> of 8 objects, while P(8,8) counts the number of ways that these 8 objects can be arranged.

Practice

9. $P(5,5) = 5! = 120$

10. $P(4,4) = 4! = 24$

11. $P(5,3) = \frac{5!}{(5-3)!} = \frac{5!}{2!} = \frac{120}{2} = 60$

12. $P(3,2) = \frac{3!}{(3-2)!} = \frac{3!}{1!} = 6$

13. $P(2,2)\cdot P(3,3) = 2!\cdot 3! = 2\cdot 6 = 12$

14. $P(3,2)\cdot P(3,3) = 6\cdot 6 = 36$

15. $P(5,3) = \frac{5!}{(5-3)!} = \frac{5!}{2!} = 60$ and

$\binom{5}{4}\binom{5}{3} = \frac{5!}{4!1!}\bullet\frac{5!}{3!2!} = \frac{5}{1}\bullet\frac{5\bullet 4}{2} = 5\bullet 10 = 50$ Thus, $\frac{P(5,3)}{P(4,2)} =$ $60/12 = 5$.

16. $P(6,2) = \frac{6!}{(6-2)!} = \frac{6!}{4!} = 30$ and

$P(5,4) = \frac{5!}{(5-4)!} = 5! = 120$. Thus, $\frac{P(6,2)}{P(5,4)}$ $=30/120 = 1/4$.

17. $\frac{P(5,3)\bullet P(7,3)}{P(5,1)} = \frac{\frac{5!}{(5-3)!}\bullet\frac{7!}{(7-3)!}}{\frac{5!}{(5-1)!}} = \frac{\frac{5!}{2!}\bullet\frac{7!}{4!}}{\frac{5!}{4!}} =$

$\frac{60\bullet 210}{5} = 12(210) = 2520$

18. $\frac{P(8,3)}{P(5,3)\bullet P(4,3)} = \frac{\frac{8!}{(8-3)!}}{\frac{5!}{(5-3)!}\bullet\frac{4!}{(4-3)!}} = \frac{\frac{8!}{5!}}{\frac{5!}{2!}\bullet\frac{4!}{1!}} =$

$\frac{336}{60\bullet 24} = \frac{7}{30}$

19. $C(5,3) = \frac{5!}{3!2!} = \frac{5\bullet 4\bullet 3\bullet 2\bullet 1}{3\bullet 2\bullet 1\bullet 2\bullet 1} = \frac{5\bullet 4}{2} = 10$

20. $C(5,4) = \frac{5!}{4!1!} = 5$

21. $\binom{6}{3} = \frac{6!}{3!3!} = \frac{6\bullet 5\bullet 4}{6} = 20$

22. $\binom{6}{4} = \frac{6!}{4!2!} = \frac{6\bullet 5}{2} = 15$

23. $\binom{5}{4}\binom{5}{3} = \frac{5!}{4!1!}\bullet\frac{5!}{3!2!} = \frac{5}{1}\bullet\frac{5\bullet 4}{2} = 5\bullet 10 = 50$

24. $\binom{6}{5}\binom{6}{4} = \frac{6!}{5!1!}\bullet\frac{6!}{4!2!} = 6\bullet 15 = 90$

25. $\dfrac{C(38,37)}{C(19,18)} = \dfrac{\frac{38!}{37!1!}}{\frac{19!}{18!1!}} = \dfrac{38}{19} = 2$

26. $\dfrac{C(25,23)}{C(40,39)} = \dfrac{\frac{25!}{23!2!}}{\frac{40!}{39!1!}} = \dfrac{\frac{25\cdot 24}{2}}{40} = \dfrac{300}{40} = \dfrac{15}{2}$

27. $C(12,0)\cdot C(12,12) = 1\cdot 1 = 1$

28. $\dfrac{C(8,0)}{C(8,1)} = \dfrac{1}{8}$

29. $C(n,2) = \dfrac{n!}{(n-2)!2!}$

30. $C(n,3) = \dfrac{n!}{(n-3)!3!}$

31. $(x + y)^4 =$
$\binom{4}{0}x^4 + \binom{4}{1}x^3y + \binom{4}{2}x^2y^2 + \binom{4}{3}xy^3 + \binom{4}{4}y^4 =$
$x^4 + 4x^3y + 6x^2y^2 + 4xy^3 + y^4$

32. $(x - y)^2 = [x + (-y)]^2 =$
$\binom{2}{0}x^2 + \binom{2}{1}x(-y) + \binom{2}{2}(-y)^2 = x^2 - 2xy + y^2$

33. $(2x + y)^3 = [(2x) + y]^3 =$
$\binom{3}{0}(2x)^3 + \binom{3}{1}(2x)^2y + \binom{3}{2}(2x)y^2 + \binom{3}{3}y^3 =$
$8x^3 + 3(4x^2)y + 3(2x)y^2 + y^3 =$
$8x^3 + 12x^2y + 6xy^2 + y^3$

34. $(2x + 1)^4 = [(2x) + 1]^4 =$

$\binom{4}{0}(2x)^4 + \binom{4}{1}(2x)^3 1 + \binom{4}{2}(2x)^2 1^2 + \binom{4}{3}(2x)1^3 + \binom{4}{4}1^4$
$= 16x^4 + 4(8x^3) + 6(4x^2) + 4(2x) + 1 =$
$16x^4 + 32x^3 + 24x^2 + 8x + 1$

35. $(3x - 2)^4 = [(3x) + (-2)]^4 =$

$\binom{4}{0}(3x)^4 + \binom{4}{1}(3x)^3(-2) + \binom{4}{2}(3x)^2(-2)^2 +$
$\binom{4}{3}(3x)(-2)^3 + \binom{4}{4}(-2)^4 = 81x^4 +$
$4(27x^3)(-2) + 6(9x^2)(4) + 4(3x)(-8) + 16 =$
$81x^4 - 216x^3 + 216x^2 - 96x + 16$

36. $(3 - x^2)^3 = [3 + (-x^2)]^3 =$
$\binom{3}{0}3^3 + \binom{3}{1}3^2(-x^2) + \binom{3}{2}3(-x^2)^2 + \binom{3}{3}(-x^2)^3$
$= 27 + 3(9)(-x^2) + 3(3)x^4 - x^6$
$= 27 - 27x^2 + 9x^4 - x^6$

37. The 4th term of $(x - 5y)^5 = [x + (-5y)]^5$ has $(-5y)^{4-1} = (-5y)^3$ and hence x^2. The coefficient is $\binom{5}{3} = \dfrac{5!}{3!2!} = 10$. The term is $10x^2(-125y^3) = -1250x^2y^3$.

38. The 3rd term of $(2x - y)^5 = [(2x) + (-y)]^5$ has $(-y)^{3-1} = (-y)^2$ and hence $(2x)^3$. The coefficient is $\binom{5}{2} = \dfrac{5!}{2!3!} = 10$. The 3rd term is $10(2x)^3(-y)^2 = 80x^3y^2$.

39. The second term of $(x^2 - y^3)^4 = [x^2 + (-y)^3]^4$ has $(-y^3)^{2-1} = (-y^3)^1$ and hence $(x^2)^3$. The coefficient is $\binom{4}{1} = \dfrac{4!}{1!3!} = 4$. The 2nd term is $4(x^2)^3(-y^3) = -4x^6y^3$.

40. The 4th term of $(x^3 - y^2)^4 = [x^3 + (-y^2)]^4$ has $(-y^2)^{4-1} = (-y^2)^3$ and hence $(x^3)^1$. The coefficient is $\binom{4}{3} = \dfrac{4!}{3!1!} = 4$. The 4th term is $4(x^3)(-y^2)^3 = -4x^3y^6$.

Applications

41. She has $5\cdot 7 = 35$ choices for her evening.

42. She has $5\cdot 3\cdot 4 = 60$ ways to make the trip.

43. The number of 6 digit license plates is $10^6 =$ 1,000,000.

44. The number of 6 digit license plates with no digit repeated is $10 \cdot 9 \cdot 8 \cdot 7 \cdot 6 \cdot 5 = 151{,}200$.

45. The number of 6 digit license plates with no digit repeated and not starting with zero is $9 \cdot 9 \cdot 8 \cdot 7 \cdot 6 \cdot 5 = 136{,}080$. (The first digit can be anything but 0, the second digit anything but 1.)

46. The number of license plates with 2 letters followed by 4 digits is $26^2 \cdot 10^4 = 6{,}760{,}000$.

47. The number of 7 digit phone numbers not starting with a 1 or 0 is $8 \cdot 10^6 = 8{,}000{,}000$.

48. The number of 10 digit phone numbers not beginning with a 0 or 1 and not allowing 000 or 911 area codes is the number of possible area codes·8,000,000 (see exercise 47 above). The total number of usable area codes is 10^3 minus the two that are omitted = 998. The total is thus $998 \cdot 8{,}000{,}000 = 7{,}984{,}000{,}000$.

49. Six girls can be lined up in $6! = 720$ ways.

50. Seven books can be arranged in $7! = 5040$ ways.

51. The 4 novels can be arranged in $4! = 24$ ways and the 5 biographies in $5! = 120$ ways. Thus, novels followed by biographies can be done in $24 \cdot 120 = 2880$ ways.

52. The candidates for mayor can be arranged in $6! = 720$ ways and those for the board in $4! = 24$ ways. Thus, the candidates for mayor followed by those for the board can be arranged in $720 \cdot 24 = 17{,}280$ ways.

53. The lock has $25 \cdot 24 \cdot 23 = 13{,}800$ possible combinations.

54. The lock has $50 \cdot 49 \cdot 48 = 117{,}600$ possible combinations.

55. She can fill the appointments in $10 \cdot 9 \cdot 8 = 720$ ways.

56. The number of words is 2^{32}, since there are 32 choices to make, and two possibilities for each choice.

57. The number of 5 digit palindromes is $9 \cdot 10 \cdot 10 \cdot 1 \cdot 1$ (the last two choices are forced once the first two choices are made) = 900.

58. There are $2 \cdot 26 \cdot 26 = 1352$ possible 3-letter radio stations and $2 \cdot 26^3 = 35{,}152$ possible 4-letter radio stations for a total of 36,504 possible radio stations.

59. The number of possible committees is

$$C(14,3) = \frac{14!}{11!3!} = \frac{14 \cdot 13 \cdot 12}{6} = 364 \; .$$

60. Jeffrey has $C(15,3) = \dfrac{15!}{12!3!} = \dfrac{15 \cdot 14 \cdot 13}{6} = 455$ choices.

61. If $C(n,3) = 10$, then $\dfrac{n!}{3!(n-3)!} = 10$ or

$\dfrac{n!}{(n-3)!} = 10 \cdot 3! = 60$. But $n! = n(n - 1)(n - 2)(n - 3)!$. Thus $n(n - 1)(n - 2) = 60$. Multiplying out the left hand side and subtracting 60 yields $n^3 - 3n^2 + 2n - 60 = 0$, which in factored form is $(n - 5)(n^2 + 2n + 12) = 0$. Since $n^2 + 2n + 12$ is never 0, it must be that $n - 5 = 0$ or $n = 5$. There are 5 people in the group.

62. If $C(n,3) = 20$, then $\dfrac{n!}{3!(n-3)!} = 20$ so that

$\dfrac{n!}{(n-3)!} = 20 \cdot 3! = 120$. Canceling the $(n - 3)!$ gives $n(n - 1)(n - 2) = 120$ or $n^3 - 3n^2 + 2n - 120 = 0$. In factored form this becomes $(n - 6)(n^2 + 3n + 20) = 0$. Since the second factor is never zero, $n - 6 = 0$ or $n = 6$ and so there are 6 people in the group.

63. The number of choices (possible lottery tickets) is C(100,6) = 1,192,052,400.

64. The questions to answer on the test can be chosen in C(15,10) = 3003 ways and the 1 question for homework in C(5,1) = 5 ways. Thus, there are 3003·5 = 15,015 ways to choose both sets of questions.

65. The two boys can be chosen in C(3,2) = 3 ways and the three girls in C(4,3) = 6 ways. Then the committee can be chosen in C(3,2)·C(4,2) = 3·6 = 18 ways.

66. The three boys can be chosen in C(5,3) = 10 ways and the two girls in C(3,2) = 3 ways. Thus the committee can be chosen in C(5,3)· C(3,2) = 10·3 = 30 ways.

67. The two shirts can be chosen in C(12,2) = 66 ways and the three ties in C(10,3) = 120 ways. The two items together can be chosen in 66·120 = 7920 ways.

68. The dresses can be chosen in C(9,5) = 126 ways and the coats in C(3,2) = 3 ways. The two items together can be chosen in 126·3 = 378 ways.